## DATE DUE

| | | | |
|---|---|---|---|
| ~~AP 25 '98~~ | | | |
| ~~JY 28 '99~~ | | | |
| | | | |
| ~~AP 22 '02~~ | | | |
| AP 28 '09 | | | |
| JE - 3 '09 | | | |
| | | | |
| | | | |
| | | | |
| | | | |
| | | | |
| | | | |
| | | | |
| | | | |
| | | | |
| | | | |
| | | | |
| | | | |

DEMCO 38-296

Biology has been advancing with explosive pace over the last few years and in so doing has raised a host of ethical issues. This book, aimed at the general reader, reviews the major advances of recent years in biology and medicine and explores their ethical implications. From birth to death the reader is taken on a tour of human biology – covering genetics, reproduction, development, transplantation, aging, dying and also the use of animals in research and the impact of human population on this planet. In each chapter there is a sketch of the relevant field's most recent scientific advances, combined with discussions of the ethical and moral principles and implications for social frameworks and public policy raised by those advances. Anybody interested or concerned about the ethical dilemmas caused by advances in science and medicine should read this book.

# Birth to Death

# Birth to Death

## Science and Bioethics

EDITED BY

**David C. Thomasma**

*Loyola University Chicago Medical Center*

and **Thomasine Kushner**

*University of California at Berkeley*

**CAMBRIDGE**
UNIVERSITY PRESS

Published by the Press Syndicate of the University of Cambridge

The Pitt Building, Trumpington Street, Cambridge CB2 1RP

40 West 20th Street, New York, NY 10011–4211, USA

10 Stamford Road, Oakleigh, Melbourne 3166, Australia

First published 1996

Printed in Great Britain at the University Press, Cambridge

*A catalogue record for this book is available from the British Library*

*Library of Congress cataloguing in publication data available*

ISBN 0 521 46297 5 hardback
ISBN 0 521 55556 6 paperback

This book is dedicated to our parents, living and deceased, Charles and Rosemary Olma Thomasma and William and Mary Louise Kimbrough. They have given us more than life.

# Contents

*List of contributors* /xiii

## INTRODUCTION /1

DAVID C. THOMASMA, PhD *Loyola University Chicago Medical Center*

THOMASINE KUSHNER, PhD *University of California, Berkeley*

## GENETICS

*Science*    **Genetics: A scientific sketch** /5

KAREN DAWSON, PhD *Monash University, Australia*

*Bioethics*    **The genetic revolution** /13

DANIEL CALLAHAN, PhD *Hastings Center, New York*

**Genetic knowledge: Some legal and ethical questions** /21

ROBERT SCHWARTZ, JD *School of Law, University of New Mexico*

## REPRODUCTIVE TECHNOLOGIES

*Science*    **The "ART" of medically assisted reproduction: An embryo is an embryo is an embryo** /35

MICHAEL E. McCLURE, PhD *National Institute of Child Health and Human Development, Bethesda*

*Bioethics*    **"O brave new world": Rationality in reproduction** /50

ALBERT R. JONSEN, PhD *School of Medicine, University of Washington*

**Reproduction, abortion, and rights** /58

ROSAMOND RHODES, PhD *Mt Sinai Medical Center, New York*

## CHILDREN AND WOMEN IN HEALTH CARE

*Science*    **The critically ill neonate /71**

> JAMES M. ADAMS, MD *Baylor College of Medicine, Houston*

*Bioethics*    **Medical technology and the child /79**

> AMNON GOLDWORTH, PhD *Department of Philosophy, San Jose State University*

**On caring for children /85**

> MARY B. MAHOWALD, PhD *Pritzker School of Medicine, University of Chicago*

## TRANSPLANTATION

*Science*    **Clinical transplantation /99**

> ROBERT A. SELLS, MB, FRCS, FRCS(Ed) *Royal Liverpool University Hospital*

*Bioethics*    **Transplantation and ethics /106**

> RAANAN GILLON, BA, MB, BS, FRCP *Imperial College of Science, Technology and Medicine, University of London*

**Legalizing payment for transplantable cadaveric organs /119**

> JAMES F. BLUMSTEIN, LLB *Vanderbilt University School of Law, Nashville*

## AGING

*Science*    **Scientific advances in aging /133**

> JOHN E. MORLEY, MB, BCh *St Louis University Health Sciences Center*

*Bioethics*    **Ethics and aging /142**

> GEORGE J. AGICH, PhD *Southern Illinois University School of Medicine*

**People with dementia: A moral challenge /154**

> STEPHEN POST, PhD *School of Medicine, Case Western Reserve University*

## PROLONGING LIFE/DEATH

*Science*    **Personal dying and medical death /163**

> STEVEN MILES, MD *University of Minnesota Center for Biomedical Ethics*

*Bioethics*    **Stopping futile medical treatment: Ethical issues /169**

> NANCY S. JECKER, PhD *School of Medicine, University of Washington* and LAWRENCE J. SCHNEIDERMAN, MD *School of Medicine, University of California, San Diego*

**The Sorcerer's Broom: Medicine's rampant technology /177**

> ERIC J. CASSELL, MD *Cornell University Medical College*

## CARE OF THE DYING

*Science*   Modern technology and the care of the dying /191

    RONALD E. CRANFORD, MD *Hennepin County Medical Center, Minneapolis*

*Bioethics*   Care of the dying: From an ethics perspective /198

    T. PATRICK HILL, MA *Consultant in Ethics, New Jersey*

## EUTHANASIA AND PHYSICIAN-ASSISTED SUICIDE

*Science*   Euthanasia and assisted suicide /207

    PIETER ADMIRAAL, MD, PhD *Laureate of the Academy of Humanism*

*Bioethics*   Physician-assisted suicide: Progress or peril? /218

    CHRISTINE K. CASSEL, MD *Mount Sinai Hospital, New York*

"I will give no deadly drug": Why doctors must not kill /231

    LEON R. KASS, MD *Committee on Social Thought, University of Chicago*

Voluntary euthanasia and other medical end-of-life decisions: Doctors should be permitted to give death a helping hand /247

    HELGA KUHSE, PhD *Centre for Human Bioethics, Monash University*

## HUMANS AS RESEARCH SUBJECTS

*Science*   Humans as research subjects /259

    HERMAN WIGODSKY, MD *University of Texas Health Sciences Center at San Antonio*

    and SUE KEIR HOPPE, PhD *University of Texas Health Sciences Center at San Antonio*

*Bioethics*   Research involving children as subjects /270

    ROBERT J. LEVINE, MD *School of Medicine, Yale University*

Future challenges of medical research review boards /283

    CHARLES R. MacKAY, PhD *National Institutes of Health, Bethesda*

## USING ANIMALS IN RESEARCH

*Science*   Animals in research /301

    FRANKLIN M. LOEW, DVM, PhD *College of Veterinary Medicine, Cornell University*

*Bioethics*   Taking duties seriously: Medical experimentation, animal rights, and moral incoherence /313

    DANIEL A. MOROS, MD *Mount Sinai School of Medicine, New York*

Animal rights and social practices /324

    TED BENTON, PhD *Department of Sociology, University of Essex*

## THE ENVIRONMENT

*Science*   **The science of the environment /339**

    ANDREW PULLIN, PhD *School of Biological Sciences, University of Birmingham*

*Bioethics*   **Environmental ethics /348**

    ANDREW DOBSON, PhD *Politics Department, University of Keele*

**Human activity and environmental ethics /357**

    ANDREW L. JAMETON, PhD *University of Nebraska Medical Center College of Medicine*

## POSTSCRIPT /369

    DAVID C. THOMASMA, PhD *Stritch School of Medicine, Loyola University Chicago*

    and THOMASINE KUSHNER, PhD *Health and Medical Sciences Program, University of California, Berkeley*

*Index* /377

# Contributors

JAMES M. ADAMS, MD
Associate Professor of Clinical Pediatrics
Baylor College of Medicine
One Baylor Plaza
HOUSTON, TX 77030–3498, USA

PIETER ADMIRAAL, MD, PhD
Koniginnelaan Sq
2201 HB Rijswijk
THE NETHERLANDS

GEORGE J. AGICH, PhD
Professor of Medical Humanities and
  Psychiatry
Director: Medical Ethics Program
Department of Medical Humanities
Southern Illinois University School of
  Medicine
PO Box 19230
SPRINGFIELD, IL 62794–9230, USA

TED BENTON, PhD
Professor, Department of Sociology
University of Essex
Wivenhoe Park
COLCHESTER CO4 3SQ, UK

JAMES F. BLUMSTEIN, JD
Professor of Law
Vanderbilt University School of Law
21st Avenue South
NASHVILLE, TN 37240, USA

DANIEL CALLAHAN, PhD
President: Hastings Center
255 Elm Road
BRIARCLIFF MANOR, NY 10510, USA

CHRISTINE K. CASSEL, MD
Professor and Chairman
The Henry Dow Schwarz Department
  of Geriatrics and Adult Development
Mt Sinai Medical Center
Box 1070
NEW YORK, NY 10029–6574, USA

ERIC J. CASSELL, MD
Professor of Public Health at
  Cornell University Medical
  College
(Office address)
1550 York Avenue
NEW YORK, NY 10028, USA

RONALD E. CRANFORD, MD
Department of Neurology
Hennepin County Medical Center
701 Park Avenue, South
MINNEAPOLIS, MN 55415, USA

KAREN DAWSON, PhD
Centre for Early Human Development
Monash University
Level 5–246 Clayton
Clayton 3168,
VICTORIA, AUSTRALIA

ANDREW DOBSON, PhD
Chairman, Politics Department
Keele University
KEELE, STAFFORDSHIRE ST5 5BG, UK

RAANAN GILLON, BA, MB, BS, FRCP
Professor of Medical Ethics,
Imperial College of Science, Technology
    and Medicine
Exhibition Road
LONDON SW7 2AZ, UK

AMNON GOLDWORTH, PhD
Professor of Philosophy
Department of Philosophy
San Jose State University
SAN JOSE, CA 94583, USA

T. PATRICK HILL, MA
Ethics Consultant
The Park Ridge
211 E. Ontario
Suite 800
CHICAGO, IL 60611-3215, USA

SUE KEIR HOPPE, PhD
Associate Professor and Chief
    Division of Sociology
The University of Texas Health Science
    Center at San Antonio
7703 Floyd Curl Drive
SAN ANTONIO, TX 78284–7792, USA

ANDREW L. JAMETON, PhD
Associate Professor and Section Head
Section on Humanities and Law
Department of Preventive and Societal
    Medicine
University of Nebraska Medical
    Center
College of Medicine
600 South 42nd Street
OMAHA, NE 68198, USA

NANCY S. JECKER, PhD
Associate Professor
Department of Medical History
    and Ethics
School of Medicine SB-20
University of Washington
SEATTLE, WA 98195, USA

ALBERT R. JONSEN, PhD
Professor of Ethics in Medicine
Chairman: Department of Medical
    Humanities and Ethics
School of Medicine SB-20
University of Washington
SEATTLE, WA 98195, USA

LEON R. KASS, MD
Addie Clark Harding Professor
The Committee on Social Thought
    and the College
The University of Chicago
1130 E. 59th Street
CHICAGO, IL 60637, USA

HELGA KUHSE, PhD
Director: Centre for Human Bioethics
Monash University
Clayton, 3168
VICTORIA, AUSTRALIA

THOMASINE KUSHNER, PhD
Associate Clinical Professor
Health and Medical Science Program

University of California at Berkeley
BERKELEY, CA 94965, USA

ROBERT J. LEVINE, MD
Professor of Medicine
Yale University
The School of Medicine
Room IE-46 SHM
333 Cedar Street
PO Box 208010
NEW HAVEN, CT 06520–8010, USA

FRANKLIN M. LOEW, DVM, PhD
Dean; College of Veterinary Medicine
Cornell University
Room 4017
ITHACA NY 14853–6401, USA

CHARLES R. MAcKAY, PhD
Project Clearance Office
Building no. 31, Room 5B33
National Institutes of Health
BETHESDA, MD 20892, USA

MARY B. MAHOWALD, PhD
Associate Professor: Department of
 Obstetrics and Gynacology
Center for Clinical Medical Ethics
University of Chicago
Pritzker School of Medicine
5841 South Maryland Avenue
Box 72
CHICAGO, IL 60637–1470, USA

MICHAEL E. McCLURE, PhD
Chief: Reproductive Sciences
 Branch
Center for Population Research
National Institute of Child Health
 and Human Development
National Institutes of Health
Department of Health and Human
 Services
BETHESDA, MD 20892, USA

STEVEN MILES, MD
Associate Professor of Medicine
University of Minnesota Center for
 Biomedical Ethics
University of Minnesota, Suite 110
2221 University Avenue SE
MINNEAPOLIS, MN 55414, USA

JOHN E. MORLEY, MB, BCh
Dammert Professor of Gerontology
Director: Missouri Gateway
Geriatric Education Center
Division of Geriatric Medicine
Department of Internal Medicine
St Louis University Health Sciences
 Center
1402 South Grand
ST LOUIS, MO 63104

DANIEL A. MOROS, MD
Associate Professor of Neurology
Mt Sinai School of Medicine
One Gustave L. Levy Place
NEW YORK, NY 10029–6574, USA

STEPHEN POST, PhD
Associate Professor
Center for Biomedical Ethics
Case Western Reserve University
School of Medicine
10900 Euclid Avenue
CLEVELAND, OH 44106–4976, USA

ANDREW PULLIN, PhD
School of Biological Sciences
The University of Birmingham
Edgbaston
BIRMINGHAM B15 2TT, UK

ROSAMOND RHODES, PhD
Director of Bioethics Education
Associate Professor Department of
 Medical Education
Mt Sinai Medical Center

Box 1108
One Gustave L. Levy Place
NEW YORK, NY 10029–6574, USA

LAWRENCE J. SCHNEIDERMAN, MD
Professor of Community and Family
   Medicine
School of Medicine 0622
University of California, San Diego
LA JOLLA, CA 92093–0622, USA

ROBERT SCHWARTZ, JD
Institute of Public Law
School of Law
University of New Mexico
1117 Standford NE
ALBUQUERQUE, NM 87131, USA

ROBERT A. SELLS, MB, FRCS,
   FRCS(Ed)
Director, Renal Transplant Unit

Royal Liverpool University Hospital
Prescot Street
LIVERPOOL L7 8XP, UK

DAVID C. THOMASMA, PhD
Professor of Medical Ethics
   and Director Medical Humanities
   Program
Stitch School of Medicine
Loyola University Chicago
2160 South First Avenue
MAYWOOD, IL 60153, USA

HERMAN WIGODSKY, MD
Emeritus Professor
The University of Texas Health
   Science Center at San Antonio
7703 Floyd Curl Drive
SAN ANTONIO, TX 78284–7792,
   USA

# Introduction

DAVID C. THOMASMA and THOMASINE KUSHNER

This book is written with you in mind. Bioethics is not something necessarily esoteric. When you and your loved ones encounter modern medicine and modern care facilities, you meet directly what C. P. Snow articulated as the "Clash of two cultures." Scientific culture and human values culture compete for our loyalties daily, but never more so than when we or someone we care for becomes seriously ill or is dying, and needs the assistance of the health care system.

In the personages of physicians, nurses, social workers, therapists, admitting clerks, nursing home administrators, case managers, and all the others we encounter individuals who not only aim for the best interests of the sick, but also represent the scientific culture of modern medicine. Due to the rapid rise of scientific and technological advancement everywhere, and especially in medicine, enormous changes must take place, willy-nilly, in our human values culture, as we call it, whether we like those changes or not. Furthermore, science and technology require rethinking cherished values about the moral status of animals, children, the dying, the mentally compromised, and even those who are healthy and must support the medical care system. Rethinking these cherished values is important because they lead us to examine even more profound assumptions about ourselves. These profound assumptions we can call "second-order" considerations, questions such as what counts as a person either when we have frozen embryos in storage or in parents with advanced Alzheimer's disease, what is personal identity, can animals actually have rights, is there a moral difference at all between animals and humans, what will happen to social values if we try to alter the gene pool, how should we treat "marginal" people, are there duties that cut across generations, and what about the environment in all this?

When we begin to root around in these questions, we enter the realm of what we call "third-order" considerations as well. These lead us to question what sort of ethical system we ought to employ: can we use the virtues we were taught, is a rights-based or principle-based ethics best, are there inalienable rights that must never be violated, do ends justify the means, should utilitarianism be used to resolve conflicts, is there some other ethical theory that might be better, are we to hold conflicting theories together like a collage, using whichever works at the moment (this is called Post-modernism)?

We do not need a degree in medicine, philosophy, or theology to grapple with these issues. After all, what Plato or Aristotle might have to say on any of these topics is interesting, but does not determine to any great extent what behaviors we exhibit as our grandmother slowly loses her edge, or as a child is born prematurely and must be placed in an intensive care unit, or as we invest in the stock of a private nursing home corporation, or as we survey dogs in a pound that, if not chosen by a family, may wind up being used in cardiology research.

In other words, when we confront the medical culture in our daily lives, we are involved in bioethics. We participate in the decisions that mesh our values with those of scientific culture. It is these decisions that are more important than theoretical reflection, since they shape our understanding of other values, and, by doing so, shape us as well.

Our book is called *Birth to Death: Science and Bioethics* because we trace current challenges to our values by biological discoveries in science and in medicine from before birth, through genetics, to our deaths, sometimes despite medical technology! These current challenges are collected in sections, most of which have three chapters. All chapters are written by experts for educated readers. Each combines a sketch of the most recent advances in the particular field, say understanding fetal growth and development, with details about how the newest scientific and technological understanding of human growth, development, aging, and eventual decline challenge our ethical and moral principles, social frameworks, and public policy about the subject of the chapter. Accordingly, the chapters contain ideas from the past, concepts of the present, and future challenges. In this way, you can gain both an enthusiastic update of the most recent thinking in biology and bioethics as well as glimmers of subsequent challenges we all may face in the future. The concerns of bioethics and the way you confront these issues will be faced every time you talk about them with friends or read them in the newspaper or see them on television. Preceding each chapter is a short Editors' Summary of the chapter so you can determine the major points of the chapter ahead of time.

The first chapter in a section describes the past, present, and future of a current scientific advancement that challenges our values. For example, we include sections on the following: genetics, which explores a fundamental basis of disease that will ultimately change the way we treat sick people; reproductive technologies, which allows us almost God-like powers over the creation and manipulation of life; newborn intensive care, which rescues children some people think are nature's mistakes; organ transplant technology, which causes problems in objectification and ownership of organs; geriatric science, which can help people live longer and with a greater quality of life; on life-prolonging technology that sometimes is used inappropriately on the dying; research on animals to benefit human beings, which causes controversies for the moral status of animals; and finally environmental science, which has a greater and greater impact on human health and well-being and increasingly requires our attention.

Each of these short sketches about the scientific advancements mostly during this century is followed by ethical commentaries. Sometimes these commentaries represent opposing views on a critical subject engendered by scientific advancement, e.g. arguments for and against using animals for medical research, arguments for and against euthanasia and physician-assisted suicide[1]. At other times, the ethics chapters reflect on complementary issues, such as ethical issues arising from caring for premature children, and then ethical issues in caring for children in general, and their increasing moral status during the past half-century, or an ethical examination of a number of issues created by increasing longevity, and a complementary chapter about how we should treat people with Alzheimer's disease, or two different views about how we might go about buying and selling organs for transplantation. All the chapters have been commissioned by the editors as fresh contributions. Several were adapted by the authors from earlier work. There is one exception. Leon Kass' contribution on "Why doctors must not kill" appears with only minor changes from an original that was published elsewhere. We included his contribution because it so strongly contrasts with other points of view expressed in the section on euthanasia.

Only the most fundamental disagreements are highlighted, so that you may come away from the book with a mind uncluttered by minor academic trivialities. We have included a List of Contributors that provides the names and addresses of all the contributors to this volume, in case you would like to contact any of them. Most chapters contain a short selection of Further Reading for your convenience, should you wish to delve further into the issues.

At the end of the book you will meet us again, where we will encapsulate the most critical issues raised by all the scientific advancements covered in the book. Right now it gives us great pleasure to thank all of our experts for their time and devotion to this project, for it is hard to collect one's thoughts about complex matters in so short a space, and write them clearly for a general audience. In editing we tried hard to maintain the author's style and interests accompanied by the clearest possible prose. Hence, the pathways taken by the chapter authors are those they chose themselves, knowing the difficulties they left behind in their choices. Nonetheless, it is like having at your fingertips the opinions and positions of noted experts in a variety of health care fields about extremely important issues. As a matter of course, then, this book could not be comprehensive, but it is suggestive of the enormous range of concerns we all must face together as we move into twenty-first century.

We wish to thank Robert Harington, PhD, at the time Comissioning Editor for the Life Sciences at Cambridge University Press, for suggesting this book to us, and encouraging its development, Tim Benton, PhD, later Popular Science Editor of Cambridge University Press, Doris Thomasma and Maggie Hall for their help in editing and corresponding with our contributors. Special thanks, too, for Patricia Marschall for her care in catching details in the final manuscript. Finally, our gratitude extends beyond limits to Sandi Irvine, our copyeditor, and Robbin Hiller, our research assistant in the Medical Humanities Program. They both spent many hours making the text as accurate and readable as possible. Thank you all!

NOTE

1. "Physician" is used throughout this book in the sense of "doctor". It does not exclude surgeons although specialties are usually specified.

# Genetics: A scientific sketch

KAREN DAWSON

EDITORS' SUMMARY

Karen Dawson, PhD, adroitly summarizes the nature of genes, the history of their discovery, and their impact on human diseases. She discusses the differences between diseases that may be caused either by numerical abnormalities (in which more genes are present than is normally the case) or structural abnormalities (in which genes are transformed, altered or misplaced). Mapping the human genome is only the first step in understanding where the estimated 100 000 genes might be. We are also interested in the role and function of each gene. Some discoveries about this point have already been made, leading to the possibility of gene therapy for diseases such as cancer and cystic fibrosis. Because of the complexity of genetic interaction, and genes and environmental influences, many challenges about treating up to 3000 genetically based diseases still exist.

From the time we are born, and at different times throughout our lives, we are the recipients of comments such as: "You really look like your mother," "You behave just like your grandfather," and so on. What these comments really are is an acknowledgment that we are showing our genetic inheritance, or our *heredity*, either in our appearance or in our temperament. Classical genetics is the study of these patterns of inheritance. Modern genetics is concerned not only with understanding these patterns of inheritance but also with understanding the answer to the questions of "why?" and "how?" these patterns of inheritance occur.

## The beginning of genetics

Genetics is not a long-established science when compared with, say, physics or chemistry. Its beginnings as a systematic study can be traced back to the work of an Austrian Augustinian monk and botanist, Gregor Mendel, in the 1860s, who discovered that the transmission of characteristics from parents to offspring depended on a set of biological elements that behaved in a predictable way. These biological elements were later to become known as genes.

The implication of Mendel's discovery was largely overlooked until about 1900 when various scientists around the world demonstrated that inherited characteristics – the genes within each cell – were carried on structures known as chromosomes. The number of chromosomes for a given species was shown to be constant, although there was a wide variation in the number of chromosomes in the cells between different species. The chromosome number for a species was also shown usually to be an even number, with half of the chromosomes coming from the sperm, or male gamete, and the other half coming from the egg, or female gamete.

The chromosomes of a species were shown to undergo a longitudinal division at the time of cell division when the organism was involved in growth. It was also shown that the chromosomes underwent a special cycle of reduction division in the formation of the male and female gametes that would participate in sexual reproduction. This special division is essential to maintaining the constant chromosome number over the generations for any sexually reproducing species.

As members of the human species, we share these features of genes and chromosomes with the many other sexually reproducing animal species. Our species has a chromosome number of 46; the chromosomes vary in size and occur in pairs; 23 of these chromosomes are inherited from each of our father and mother. In the human karyotype, an arrangement of the chromosomes in a cell constructed from a photograph, it can be seen that there are 22 chromosome pairs, the autosomes, and two sex chromosomes that are the same in a female (i.e. XX) or different in a male (i.e. XY) (see Figure 1).

The beginning of genetics provided us with some knowledge that is used widely in medicine today.

## Detecting chromosomal abnormalities

Sometimes problems can occur in the formation of the sperm and egg that combine to form a new generation. It is possible to detect these errors as chromosomal abnormalities in the fetus before birth, using karyotyping and prenatal diagnosis techniques such as amniocentesis or chorionic villus sampling (CVS).

Chromosomal abnormalities can be of two main types: changes in chromosomal

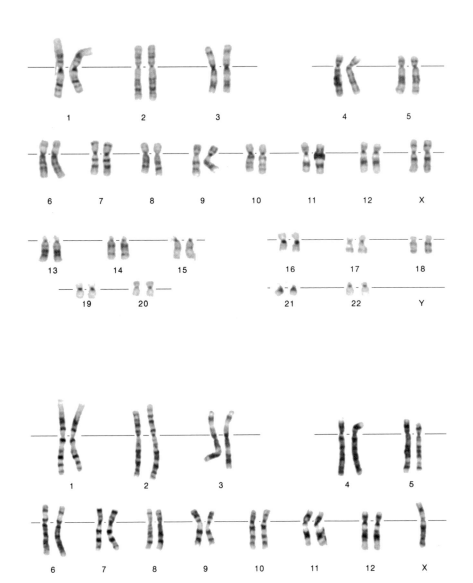

Figure 1. Chromosomes of the human female (XX) (*top*) and male (XY) (*bottom*) arranged in a karyotype.

number (numerical abnormalities) or changes in chromosomal structure (structural abnormalities) and the abnormality can involve either the autosomes or the sex chromosomes.

## Numerical chromosomal abnormalities

Numerical chromosomal abnormalities can be either an addition or a reduction in the normal chromosome number. Numerical abnormalities in autosomes that are compatible with life tend to be additions to the chromosome number.

The most common chromosomal abnormality of the autosomes that can result in a live birth can be found in people with Down syndrome. These people have 47 chromosomes, with three copies of chromosome 21 being present instead of the usual two. Hence this disorder is sometimes also referred to as trisomy-21.

Down syndrome can result in both physical and mental handicaps in a child, and there is no known cure. Its incidence is strongly related to the age of the mother at the time of conception, increasing as maternal age increases. It is believed that the chromosomes in the egg cell become less likely to separate properly during gamete formation as a women gets older. The incidence of Down syndrome in relation to maternal age is presented in Table 1.

For a female, numerical chromosomal abnormalities of the sex chromosomes can be either an increase or a reduction in chromosome number. About 1 in 2000 girls are born with Turner syndrome, the loss of all or part of an X-chromosome. There are often only 45 chromosomes in the karyotype of girls affected by Turner syndrome. The effects of this abnormality are that the girls are short in stature and have delayed sexual maturity.

For a male, changes in sex chromosome number that are compatible with life are limited to increases in chromosome number. The most common of these conditions is Klinefelter syndrome, where the male has an extra copy of the X-chromosome, i.e. XXY. This disorder, estimated to occur in about 1 in 2500 males, can have serious effects on fertility.

## Structural chromosome abnormalities

Structural chromosome abnormalities can result from a small part of a chromosome being deleted or duplicated, or part of a chromosome becoming translocated (attached) to another chromosome. Whether these changes will result in an abnormality largely depends on the amount of chromosomal material involved and whether or not the change results in the net gain or loss of genetic information.

Table 1. *The chances for women of different ages having a live-born baby with Down syndrome or some other chromosomal abnormality*

| Mother's age at expected date of delivery | Chance of having a live-born baby with Down syndrome | Chance of having a live-born baby with another chromosomal abnormality |
|---|---|---|
| 35 | 1:384 | 1:179 |
| 36 | 1:307 | 1:149 |
| 37 | 1:242 | 1:124 |
| 38 | 1:189 | 1:105 |
| 39 | 1:146 | 1:81 |
| 40 | 1:112 | 1:64 |
| 41 | 1:85 | 1:49 |
| 42 | 1:65 | 1:39 |
| 43 | 1:49 | 1:31 |
| 44 | 1:37 | 1:24 |
| 45 | 1:28 | 1:19 |

From: *1993 Directory of Genetic Support Groups and Genetics Services Australia & New Zealand. Canberra: New South Wales Genetic Education Program.*

## The development of biochemical genetics

The study of the biochemical effects of genes can be dated from the work of British physician Archibald Garrod in 1902. He studied a human disease, alkaptonuria, and concluded that this disease was inherited and resulted from a blockage in the metabolic pathway that led to urea being excreted as a final break-down product in the urine. What Garrod had demonstrated was that abnormal gene products such as enzymes – the proteins that control metabolic pathways – can cause a blockage at a definite point in a metabolic pathway. He also showed that similar blockages in different pathways could result in diseases such as albinism, cystinuria and porphyria. The gene was now known to have both a structure and a function. This knowledge has been crucial in the diagnosis and treatment of many single gene disorders, genetic disorders that result primarily from a defect in one gene, today.

## Understanding the genetic code

In 1953 genetics took a giant leap forward in understanding what genes were and how they functioned. James Watson and Francis Crick discovered the structure of the genetic material – deoxyribose nucleic acid, more commonly referred to as *DNA*. DNA was shown to be a double helix molecule that was composed of only four paired chemical bases – guanine (G), cytosine (C), adenosine (A) and thymine (T). For each gene, the genetic code is read in triplets, sequences of three bases at a time to yield messenger RNA (ribonucleic acid), which is then translated within the cell to produce a protein that participates in biochemical pathways within the organism.

It is differences in the order of these bases that determine the differences between pieces of DNA, and hence the differences in the composition of the different proteins produced. An insertion, a deletion or a change in any base in a specific piece of DNA – a mutation – can disrupt the production of a protein usually coded for, stop the protein from being produced, or upset the normal function of the protein produced.

A gene can consist of hundreds of thousands of base-pairs of DNA. Special techniques have been developed to enable the position of a particular gene to be determined on a chromosome (i.e. mapped) and the precise sequence of a gene to be determined. These techniques are the basis of the most ambitious research project yet to be undertaken by geneticists (see next section).

## The Human Genome Project

In 1990 scientists began work on a controversial large-scale internationally collaborative project to map and sequence the entire human genetic complement or genome. The human genome consists of 300 000 000 base-pairs of DNA that are estimated to contain about 100 000 genes. There are two major goals of the Human Genome Project: first, to create a physical map of the chromosomes that comprise the human genome; and, second, to define the function of each of the genes present on each chromosome.

The first goal of the project has almost been completed due to the amazing developments that have occurred in sequencing techniques and the number of scientific teams that have been contributing data to the project. The achievement of the second goal of the project will be more time-consuming than the first, but also will be potentially more valuable because it will provide more medically useful information. It will change the way we treat many human illnesses.

Understanding the sequence of the bases in human genetic material is the essential first step in understanding the genetic basis of inherited diseases and chronic illnesses, and the future possible development of the prevention, diagnosis or improved treatment of these diseases.

Application of the knowledge coming from the Human Genome Project has already yielded some of these benefits. For instance, it is now possible to diagnose before and after birth, with increased accuracy, individuals who are affected by, or who are carriers of, *fragile-X syndrome*. Fragile-X syndrome is now known to result from the duplication of a small section of genetic material on the X-chromosome. It is one of the most common chromosomal abnormalities and can result in mental retardation, delayed development and subtle behavioral deficits. This disorder affects predominantly males because of their single X-chromosome. Females, normally with two X-chromosomes, can carry the disorder on one X-chromosome but be largely unaffected by its presence.

Other information from the Human Genome Project has led to the identification of genes involved in some forms of cancer. Strategies are being devised to counter the effects of these altered genes through manipulating the sufferer's immune system. Some strategies for treating different types of cancer using gene replacement, known as gene therapy, are entering the clinical trial stage. In this instance, our increased knowledge of the human genome has allowed a more aggressive approach to be adopted in the battle against some diseases.

It is also possible to identify, from a knowledge of the human genome, people who have a susceptibility toward developing diseases such as diabetes, arthritis, high cholesterol levels and some forms of heart disease. These diseases are polygenic, i.e. caused by the action of many genes, but a person's susceptibility can be determined by identifying only several sequences in their DNA. The identification of people who are likely to develop these types of disease later in life as an outcome of genetic preconditioning has immense consequences for the introduction of preventative programs to minimize the chances of the disease susceptibility becoming realized.

## Future possibilities

About 3000 genetic disorders that can affect our health and well-being to different extents have been identified. Progress on understanding where these genes are located, how these disorders occur, and how they might be corrected is occurring rapidly as more becomes known about the human genome. The rate of progress is encouraging, but there are many more aspects about the functioning of our genome and about individual genes that have yet to be understood. If these secrets can be unraveled, human medicine will become incredibly powerful in diagnosis, prevention measures, treatment possibilities that might be adopted.

These prospects are exciting and perhaps a cause for some concern. Scientists have the potential to make these discoveries, but some guidance as to how this new knowledge might be used and for what purposes, is required. It must come from informed discussion within the community.

## Suggestions for further reading

Dawkins, R. *The Selfish Gene*. Oxford: Oxford University Press, 1989.

Dawkins, R. *The Extended Phenotype: The Long Reach of the Gene*. Oxford: Oxford University Press, 1989.

Gould, S. J. *Bully for* Brontosaurus: *Further Reflections on Natural History*. London: Penguin, 1992.

Gould, S. J. *Eight Little Piggies: Reflections on Natural History*. New York: Norton, 1993.

Nossal, G. J. V. and Coppel, R. L. *Reshaping Life: Key Issues in Genetic Engineering*, 2nd ed. Melbourne: Melbourne University Press, 1989.

Suzuki, D. T. *Genethics: The Ethics of Engineering Life*. Cambridge, MA: Harvard University Press, 1990.

# The genetic revolution

### DANIEL CALLAHAN

EDITORS' SUMMARY

Daniel Callahan, PhD, focuses on the challenges to society created by our increase in genetic knowledge. First and foremost, such knowledge will only create additional uncertainties, he thinks. Should society not handle these properly, the genetic knowledge we obtain will only make us more obsessive about trying to control disease, death, and deviant behavior, without actually contributing to our personal and social happiness and well-being. This warning is in keeping with his life-long concern about what sort of society we ought to be and become, and a healthy wariness about the capabilities of science and medicine in forging a good society.

Not long ago I visited the city of Brno, in the Czech Republic. There I gazed with awe on the small garden patch – no more than 3m by 6m – where the botanist monk Gregor Mendel in the 1860s worked out the first truly scientific understanding of genetics. No one paid much attention to him at first, but the world has now definitely caught on. For well over a century human genetics has been the stuff of nightmares and dreams.

The nightmares have a basis in history. The Nazi devotion to racial purity, based on faulty genetic knowledge harnessed to ancient prejudices, led to the greatest crime in human history. Somewhat more benign forms of the same nightmare appeared in early twentieth-century American laws allowing sterilization of the retarded, and in widespread beliefs that physical defects, mental abnormalities, and undesirable, socially harmful behaviors could be effectively excised with better genetic knowledge. Genetics has thus been the source of an enormously seductive, often dangerous temptation: to use genetic knowledge to improve on human nature, to give us more choice about our human traits, and decisively to get rid of genetically based disease (which may turn out to be most disease), or, failing that, even the people with disease.

The dreams are simply the other side of the nightmare coin. Why should we not use genetic knowledge to improve the human condition? Why should we assume that we will simply repeat all the errors and corruptions of the Nazis? Why cannot we, with a better scientific basis, develop forms of genetic knowledge and application that will mainly do good and minimally do harm? And who among us can doubt the advantage that our more recent genetic knowledge has already given us: improved family and reproductive counseling, prenatal diagnosis and the identification of the markers for many diseases that will appear only later in an individual's life?

If genetic knowledge has been the stuff of dreams and nightmares, it is also the stuff of something even more anxiety-producing and perhaps more insidious. Much future genetic knowledge will not give us absolute certainty: it will give us only probabilities and possibilities, leaving it up to individuals and society to know what to do with much ambiguous knowledge.

I focus on the uncertainty because it will make a great deal of practical difference. Already, as part of the great scientific effort to map the human genome, there is increased likelihood that science will, in the not too distant future, be able to discover all kinds of information about our genetic make-up and our genetic expectations. It is knowledge that we may (or may *not*) want as individuals, but surely knowledge that our employers and our health insurers will want. When they hire us, our employers will want to know what are our chances of getting sick, and our health insurers will want to know how likely it is that they will have an expensive patient to pay for some day. That knowledge of our possible medical fate will be available, but to whom and under what conditions? How can we maintain the confidentiality of knowledge that others will want?

It is no less likely that those who still hanker after the use of genetics to manipulate, change, and improve the human condition will be eager to have this knowledge also. Perhaps it is the case, as some speculate, that there is a genetic predisposition to violence in some people. If we could identify the source of that trait, perhaps we could change or eliminate it. But almost certainly there will be a long period when much of our knowledge will be fragmentary, and tantalizingly so; and a still longer period after that where the knowledge will be uncertain, ambiguous and probabilistic. *Some* people may, under *some* environmental conditions, have a predisposition to violence in *some* circumstances *some* of the time.

In short, we will not often know just what to make of the "knowledge" we have, much less what to do about it. What if we can find out, early in life, that some of us will have a predisposition to Alzheimer's disease later in life – but a predisposition only, not a certainty of getting it? And what will such knowledge mean when there is no cure for Alzheimer's disease or any way to avoid getting it? How could we tolerate such knowledge without it casting a dark shadow over those many years of life when we were perfectly healthy? Could we perhaps vitiate 30 or 40 or 50 years of our life out of anxiety for the 5–10 years after 80 when the disease may finally bring us down?

Take still another form of uncertainty, that generated by advances in prenatal diagnosis. More and more parents will have a choice over the genetic traits of their children, becoming able to decide whether they can tolerate certain defects. Surely they can be seduced into trying to guarantee certain traits, such as height, or intelligence, or eye color. We can be certain that some parents, perhaps many, will want such choices. We can be no less certain that the outcome will be uncertain, for the children and for the parent. Will those parents be happier with children whose traits they have specified in advance? Will the children be happier? Can we have any certainty whatever that we will, in the long run, produce better children – whatever "better" might mean in this context?

And what of those other parents, the luckless ones who chose to forego the genetic manipulations, trusting their fate to nature, that cruel nature that then gave them a damaged child? Could others not say "I told you so?" clucking their tongues in self-righteous pity? Is it not almost certain that they would, such being the well-known fate of those who have the hubris to turn their back on the deliverances of medical science? Will they be stigmatized, along with their children?

If uncertainty is one great and looming problem, another is that of putative personal choice and actual state (or social) coercion. All new genetic technologies have been introduced as means of expanding individual choice. There is now a powerful abhorrence of state-imposed genetic technologies. The Nazi experience remains fresh in the scientific mind. The great espoused value is that genetic knowledge should always be discretionary knowledge, to be used or not according to the moral values of individuals.

Yet there is a problem here. New medical technologies rarely remain discretionary for long. If they are not legally imposed on people, something hard to do in our western society, they can just as effectively be imposed by social pressure. When prenatal diagnosis was being introduced in the late 1960s, every assurance was given that no woman would be forced to have such a diagnosis, much less be forced into an abortion if the diagnosis turned up a terrible genetic defect. The great anxiety focused on government coercion of those who would be judged unfit mothers.

The worry about direct coercion was misplaced, but that did not mean influence was absent. Not at all. Social pressure and developing mores on prenatal diagnosis soon turned it into a routine "medically indicated" procedure, as common and standard for women over the age of 35 as taking blood pressure. It takes a tough and determined woman these days to exercise the "free choice" of not choosing prenatal diagnosis, and a still tougher and more determined woman not to have an abortion if a serious defect is discovered in her fetus after that diagnosis. Here as elsewhere, social pressure can accomplish quite nicely and smoothly what could never be legislated directly.

In the eyes of its most enthusiastic supporters, genetic knowledge and application are expected to become the greatest of all scientific breakthroughs in the history of medicine. We will, finally, understand the deepest biological basis of illness and disease. Even those who are something less than true believers are impressed with the possibilities, for both good and ill. Yet in the long run, what is likely to make the greatest difference is not the genetic knowledge as such, but what kind of soil that knowledge grows in. It was not just bad genetic knowledge that led the Nazis astray: it was their culture of racism and anti-semitism that allowed that knowledge to flourish and take root.

In our day, the danger is not likely to be so evident or so crass. Much will depend upon whether we think the world can tolerate some disabled, genetically diseased people. It will depend upon whether we are willing to tolerate the long-standing human fact that human beings will and must die of some disease, even genetic disease. It will depend upon whether we are willing to bear children who are less than perfect, other than their designing parents might want them to be. It will depend upon whether we are willing to continue trusting the genetic lottery of nature, not perfect but not always so bad either. It will depend upon whether we believe we must socially coerce people for their good, and for our good.

Follow carefully the development of genetic knowledge. Watch in particular for the evidence about the uncertainty of that knowledge, and for the kinds of social manipulation it is likely to draw us into. But, above all, watch out for the kind of society we could become and what that could mean as a context for the genetic developments. As the geneticists always like to say, how a gene expresses itself will be affected by its environment. That is no less true of genetic knowledge itself. A good society is not likely to produce a bad genetics. But even a good genetics cannot be saved from the influence of a bad society.

We need, in short, to ask the following question: what sort of society, with what values, will be the safest into which to introduce new genetic possibilities? And what kind of society is likely to be the most hazardous for those possibilities? Let me now aim at the last question, centering my remarks on potentially hazardous societies. Three societal patterns seem to me most fraught with danger: (a) those that demonize illness and biological death; (b) those that, finding ordinary forms of social control insufficiently powerful, want to find biological solutions; and (c) those that think there is no common social good, only a plurality of individual goods.

## Demonizing illness and death

One of the great paradoxes of twentieth-century medicine is that, in parallel with its success in alleviating illness and forestalling death, it has seemingly increased and exacerbated the fear of disease and death. The more we have gained, the more we seem to want. The more we have reduced the risk of disease, the lower still we want that risk. The more we have succeeded in pushing death off into old age, the more insistent we have become that everyone live to old age; and, once into old age, to extend life even further. Despite the fact that life expectancy has increased by over 30 years in the twentieth century, and infant mortality has been radically reduced, the thrust of biomedical research (with full public backing) has been to keep its momentum going, now targeting all those diseases and medical conditions still left to be conquered. We have come close to demonizing illness and death, turning them into the greatest human enemies, and worth whatever resources it takes to root them out, down to the last illness and the last death.

Two great approaches to the further reduction and elimination of disease now compete with each other. One of them is oriented toward changing those hazardous health behaviors known to be major causes of illness and mortality. Here the target is the way we live our lives, changing us from the outside (so to speak), hoping that by doing so we can take advantage of the body's natural capacity to renew and cure itself. The other great approach is from the inside (again, so to speak), looking to the genetic bases of disease and mortality and to change and manipulate them to improve our health. Each of these approaches, however, can be pushed too far. The behavioral approach can lead to a kind of health fascism, coercing everyone to live healthy lives whether they want to or not, running roughshod over our individual values and desires. No one seems too much worried about the growing trend to limit the places where people can smoke, forcing them out of doors in the end, but the same practice directed toward our eating and exercise habits might be a different matter. Will we want the law standing next to our refrigerator and monitoring the time we spend on the rowing machine?

The genetic approach has more disturbing possibilities. It will, in fact, be hard to carry the coercion of our life styles too far, simply because of the difficulty of policing them. With genetic knowledge, we will have at hand a more potent tool. For its promise will be, by means of the short cuts of science, to get to the bottom of disease, and to extirpate it at its source. It will be hard to resist the attraction of genetic screening, gene therapy *in utero*, and genetic profiles that will lay out for us – and everyone else potentially – what our genetic fate and health history may be over a life time. This will be all the more the case if we treat sickness and death as a great, totally unacceptable part of our biological life. We have for decades in this country carried out "wars" against disease, and death has been made the enemy by scientific medicine. To the extent that we continue the trend of making illness and mortality increasingly unacceptable to an ambitious science, to that extent we will be setting the stage for a fierce, unforgiving, and uncompromising use of genetic knowledge to perfect the human condition. We will then have no defense against a genetic juggernaut. It is difficult enough as it is to accept the handicapped, those who are not in perfect health. It will not be easy to keep alive that acceptance when the means will be at hand to make certain they do not exist.

## The drive for social control

There can be little doubt that modern societies are marked by a wide and disturbing range of social pathologies, most notably violence and the breakdown of human relationships. This is most acutely the case in the USA. There can also be little doubt that those countries most afflicted with such pathologies are frustrated, and often angry, about their inability to bring the social situation under control. Nothing seems to work well, and fear often stalks the streets.

No wonder, then, that genetics offers an attractive lure to deal with such problems. We have at least some reason to suspect that a predisposition to violence may have a genetic basis, as does a predisposition to alcohol and drug abuse. We also know that many children are born into the world burdened by handicaps of one kind or another that add still a further social burden. Why not, then, look to genetics to see whether we can find a scientific way out? Why not employ it to circumvent the usually cumbersome and ineffective means we now use to control social misbehavior and deviancy? Why not use genetics to see whether we can, over the coming years, develop a more kindly, sociable human being, one less marked by the destructive traits so evident in so many of our fellow citizens?

These questions are so obvious, and the attraction of answering them in the affirmative so beguiling, that it is difficult to resist being drawn into them, prepared to go the next step. Yet we know that much social deviancy is a result of social conditions, not genetic determinism. There are now, and have always been, some peaceful societies. There are now, and have always been, some societies that are not drug dependent. Even if there are genetic predispositions to socially harmful behavior, that does not itself mandate the value of genetic manipulation to rid ourselves of that behavior. Nor is it likely we will know enough in any even distant future to know how best to use genetic knowledge to manage large-scale social problems. But the real hazard is that, growing desperate, we will be tempted to "medicalize" social deviancy, and then use the big guns of genetics to cure them. Once again, the danger lies less in the genetics than in the kind of social desperation that would lead us to look to genetics for a decisive way out. Our difficulties in controlling guns, the response of a fearful society, ought to be warning enough: the garrison-state mentality is full of dangers, and there is no reason to believe that genetics offers a good way of managing them.

## Private ends and common good

The most popular argument for taking advantage of genetic knowledge and power will be that we ought as individuals to be free to do so. We look with great wariness in our society upon arguments stressing the common good. We seem to believe, in fact, that there is no common good, only an aggregation of our private goods. And – short of patent harm to others – we seem increasingly to believe that we neither can, nor should, pass judgment on the private goals and lives of others.

That is a situation ripe for genetic exploitation. We are left with little that can be used to stand in the way of the most radical use of genetic knowledge to serve private ends: to have the kind of health we want, the children we want, and the genetically fit adults we want. It will always in the nature of the case be difficult to show long-term social harm from genetic manipulation; thus the advantage will always lie with those individuals who want to move ahead, to pursue their own ends, and nowhere more forcefully than in the name of health.

Yet here is a perfect arena to ask about the cumulative, aggregate impact of private choice. When all our free choices are put together, what kind of a society will we get? Why should we think that the sum total of our self-interested genetic choices will result in a decent society? There is, actually, no challenge more important than trying to ask, as a community task, what would serve our common and collective welfare in the future, to make use of genetics to improve our general well-being, not just to fulfill our personal wishes. But to be able to do this requires that we be willing to talk the language of the common good – recognizing that, if the genetic predictions turn out to be true, we will be creating a different kind of society. The question is not just what we will do with the new choices that genetics forces on us; more importantly, we must first ask what kinds of choice we should have in the first place. For by the time we are ready to exercise a genetic choice, our hand will already have been forced by having to deal with such a choice. What kinds of choices will be good for us as a whole, not just one by one?

## Conclusion

I have focused my comments on what a hazardous society might look like for the development and the context-setting of genetic developments. We need not have that kind of society. We can have a society that does not become obsessed with disease, illness, aging, and death – a society that understands that it is now, and will remain, part of the human condition. Not by any means a good part, but a part, to be lived with and accepted, doing something about it when possible, but taking care not to let itself do moral harm in the name of reducing the harms of the body.

We can have a society that does not look to medicine to clean up a society marked by dangerous behaviors whose causes are far more likely to be environmental and social than genetic. We can have a society that wants always to know what will give us a better society and not just an expanded range of individual choice. We can have a society, in short, that takes genetic knowledge and possibilities in its stride, hopeful about some of the more benign possibilities, suspicious about the potential harms that could be done in its name, and skeptical that it will transform the human condition. We know that genetics can do good for us, and can do harm to us. But we should also know that it is not genetics as such that is the problem. It is the values and beliefs and practices we bring to bear on genetic knowledge that are crucial. It is the kind of cultural environment we create that will, in the end, determine whether the societal expression of genetic knowledge will harm or hinder our common life.

### Suggestions for further reading

Duster, T. *Backdoor to Eugenics*. New York: Routledge, Chapman, and Hall, 1990.

Lippman, A. Prenatal genetic testing and screening: Constructing needs and reinforcing inequities. *American Journal of Law and Medicine* (1991), **17** (1&2).

Ramsey, P. *Fabricated Man: The Ethics of Genetic Control*. New Haven, CN: Yale University Press, 1970.

Wertz, D. C. Ethical and legal implications of the new genetics: Issues for discussion. *Social Science and Medicine* (1992), **35** (4).

Wertz, D. C. and Fletcher, J. (eds.) Proposed: An international code of ethics for medical genetics. *Clinical Genetics* (1993), **44**, 37–43.

Wertz, D. C. and Fletcher, J. (eds.) *Ethics and Human Genetics: A Cross-Cultural Perspective*. New York: Springer-Verlag, 1989.

Wilkie, T. *Perilous Knowledge: The Human Genome Project and Its Implications*. San Francisco University of California Press, 1993.

# Genetic knowledge: Some legal and ethical questions

ROBERT SCHWARTZ

EDITORS' SUMMARY

Robert Schwartz, JD, details the almost overwhelming problems that occur when new genetic knowledge is gained. First, he warns of the dangers of social eugenic policies, particularly since there is no perfect genome map, no standard against which one can judge what is health and what is disease (those standards are socially determined, not just medically). He then discusses the legal and moral problems that arise when clinical judgments would be made on the basis of such an "imperfect" science. The lay public as well as public policy figures would have to become used to dealing with statistical data rather than an exact science. Expectations and agreements on language would require intense effort on everyone's part. Most telling are the questions he poses about parental judgment for fetuses and children who might suffer from a genetic disease that does not express itself immediately, but rather later in life. Should such individuals be discriminated against, either by not selecting them for implantation after fertilization of the egg, or through abortion after the embryo is implanted? How does that "discrimination" differ from that of social eugenics itself?

Over the past century, the history of the use of genetic knowledge has given us reason to worry. A century ago, long before Proposition 187 was a twinkle in the California electorate's eye a surge of interest in human genetics coincided with a fear of new waves of immigrants and an interest in developing a new American immigration policy. Among the various arguments raised against the Irish, Italian, Polish, Jewish, Chinese and other immigrant groups was the argument that the addition of such base and uncultured ethnicities to our body politic would pollute the American gene pool. The nineteenth-century anti-immigrant movement gladly adopted whatever it could from the growing scientific eugenics movement that had been established throughout Europe and the USA. While the anti-immigrant boosters of American gene purity might have admitted that what they were doing was not very charitable, they were driven by what they saw as the inexorable march of science. They knew that the genetically stronger races would ultimately prevail and that it was useless and counterproductive to attempt to stem the tide predicted by the genetic scientists. Indeed, they viewed those opposed to their science as know-nothings. Only fools could believe that any kind of national policy could hold back scientific progress.

During the early part of this century the interest in putting science to the use of humanity continued to manifest itself in support for eugenic policies. These policies were never so dangerous as when they were also incorporated into law. Probably the greatest legal acknowledgment of genetics came in the 1930s when Justice Holmes, in an opinion that has been ever since to his shame, upheld a decision that Carrie Buck, a "feeble-minded" woman in a state institution, could be forcibly sterilized by the state of Virginia entirely for eugenic reasons. Ms Buck had her mother and a daughter in the same institution, and, in his version of the "three strikes and you're out" principle, Justice Holmes declared that "[t]hree generations of imbeciles are enough." Several decades later Ms Buck was discovered wandering the hills of Appalachia, still expressing concern that she was unable to have any more children.

Of course, the scientific eugenics frenzy reached its nadir in Nazi Germany, but the fact that the horrors of that regime have been decried by every civilized society since does not mean that the rest of the world has rejected all of the underlying eugenics principles. Justice Holmes's infamous declaration has never been overruled, and even after the Second World War the United States Supreme Court was called upon to deal with eugenics-based laws. In the 1940s in *Skinner* v. *Oklahoma* the Court addressed a state statute that provided for the eugenic sterilization of some third-offense felons. While the Court found that the statute was a violation of the United States Constitution, its reasoning was not based on the uncertainty of the scientific information or the significance of the decision to reproduce; rather, the opinion was grounded on the finding that the statute violated the equal protection clause of the Fourteenth Amendment because it provided for the sterilization of third-time offenders who had committed certain listed crimes (generally, "blue collar" crimes) and excepted those who had committed other crimes (generally, fraud-based "white collar" crimes) without any rational basis for that distinction.

While the interest in eugenics has never disappeared in Western society, it has waxed and waned and its form has changed over the course of the last century. It has reemerged this decade in the form of criticism of social programs of the past several decades that have been used, in the perspective of some eugenicists, to prop up the genetically unworthy. In America, some view the social policy of the past several decades as being as counter-scientific, as Lysenko's agricultural science proved to be. Independent of the views taken in Hernstein's *The Bell Curve* and other reviews of the genetic basis of America's increasingly class-based society, interest in human genetics has arisen also out of the Human Genome Project, the 15 year international cooperative project to map the entire human genome.

The Human Genome Project, which remains on schedule as it approaches the five-year mark, is the largest, most ambitious, and most expensive biology project in history. Those who have become particularly interested in this project have done so for a number of different reasons. Scientists, philosophers, entrepreneurs, lawyers, and others have been motivated primarily by two factors, however, in bringing their interest to this quickly developing subject: (a) the intellectual excitement of being at the cutting edge of science, where science will so clearly and certainly affect social policy; and (b) the enormous amount of money that has been flowing into this research, and that will be flowing to pharmaceutical companies and others as a consequence of the expected scientific discoveries. The Human Genome Project is the next-generation moon quest – but it is one that is taking place in an environment that encourages private entrepreneurial profit. Thus, uses of the genetic information that arise out of the Human Genome Project will face all of the ethical and legal concerns that did (or should have) informed decisions concerning the use of genetic information in earlier generations, and it will face one more issue as well – how to tame the incentives created by the possibility of utterly gigantic profits. In order to allow the ethical and legal issues to be carefully drawn and analyzed, the United States Congress has allocated 3% of the roughly $15 billion that it will spend on this project for research on the ethical, legal and social issues that are likely to arise from it; European governments have allocated even greater sums. While it remains to be seen whether the ethics and law will, in fact, keep up with the scientific developments and their likely social impacts, this move by Congress has drawn thousands of ethicists and lawyers to this general area.

## Three preliminary principles

Before looking at a catalog of the risks and promises for our social and legal institutions held by the Human Genome Project, it is worth recognizing that proper use of genetic information is premised on three assumptions: (a) there is much in our lives that is determined by environment, not genetics; (b) there is no "normal" human genome; and (c) the public must be able to understand the language of genetics – and, thus, the language of statistics – to be able to participate in real community discussion on the proper availability and use of genetic information.

First, there is much we do not know about the relationship between a person's genome and the expression of that person's genetic make-up. Most expressed genetic attributes involve at least some environmental factors, and it would be misleading to base any policy on genetics alone when there are other environmental factors – perhaps unknown to us – that may have a much greater effect on an attribute than the genetic factors. This society reveres science, and it runs the risk of attributing much more to a scientifically identified cause of an attribute than to the yet-unidentified causes that may be far more important. To what extent is alcoholism genetic, and to what extent is it environmental? As our scientific research has moved from environmental research to genetic research, our society seems to be willing to readjust its view of the "real" cause of alcoholism, which is likely to be a combination of the two. How much of an alcoholic's behavior is willed (and thus culpable) behavior? Our uncertainty here is reflected in our disagreement over how to deter alcoholics: should we send them to jail because they are morally blameworthy, or send them to the hospital because they are sick? An attempt to make these kinds of policy decision on the basis of incomplete or uncertain scientific data could result in "medicalizing" everything that has a genetic component, and that could be an ethical and legal disaster. To address questions about the use of newly developed genetic information, we must be able to talk about how much of an action is genetically determined, and how much is a matter of free will. We do not yet have enough knowledge to do this, in most cases; indeed, we have not yet even developed a vocabulary for this kind of discussion.

Second, there is no "normal" genome that is naturally expressed in a "normal" person. There is no currently available definition of the standard, no-frills, fully operational human being. If that is true, though, when does a deviation from the common gene constitute a genetic disease, and when, on the contrary, is it merely an interesting variation? When is a genetic composition "substandard"? When is a person too short? Too cancer-prone? Too likely to be sexually abusive? As George Annas said in 1990, "[W]e know that most 'diseases' and 'abnormalities' are social constructs, not facts of nature. Myopia, for example, is considered acceptable, whereas obesity is not. We won't discover a 'normal' or 'standard' human genome, but we may invent one." If we do invent one – and that may be necessary to make medical use of the information we are acquiring – it is hard to know how we can avoid describing the paradigm of human genomes, the arguably benign version of the perfect genome, the master race.

For society at large to participate in a policy debate on the use of genetic information, the public must be able to understand the language that scientists speak. In this case, the public must understand the language of statistics. It will not be easy to educate the public to talk about percentages and probabilities, not certainties, when talking about science. Particularly where we are dealing with "big science," with real scientific breakthroughs, the public expects science to be exact. We will make it to the moon or we will not. A vaccine will prevent polio, or it will not. Whereas medical researchers are used to talking about probabilities, patients are not.

Further, the public inability to deal with probabilities when dealing with scientific "facts" is exacerbated by the commercial nature of genetic testing. For-profit genetic testing ventures can be expected to encourage patients to be absolutely certain about their genetic propensity toward a condition, even if the condition is not always very closely related to the genetic trait, and even if the patient can do nothing about the condition once it is known. The competition among genetic testing firms and the huge amount of money potentially available for genetic testing (through third-party health care payers) have led to the development of a quickly growing market for these tests. Those who profit from this market have little interest in teaching prospective patients about the uncertainties and subtleties inherent in any genetic evaluation; they are better off as long as they can deliver their clients' patients a "yes, you've got it" or "no, you don't" answer, with an explanation provided by some lesser (or, at least, cheaper) professional.

Finally, the traditional medical definitions of causation are different from the definitions used in making other forms of public policy. Consider the question, "Did exposure to substance X cause this cancer?" In law (and in the minds of most people), the answer is "yes" if there is a greater than 50% chance that it was a substantial contributing cause. Since an error in either direction is just as likely to be harmful, from the point of view of law and public policy a "more likely than not" standard is all that is required. Even a 10% chance – or a 1% chance – that substance X contributes to the development of some form of cancer is likely to have a tremendous effect on the way people deal with it. Nevertheless, medicine deals with a different level of certainty. When medical researchers say that there is a causal connection between substance X and cancer, they are saying that there is a very high probability of such a causal connection – perhaps one with a probability of more than 0.95. If it is likely that there is a connection, but the connection is less than 95% certain, the medical researchers will say that there is no proven relationship between the two. Thus, doctors and public policy makers may look at the same causal data, and while the public policy makers announce that there is a causal relationship, the doctors will announce that there is none. Unless we all start using the same language, and unless we are willing to take on the tough task of talking in terms of probabilities, it will be extremely difficult to make meaningful public policy to deal with developments coming out of genetic research.

## Issues arising from the development of knowledge about the human genome

### *Clinical medicine*

Genetic testing will enable us to determine whether people (or, probably more common, fetuses) are likely to develop any of a wide range of genetic illnesses. Is there any limit that ought to be placed on the treatment of genetic conditions? Are there adverse consequences that we cannot foresee in attempting to change attributes that were provided through the genetic lottery? There is little objection to treating conditions that manifest themselves at birth and are then treatable through the use of conventional, nongenetic medicine. Society has been treating the disease phenylketonuria (PKU) for years, and the process gives rise to no ethical dilemma. We may be a bit more cautious in approving genetic screening for conditions that can be treated only through gene therapy, in which the "bad" gene is literally replaced by a better (normal?) gene. Medicine is now able to engage in gene therapy to treat severe combined immunodeficiency disease (SCID), the "bubble baby" disease that renders some children virtually without an immune system, by removing the offending gene and replacing it with the normal gene under conditions that will allow the cells with the normal gene to replicate themselves enough times to overcome the effects of the cells with the defective gene. While such treatment seems humane and entirely unobjectionable, it opens the door to treatment protocols that will allow for the replacement of other "bad" genes by "good" genes. When should the gene for height or intelligence (or eye color) be considered a sufficiently "bad" gene that gene therapy would be permitted? Further, should the use of somatic cell gene therapy, in which the genes of the functional human cells are changed, be treated differently from the use of germline therapy, in which the genes in the sperm and the eggs are changed as well? In the first case, the genetic changes affect only the patient, not the patient's offspring. In the case of germline therapy, however, the genetic composition of future generations will also be changed and the gene pool of our species will be forever (even if, in each individual case, only slightly) changed.

The problems with screening *in utero* become even more difficult when the potential genetic disease does not manifest itself until many years after birth, or when there is no treatment – genetic or conventional – currently available for the condition being screened. Should we be screening for Huntington's disease *in utero*? Is it fair for the parents to make a decision about aborting a fetus because that fetus may have a crippling and ultimately fatal disease 45 years after its birth? Do we want the child (or the parents, for that matter) to possess this potentially explosive information? How does it affect a person to know that he or she *will* suffer from this cruel disease sometime later in life? Of course, potential Huntington's disease patients know that there is a 50% risk of suffering the disease as soon as they find out that one of their parents is suffering from the disease – an event that may not take place until the child is in his or her 20s or 30s. Still, a large percentage of those who are at risk for Huntington's disease (i.e. one of their parents has it) choose not to be tested, or not to know the results. Should we applaud their courage, or condemn their failure to get the information about their health that is necessary to make appropriate life decisions? Should we require that people be tested even if they do not want to know the results? Do they have a moral obligation to know their own genetic status?

Screening decisions need not be individual decisions, either. We could engage in screening of the entire population for specific genetic traits, so every person will know of his or her risk for those genetic conditions. Where the incidence of some gene in the whole population is low, however, population screening will yield a very large number of false positives, and the financial and emotional costs of those false positives may not be worth the benefit that comes out of identifying the comparatively small number of true positives. So population screening may not be justified, even when the condition that is the subject of the screen is treatable in some way.

Even when the condition that is the subject of the population screening is not treatable, the information may be helpful to the patient in making life decisions, such as those related to reproductive plans. But if we are testing people to give them information that might be useful when they decide whether or not to become parents, who should we test, and when should we test them? Should we provide screening only to those who might have children? We could provide this screening as part of the marriage license process, although we would miss testing of a large percentage of parents. In any case, why would we be testing those who are about to reproduce? Just to provide them with information, or to do more? Would we then recommend or require fetal screening for their child? Would we recommend or require abortion if the testing revealed that the child were affected? Would we forbid the marriage of a genetically unsound couple? Would we permit the marriage only if one of the partners were sterilized? Would we offer treatment? If so, what kind of treatment? In any case, to whom would this screened information be available – to the screened person only, to potential partners, to medical practitioners, to potential health insurers and employers, and to the government as well?

There are a great many potential liability issues that arise out of the availability of genetic testing and treatment for the tested conditions. First, physicians who know of the existence of genetic screening and do not offer it to the patient may be legally liable, just as they would be liable for failing to provide a nongenetic diagnostic tool. There is both tremendous public interest and a huge amount of venture capital in the area of genetic screening; every possible genetic screen will be made available to the public, through the commercial medical community. Further, we know that if a genetic screen is made available, the public will patronize it. The tremendous expense of some kinds of genetic tests, the deceptively certain results, and the absence of treatment for some of the conditions for which screening is available may lead physicians to withhold information about the tests from their patients. Whether the law will recognize such actions as justifiable remains to be seen.

Second, physicians who know of a genetic risk (or the absence of a genetic risk) to a patient but do not tell the patient may be liable for failing to provide that information. However, where a physician should know that providing genetic information will cause emotional distress, he or she may be liable for providing that information. The obligation to provide genetic information, but only when appropriate, may put the physician in a difficult position. For example, information about the *BRCA*-1 gene, which is highly predictive of some kinds of breast cancer, may lead a patient to have more regular mammograms and get better prophylactic care, but some evidence also suggests that it may cause some patients to avoid mammograms and other forms of care because they are fearful of getting the bad news that they are suffering from the disease. A physician will have to know the patient well to know whether the patient ought to be screened, and how the patient ought to be told of the result of the screening test.

Finally, physicians and other health care workers could be liable for revealing confidential genetic information – or for not revealing it, when that is appropriate. A genetic screen may indicate that others – the patient's siblings, for example – are at high risk of some genetic condition. Should they be informed of their risk, even though this means breaching the confidentiality of the patient? If physicians tell, they may be liable for a breach of confidence; if they do not, they may be liable for a breach of obligation to third parties.

## Employment

Genetic information can be used by employers and employees in several different ways. Employers may seek to do a genetic screen of their employees in order to ensure that those who are hypersusceptible to some kinds of risk do not suffer any exposure. This protects the health of the employee while at the same time maintaining a productive work force and lowering the cost to employers, who will not be forced to pay for employee health and disability benefits. But this kind of screening may provide an excuse for an employer who does not want to provide a safe workplace for all employees. Further, the cost control benefit of employee genetic screening will be realized primarily through health care insurance premium savings, and the employer may be motivated to hire workers who have had genetic screens that have shown that they are unlikely to be subject to any illness (for which the employer might have to pay, anyway), and that their children, whether born or not, are unlikely to suffer from any genetic illness. After all, the employer and his health insurer probably provide health coverage to the employee's family, too. Ultimately, preemployment genetic screening will effectively eliminate some people from the job market altogether.

## Education

Some skills are partly – and perhaps largely – genetically based. Mathematical ability, for example, is presumed to have a substantial genetic component. Should genetic information be used to track students in schools? To determine who goes to academic high schools (or middle schools, or elementary schools) and who goes to vocational schools? How is the use of genetic information for these purposes any different from the use of IQ and other tests that now determine the special education status of children, except that the genetic tests may be more accurate? Of course, they may only appear more accurate because they appear more certain and more scientific, because the results do not change from test to test. The apparent certainty of one's genetic status may hide the subtleties and complexities (and arbitrariness) of our attempts at multigene analysis. What is the proper genetic make-up of a doctor? Do we want more scientific brilliance or compassion, assuming either of these is, at least in part, a genetic attribute? This is hardly a scientific question with an obvious answer, it is a matter of social policy. Might we track the wrong people into medical school? If we do start tracking on the basis of genetic predictions, it will certainly increase the relationship between genetic make-up and perceived intelligence, income and social class. Whether or not genetics have had any role in determining social class in the past, they will in the future – by definition.

### *Criminal law*

Genetic information could be extraordinarily valuable in the criminal justice system. Of course, genetic testing is already used for purposes of identification, but this DNA fingerprinting raises few ethical issues. Some criminal defendants may have a genetic propensity to commit certain kinds of crime, and the use of evidence of such a propensity creates substantial ethical problems. Arguably, those propensities should not be irrelevant in determining the defendant's guilt or in determining the appropriate sentence once guilt has been established. For example, there is now some support for the proposition that violence is, at least in part, a genetic attribute, and that sexual violence is, in part, the result of a variety of different genes. But how should these factors be relevant in an individual prosecution? Is the fact that one has a propensity to commit violence admissible, as evidence submitted by the prosecution, to show that it is *more* likely that a particular defendant actually did commit a crime? Or is it admissible, as evidence submitted by the defendant, to show that the defendant was forced to commit the act; that he or she did not do so voluntarily or with the *mens rea*, the mental state, required by the criminal law? Further, if the evidence is submitted by the prosecution, does it violate the Fifth Amendment to the United States Constitution by forcing the defendant to be a witness against himself or herself?

Even if genetic evidence were not admissible at the guilt stage, it may make sense to introduce it at the sentencing phase. But should the fact that one is genetically disposed to commit a crime result in a longer sentence (because the convict cannot be "rehabilitated" from his or her genetic condition, and thus may pose a greater continuing threat to those on the street), or should it result in a shorter sentence (because the genetically driven convict is less culpable for the act and thus is deserving of less punishment)?

There are further uses of genetic information within the criminal justice system. A genetic propensity might be used as one basis for denying bail, especially under the federal process that allows the judge setting bail to consider the likely conduct of the defendant during the pendency of the judicial proceeding. As a matter of general course, we do not allow preventive detention of those who might someday commit a crime because, in part, our predictions of dangerousness are so inaccurate. If genetic tools make such predictions easier, should this society be more willing to allow for pre-crime detention? Could we order a civil commitment for one who poses a risk to him/herself or others because of a genetic disposition to crime, as we do when a mental illness (which also has a substantial genetic component) renders someone a danger to him/herself or others? If we know someone is likely to commit a crime, why not arrest that person before he or she actually commits the deed? Further, within the corrections system prisoners could receive treatment, work assignments, and even living assignments based on their genetic profile. Finally, their parole time and conditions could be affected by their genetic profile.

### Insurance

Of course, both life and health insurance are affected by genetic information. What is more, it seems appropriate that genetic condition be taken into account when insurance policies are issued – at least as long as what is really at stake is true insurance. Insurance is a bet, and, to be fair, it requires that both sides have the same information. Thus, there is little reason to object to the insurance company practice of collecting genetic data on potential insureds, as long as that information is made available to those insureds as well. Conversely, to the extent that our insurance system has been drafted to serve other purposes, it may be inappropriate to allow it to discriminate on the basis of genetic condition. For example, if the American health insurance system is not really a system to insure against unpredictable and catastrophic risks, but, rather, a method for prepaying medical expenses, or a surrogate national health system, then treating those with potentially costly genetic conditions less favorably may be inconsistent with the cost-leveling function of the system. Further, because the United States health insurance system is partly employer financed, insurance company policies that provide for genetic rating will ultimately discourage employers from hiring employees who may develop a condition that is expensive to treat, or who have (or might have) a family member who may develop such conditions.

While there is thus a powerful argument that genetic (and other) conditions should not be considered relevant in the group health care insurance system, that argument may not extend to life insurance, disability insurance, and other forms of insurance. However, if life insurance is really a substitute for a state program that would provide aid to widows and orphans, and disability insurance is just a privately administered addition to a state unemployment scheme, then acknowledging genetic conditions may be inappropriate within these insurance systems also.

### Other issues

Genetic information may also be useful to courts in making child custody decisions. Child abuse may turn out to have a genetic component, and it is not totally inconceivable that we could develop a genetic profile of a "good parent." We may also wish to use genetic data in making immigration decisions. One function of immigration law is to protect the national gene pool; that is the reason that we do not admit some people with genetic diseases. How much further should we go in considering the effect that immigration will have on our gene pool? Should we admit immigrants of reproductive age who are likely to make us more competitive in the international economy? If we could, which traits should we favor, and which should we disfavor, in immigration policy?

Finally, should the government (or anyone else) be permitted to maintain a bank of genetic profiles, as they do now with fingerprints? This is a powerful form of identification, and, arguably, it would make it much easier to solve some crimes, to identify bodies, and to resolve paternity cases, to give just a few examples of its utility. But it would give the government, and whoever else would have access to it, an extraordinary amount of information about each one of us, and thus an extraordinary amount of power over us.

## Conclusion

This list of ethical, legal and social issues raised by the genetic revolution is hardly comprehensive. Newly developed information also will require society to address several difficult questions that have remained unresolved over the past decade – and the past 20 centuries. There will be a new debate on the relationship between determinism and free will, for example, as we discover that virtually no attribute is entirely independent of genetics, but that virtually no attribute is entirely genetic either. Once again we will have to address the question of the relationship between genetics and personal identity. What makes a person *the* person that he or she is? Is it genetic make-up, or something else? To the extent that individual identity is tied to genetics, does that mean than any genetic alteration yields a new person, at least in some sense?

Our society will also have to determine whether there is an ethical difference (and whether there ought to be a legal difference) between negative eugenics, where a genetic infirmity is corrected and positive eugenics, where a "normal" person is "improved" through genetic manipulation. Consider the difference between allowing a patient to die and euthanasia: this is a distinction that is much more appealing in theory, as an arbitrary limit on a potentially frightening technology, than in fact. If you live on Garrison Keillor's Lake Woebegone, where "all children are above average," the distinction between positive and negative eugenics is awfully blurry; it all depends on how the "normal" genome is defined.

We will also have to decide who is going to pay for any available genetic technology. The USA has just completed a national debate about the financing of health care, with the conclusion that, in general, except in an emergency, each person can get the health care he or she can pay for. Is there any problem with giving a rich person an opportunity to improve his or her family's genetic condition (and genetic legacy) while denying that opportunity to a poor person? There is nothing wrong with allowing richer people to buy their children nicer homes, more expensive clothes, and more teacher-intensive education than the poor can afford. Is there anything wrong with allowing the rich to buy their children more genetic intelligence or a more desirable eye color? Finally, we must worry about whether genetic make-up will become a surrogate for race in a way that will allow society subtly to perpetuate its racist history.

The Human Genome Project and newly developed genetic information offer us a potential for fabulous medical and social advances – a chance to overcome problems that have cursed our species from time immemorial. They also offer us an opportunity to perpetuate the worst that our species has developed over centuries, and to develop a true genetic caste system. We ought to design the legal and social institutions that will control the use of genetic information with the presumption that the intellectual excitement associated with genome research, and the money that will flow from it, will provide powerful incentives for both paths.

## Suggestions for further reading

Annas, G. and Elias, S. *Gene Mapping: Using Law and Ethics as Guides.* New York: Oxford University Press, 1992.

Bishop, J. and Waldholz, M. *Genome: The Story of the Most Astonishing Scientific Adventure of Our Time – The Attempt to Map All the Genes in the Human Body.* New York: Simon & Schuster, 1990.

Cranor, C. (ed.) *Are Genes Us? The Social Consequences of the New Genetics.* Brunswick, NS: Rutgers University Press, 1994.

Hernstein, R.J. *The Bell Curve.* New York: Free Press, 1994.

Kevles, D. and Hood, L. (eds.) *The Code of Codes: Scientific and Social Issues in the Human Genome Project.* Cambridge, MA: Harvard University Press, 1992.

President's Commission for the Study of Ethical Problems in Biomedical and Behavioral Research. *Splicing Life: A Report on the Social and Ethical Issues of Genetic Engineering with Human Beings.* Washington, DC: US Government Printing Office, 1982.

Weir, R. F., Lawrence, S. C. and Fales, E. (eds.) *Genes and Human Self-Knowledge: Historical and Philosophical Reflections on the Modern Genetics.* Iowa City: University of Iowa Press, 1994.

Wilkie, T. *Perilous Knowledge: The Human Genome Project and its Implications.* Berkeley, CA: University of California Press, 1993.

Wingerson, L. *Mapping Our Genes: The Genome Project and the Future of Medicine*. New York: Dutton 1990.

In addition, the National Technical Information Service in Springfield, VA, regularly publishes a bibliography on the ethical, legal and social implications of the Human Genome Project.

# The "ART" of medically assisted reproduction: An embryo is an embryo is an embryo

MICHAEL E. McCLURE

EDITORS' SUMMARY

Michael E. McClure, PhD provides a thorough overview of current, advanced, experimental, and future possibilities in medically assisted reproduction. At each stage, personal, medical, economic, and social risks are encountered, all of which contribute to the ethical dilemmas posed by such interventions. He cautions that we still do not know well enough all the effects of current standard practices, nor are their outcomes uniformly evaluated. This holds true to a greater degree for more experimental manipulations of sperm and stem cells themselves. Each step requires additional research because the technology is so new and expensive. Applications to germline therapy are also suggested.

Sexual reproduction involving the exchange and joining of cells containing the genes of two different individuals began on earth nearly 1.1 billion years ago. The theory is that it started with tiny, single-celled sea creatures who developed pores in their surface coverings. This allowed them to release sex cells into the surrounding warm ocean water. There they fused with other cells and were cultured as embryonic forms struggling to attain independent individuality. In the passage of time a latecomer to the scene, our human species, has sought through medical research to develop technology to assist infertile or sterile couples in their quest for parenthood. This is done out of profound respect for the right of each couple to choose to procreate.

In one respect, we have come full circle today from the origin of "sex in the ocean" by developing a procedure for "sex in a dish," which depends largely on human sex cells and embryos being cultured in a modified form of ocean water. The development of *in vitro* fertilization (IVF) and embryo transfer (ET) techniques between 1944 and 1977 and their successful clinical application in producing the world's first test-tube baby, Louise Brown, in 1978 totally revolutionized reproductive medicine within a decade. We are now at the point where basic assisted reproduction technology (ART) applications are now considered standard infertility treatments.

At present, the medically assisted pathways to parenthood have become many, highly variable, and anything but simple. The complexity is mind-boggling, as I will show. In the words of one colleague, the ART of medically assisted reproduction "opens a Pandora's box of ethical, moral and legal questions . . . [regarding] . . . what we *should* do, as opposed to what we *can* do." The purpose of this chapter is to discuss not only what we can do today but what we will be able to do in the very foreseeable future. The future view adds a whole new dimension to the ART of human reproduction and its ethical challenges.

## What is "ART"?

All assisted reproductive technologies share one attribute in common. They are attempts to bypass a blockade of one or more critically required functions (vulnerability points) in the biological continuum of events from sex cell rendezvous and fusion to early embryo development, transport and implantation in the uterus. Such blockages can be physical, biological or genetic in nature. The respective nature of the block, the route by which the bypass therapy is provided, and the number of individuals involved (two party or three party) in the clinical therapy design raise, in various combinations, very different sets of moral, ethical, cultural, and legal questions about selected therapy options. All ART procedures ultimately involve some form of clinically manipulating sex cells and/or embryos. For orientation to the differences in ART procedures, one must first consider the identifying term for the size, and nature, of the cellular forms being manipulated.

All of the biological materials (sex cells and embryos) being handled in ART procedures are very, very tiny – no bigger than the period at the end of this sentence. Among today's options, there is no clinical ART application that manipulates and transfers biological forms beyond single cells and early preimplantation embryos consisting usually of clusters of 2 to 8, and rarely of 16, cells. Public misperceptions about the size of the cluster contributes to controversy about manipulating embryos. There is also considerable controversy, amongst both academic professionals and the public, regarding the use of the term "pre-embryo" to attempt to differentiate the first 14 days of development as a phase preparatory to becoming an embryo. This is because the early cells in such embryos are capable of developing into a complete human being if separated off by "cloning," or more precisely speaking, "twinning." In my opinion, to paraphrase the psychologist Gertrude Stein, an embryo, is an embryo, is an embryo.

## The basic "ART" of medically assisted reproduction

One category of ART applications, low to high technology manipulations, could be called standard "treatment." This set of procedures includes fertility drugs for the woman. Such drugs cause ovarian stimulation (OS), controlled ovarian hyperstimulation (COH) or ovulation induction (OI) as a means of medically increasing the number of eggs produced, or instead, regulating the timing of their production. Although such initial efforts at assisted reproduction may rely on spontaneous intercourse as the source of the male sperm, commonly recommended options to improve a possible male infertility factor include intravaginal insemination (IVI), first reported as successful in 1790 by John Hunter for a man who used a syringe to vaginally deliver his semen, or more sophisticated variations of the procedure in which processed or unprocessed sperm are delivered into the female reproductive tract by intracervical insemination (ICI) or intrauterine insemination (IUI). For couples with unexplained male factor infertility, there is considerable controversy about which of the latter two sperm-delivery procedures is the most effective and whether the outcome of either of them is improved by the use of ovarian stimulation treatment of the normal woman. A large-scale national clinical trial is currently underway to resolve this controversy.

Although all of the above ART applications utilize the vaginal route of sperm delivery, pregnancies have been reported for procedures accomplishing sperm delivery by syringe needle injection of a processed sperm preparation through the vaginal wall and into the appropriate (intraperitoneal) region of the reproductive tract. These latter procedures, termed intraperitoneal insemination (IPI) or direct intraperitoneal insemination (DIPI) are largely considered obsolete. They presently remain uncommon. There is insufficient experience to permit any evaluation of efficacy. In all of the insemination procedures noted, the rationale has been to promote the delivery of functional sperm to the site of fertilization in the female reproductive tract. These procedures all have the disadvantage that, in the absence of a pregnancy, there is no proof that fertilization even occurred. Hence the experience does not provide clues for deciding whether to repeat the procedure or to select other ART options.

Many of the ART options available today have evolved from a combination of low technology hormonal treatments with fertility drugs to stimulate the ovaries of women and the basic method of *in vitro* fertilization combined with embryo transfer (IVF-ET). It is instructive to consider briefly the steps that comprise basic IVF-ET, since variations of these steps, such as adjustments, deletions, or additions to them, together with a new procedure name, comprise the other commonly recognized ART applications.

· First, an oocyte retrieval attempt is accomplished from spontaneous or hormonally treated menstrual cycles under timed conditions. This process requires intense scheduling attention by patients and clinicians.

· In cases where a poor response did not "cancel the cycle," oocytes are surgically retrieved, evaluated for quality and processed for fertilization attempts.

· The fertilization attempt involves inseminating the processed oocytes (eggs) in a culture and documenting fertilization success.

· The embryos, almost always multiple embryos, resulting from the successful fertilization efforts are cultured, usually to the 4- to 8-cell stage, and either transferred immediately, usually in groups of two to four, or frozen using a cyropreservation technology to permit future transfer attempts.

· The transfer in either case involves careful timing with regard to either the spontaneous or the hormone-treated status of the recipient's uterus to ensure the optimal receptivity necessary for the successful implantation of the transferred embryo.

· Some programs, in an attempt to improve the embryo's opportunity for implantation, may include an additional step termed assisted hatching (AH), which consists of using micromanipulation techniques to physically or chemically weaken the shell around the unhatched embryo and ease the embryo's effort to "hatch" from it.

· Hormonal support, either spontaneous or medically supplemented, for the continuation of the pregnancy leads in successful cases to appropriate conditions of gestation. This produces what is called a live-born, take-home baby. While standards for IVF-ET used to be aimed at developing a single protocol for best performance, an increasingly diverse patient population, medical techniques, moral and ethical problems, and cultural and economic criteria have, in more recent years, led to IVF-ET procedures being individualized for the particular couple being treated. Some other procedures include GIFT and frozen embyro transfer described below. These are still regarded among the standard therapies I am discussing.

· While the ovarian stimulation and oocyte retrieval procedures of gamete intra-fallopian transfer (GIFT) are similar to those for IVF-ET, there are no "*in vitro*" manipulations outside the body other than egg and sperm processing for their simultaneous delivery to the natural site of fertilization in the woman's reproductive tract. In addition to having multiple eggs that offer an enhanced probability for obtaining one or more take-home babies, this procedure provides the couple with an option for fertilization and embryo development occurring in the natural environment of the woman's reproductive tract, a circumstance that may relieve certain moral, ethical and/or cultural concerns held by some couples. A disadvantage of GIFT is that this procedure, unlike that for IVF-ET, involves general anesthesia and abdominal surgery. Although nonsurgical procedures are being investigated for the transvaginal delivery of the gametes into the fallopian tube, this approach remains experimental. It has, however, been reported that a successful pregnancy resulted from a transvaginal, intraperitoneal injection approach, i.e. peritoneal ovum sperm transfer (POST). The experience with this approach remains too limited to evaluate properly.

· A common variation of the basic IVF-ET approach includes the additional option of subsequent frozen embryo transfer (FET) cycles, which permit the opportunity for a couple to space their medically assisted pregnancies. For example, the initial oocyte retrieval cycle could provide nine embryos, of which three are immediately transferred and the other six are frozen and stored for transfer years into the future. A couple could theoretically have two children from the first three embryos transferred, two years later have a third child from a second transferred group of three embryos and two years later still have a fourth child from the last group of three embryos. One year after the birth of the last child, the family would have one-year-, three-year- and five-year-old children all derived from the same group of embryos concurrently conceived from simultaneously retrieved and fertilized oocytes. It is interesting to consider the relationship of the children in such a case. If the first two are twins, what are the last two?

· Other ART options derived from basic IVF-ET involve a combination of the IVF-ET and GIFT procedures described earlier. The primary difference between some of these options is the stage of development of the embryo being transferred, e.g. the 1-cell stage (pronuclear stage transfer or PROST, and zygote intrafallopian transfer or ZIFT) or 2- to 4-cell stage embryos (tubal embryo transfer or TET). These tubal procedures do have at present the added disadvantage of general anesthesia and surgery.

## The advanced "ART" of medical assisted reproduction

Well beyond the more general ART applications discussed above are the series of more recently developed advanced ART options that represent the most demanding and complex of the ART clinical applications. Animal research has recently provided insights into truly remarkable future applications involving advanced ART techniques that were not even considered realistic as recently as a year ago. All of these advanced ART options involve the use of specially engineered machines that permit microscopic observation by the maneuvering of tiny microneedles, often no thicker than a hair width, and the hollow, straw-like, glass micropipettes used to perform extremely delicate procedures on sex cells and/or embryos. These procedures differ only in the purpose for which micromanipulation is being done and how the microtools are used to accomplish this purpose.

These newer procedures are often used to remedy male infertility problems. One of the most difficult of the infertility circumstances to treat has been severe male factor infertility. In an effort to provide clinical ART applications to relieve instances of this problem, a number of variations of micromanipulation techniques have been devised to provide fertilization assistance. The technique of zona drilling (ZD) or partial zona dissection (PZD) involves micromanipulation procedures to thin or penetrate the shell of the material called zona pellucida coating the egg. Sometimes when women are infertile, this shell may be abnormally thick or prematurely hardened. When male infertility is the problem, and the woman's egg is normal, the shell may still be unable to be penetrated by the weakened or dysfunctional sperm produced by the male. Early ZD/PZD procedures used a micropipette-delivered stream of mildly acidic culture fluid to thin the shell and reduce its surface strength. While this treatment enhanced the odds of penetration by sperm placed next to an egg so treated, it also bypassed the natural biological mechanism that permits only a single sperm to fertilize an egg and introduced a chance of unwanted damage to the egg itself.

This technique raises serious concerns about genetic consequences called "polyploidy," which means that too many sperm have penetrated an egg. It requires careful monitoring of the fertilization process and the genetic status of the resultant embryo. While the detection of polyploidy at the *in vitro* embryo stage is distressing, it occurs before a pregnancy has occurred and the dilemma it poses for the couple is quite different from the far heavier one posed by detection of polyploidy in fetal cells after a pregnancy is established. In the former case, the couple would choose not to implant the embryo; in the latter instance, abortion would be contemplated.

A variation of assisted fertilization more advanced than ZD/PZD is a technique in which the the shell is penetrated with a sharpened micropipette and individual sperm are delivered through the hollow in the micropipette into the area between the inner shell and the outer egg membrane (perivitelline space). This application, termed subzonal insertion (SZI or SUZI), improves control of the process but raises the dilemma of how many sperm to deliver to maximize the chance of fertilization while minimizing the genetic risk of polyploidy. Studies have been reported indicating that when more than eight sperm were inserted, the polyploidy rate was 100%. This demonstrates the need for monitoring and for additional research.

The most advanced ART application developed to date for assisted fertilization by micromanipulation procedures is termed intra-cytoplasmic sperm injection (ICSI). It is currently considered a major breakthrough for treating infertile couples suffering from severe male factor infertility in which the underlying cause in the male is the production of extremely low numbers of sperm or sperm forms incapable of spontaneously fertilizing an egg. In this technique, a very demanding micromanipulation technology involves the injection of a single sperm into the egg, using a sharpened micropipette holding a single sperm in its center. To be technically successful, the microinjection process must delicately deliver the sperm directly into the cytoplasm of the egg without damaging the egg in a manner causing it to die. The procedure clearly reduces the danger of inducing genetic polyploidy found in earlier variations of this specific fertilization approach. It has the disadvantage of high costs and limited success rates.

The use of ICSI combined with either the microscopic epididymal sperm aspir-ation (MESA) procedure developed earlier or the very recently developed round spermatid nuclei injection (ROSNI) procedure offers a powerful ART plan for relieving the infertility of men with certain types of severe male factor infertility. The MESA procedure requires the surgical retrieval of sperm from the ducts (epididymides) adjacent to the testes. A processing procedure is then used to concentrate the sperm for use in either therapeutic insemination or IVF-ET appli-cations. In addition to the concern that the sperm obtained in this fashion is not of the highest performance quality, the low success rates and restrictions on repeated microsurgical episodes for the collection of sperm are disadvantages that have made this approach a rare ART option. The advent of the ICSI procedure could, however, make epididymal sperm aspiration a more viable option by dramatically reducing the number of surgical episodes. Theoretically, a single collection surgery with cryopreservation storage of small groups of the sperm collected could permit repeated future efforts to establish a pregnancy by ICSI combined with IVF-ET. The ROSNI procedure is perhaps among the more astonishing developments in the evolution of ART applications. In 1995, it was reported that the ROSNI procedure had successfully established human pregnancies involving males with conditions that ordinarily totally prevent the male from fathering children by any means. The approach used a very immature underdeveloped form of sperm rather than sperm cells themselves as the genetic source of the male chromosomes. In this work, immature round spermatid stage nuclei were isolated by a microsurgical procedure from the testes of men medically diagnosed as azoospermic (lacking sperm production). The male sex cell nuclei obtained were then used to "fertilize" an egg with an ICSI-like procedure in which the male round sperm cell nuclei were microinjected into the cytoplasm (ooplasm) of the egg and the egg was then stimulated (mechanically or chemically) to activate development. The resultant embryos were transferred similarly to an IVF-ET procedure. In the light of the extreme nature of the procedure, its low success rate in establishing pregnancies and its potential for inducing broader degrees of genetic risk, these reports have stirred considerable ethical controversy and debate.

Another potential ART application, not yet attempted in humans and very recently reported as feasible, is called stem cell transplantation (SCT). It will undoubtedly become a major source of controversy in the near future. This application involves the microsurgical isolation of primitive male germ cells (stem cells) capable of dividing and replenishing themselves. In the animal experiments conducted so far, stem cells obtained from genetically unrelated donor mice were microinjected into the tubes in the testes of recipient mice, where they colonized the tubes and generated functional sperm production. The recipient mice, previously sterile, were now able to produce offspring by natural matings, which they did. In humans, for example, this approach could be used to provide donor stem cells to replace the dysfunctional or genetic disease carrying stem cells present in a male with severe male factor infertility, or even to restore fertility in male cancer patients, whose cancer therapy normally causes sterility. They would store their own stem cells prior to the cancer therapy. Later, after therapy, these would be returned to the man's own testes when it was safe for their development.

## The quandaries of "ART"

The personal, ethical, medical, and social quandaries of ART include how commonly it is used, how successful it is and could be, how costly it is, and how risky it is. A comparison of basic IVF-ET outcomes is enlightening in several respects. First, while it is true, as has been said, that somewhere in the world a "test-tube" baby is born every day, in the USA IVF baby births are uncommon occurrences (about 0.2%) with respect to the overall annual birth rate (over 3 million). Second, the success rate for live births remains lower than one would like and has not improved much during the last five years. Recent data show that for every 6.5 treatment cycles, a live birth was obtained. In US surveys, depending on the ART procedure used, 16.9 to 27.9% of each group of eggs collected resulted in births. The age of the mother and the presence of a male infertility factor remain major negative factors. ART options remain an emotionally intense experience for the couple with little overall improvement in the live birth success rates compared to summaries in earlier years.

Informative answers regarding the success rates for the common high technology ART procedures are now available. Answers on success rates for the advanced ART applications, however, are not so easily found or interpreted when they are found. Understandably, these very advanced ART options are only a small proportion of the ART procedures performed each year. In addition, their early stage of development contributes to a lack of standards and measures for success. One of the quandaries encountered by an infertile couple weighing their options is how to determine what the various "success rates" they read about mean in regard to their goal of achieving a take-home baby. The rates encountered in some studies variously count the following as "success": oocyte recovery, fertilization itself, implantation, chemical pregnancy, clinical pregnancy, full-term gestations, deliveries (one or more babies) with or without ongoing pregnancies (expected deliveries), and ultimately, take-home baby rates. While this range has academic significance to the scientists and clinicians that use them, the only truly meaningful success rate to the infertile couple is the last one. Virtually none of the complicated advanced ART options, especially those involving sperm microinjections, have easily understood success rates allowing a simple evaluation regarding take-home babies.

Another quandary faced by the infertile couple considering ART therapy is the cost involved. The author receives, either directly or routed through congressional offices, numerous letters every year from couples desperately seeking funding to permit them to pursue IVF-ET or related ART procedures. Just how costly is ART? In a recent study of the costs of IVF in the USA it was estimated that the cost for a couple achieving a successful delivery ranged from $44 000 to $211 940. With each failed cycle, the chance of success declined and the costs increased. The average actual cost per delivery observed for a first cycle success in young women with no male factor infertility increased from $66 667 to $114 286 for couples undergoing a sixth cycle attempt. For older women of a couple with male factor infertility, the cost per delivery rises from $160 000 (first cycle) to $800 000 (sixth cycle). These cost figures dramatically illustrate why couples with no or limited health insurance and limited personal resources write in desperation. It is because they are being excluded from the health care services they want. Such an exclusion, based on economics, inequitably discriminates most heavily against low and modest income couples in our society. This issue remains unresolved for most nations, since many procedures are still considered experimental. How risky is ART therapy? The data accumulated for common ART procedures has shown that general risks of surgery and the occurrence of the ovarian hyperstimulation syndrome in the female member of the couple have been minimized with careful medical practice. In terms of the children born of such therapy, the birth defects that have been observed occur in frequencies and types not significantly different from that found in the general population. These comforting results are, however, all short term. Our experience with the associated long-term risks is far less informative. It has been only 17 years since the first IVF-ET baby was born and the population of such babies born, even world wide, remains too small (about 40 000) to conduct statistical studies about long-term effects.

There are, however, some outcome aspects that suggest caution should not be relaxed in regard to concerns for the potential risks of ART therapies. Perhaps the most striking of these is the risk of premature birth associated with ART procedures that promote multiple embryo implantations and gestations. It has been well established that multiple gestations are associated with increased rates of premature delivery, perinatal mortality, long-term morbidity, and higher health care costs. A study of multiple gestation pregnancies recently completed at one major United States hospital evaluated 13 206 pregnancies with respect to hospitalization costs and showed that 2% of singleton, 35% of twin, and 77% of higher-order multiple gestation pregnancies were associated with the use of assisted-reproduction techniques. Hospital charges for the delivery of triplets were 11 times higher than for a singleton delivery with escalating total charges to the family for a singleton ($9845), twin ($37 947) and triplet ($109 765) delivery.

More importantly, the dramatically higher costs noted are but a secondary issue. The primary issue is the more serious matter of the mortality and morbidity associated with higher-order, multiple gestation births. A study of more than 7.5 million births in the USA that was published in 1995 concluded that prematurity itself, i.e. being born too soon, is the principle factor linked to the high infant mortality in the USA. This concern was very recently heightened by a report on long-term outcomes for premature babies born weighing less than 750 grams, who survive to school age. The findings showed that up to 20% of these children have one or more severe functional disabilities and up to one-half have one or more less serious, but significant, functional disabilities. The investigators authoring the report concluded that newborns with birth weights less than 750 grams who survive are at high risk for neurobehavioral dysfunction and poor school performance. Such knowledge supports the need for improved ART success rates that allow establishing singleton pregnancies by the induction or transfer of one or as few embryos as possible.

For the female member of the infertile couple or for women considering whether to serve as a third-party egg donor, evidence linking fertility drug use to a 2.5-fold increase in the risk of ovarian cancer must be considered. Studies reported over the last two years from a national research program (the Collaborative Ovarian Cancer Group) sponsored by the National Institutes of Health (USA) have suggested that treatment with infertility drugs causing ovarian stimulation may, under certain circumstances, significantly increase the chance of the woman developing ovarian cancer in the future. From the ongoing controversy that arose from these reports has come the view that the greatest risk is associated with women who experience 12 or more ovulatory cycles stimulated by fertility drug treatments. Women with less exposure to the fertility drugs who have no family history of ovarian cancer are not considered to be at high risk. In consideration of the balance of the risks and benefits of exposing women to fertility enhancing drugs, it has been pointed out by several authors that the medical benefits under consideration do not apply to the increasing number of women taking fertility drugs for the sole purpose of serving as an egg donor. The uncertainty generated by the reports noted above clearly counsels conservative caution in designing ART treatment plans.

Little is known at present about the long-term potential risks for ART procedures introducing genetic defects that are manifested at some later time in the life span of the child. As noted earlier, certain techniques such as zona drilling, partial zona dissection and subzonal insertion are associated with significant risks for inducing a genetic state in which extra sets of chromosomes are present (polyploidy) in the fertilized egg and early embryo. In the case of ICSI, there is concern that the nature of the conditions in which intact or membrane-denuded sperm heads are manipulated and forced into the cytoplasm (ooplasm) of the egg may expose the male genetic material to unanticipated gene damage at the molecular level. For the ROSNI procedure, the genetic material in the nuclei being transferred is in an even less protected state and represents an immature stage in the development of the sperm in which ongoing modifications to the genetic material have not been completed. The high fertilization success rates and contrasting low implantation rates reported in some series may indicate why so many embryos die. Such concerns have underscored the active debate currently underway between those who ask the question "How can you prove it is safe in humans without doing it?" and those that counsel that rushing into the use of micromanipulator-assisted fertilization procedures is a "recipe for disaster" that represents "an ethical time-bomb." These latter concerns have been heightened by the realization that even the common IVF procedures with which we are more scientifically comfortable recently have been shown to possibly have subtle longer-term genetic effects in animal studies.

For example, in recent studies conducted with bovine embryos, fetuses from embryos produced outside the body (*in vitro* embryos) and embryos produced inside the body (*in vivo* embryos) showed no differences in high grade embryo survival upon transfer. Comparison of the fetuses derived from each type at seven months of gestation, however, showed that fetuses from embryos produced by *in vitro* manipulations were heavier and had skeletal measurements that were disproportionate to weight.

The effect of cryoinjury from embryo freezing (cryopreservation) is thought to be the reason that only 50–60% of frozen human preimplantation embryos survive the freezing and thawing process. While there has not been a sufficient number of human live births from frozen embryos so far to be able to study the question of whether there are delayed effects produced in children resulting from such technology, the circumstance is otherwise in animal research reported in 1995. The results showed that progeny derived from frozen mouse embryos had increased adult and senescent body weights unrelated to maternal weight or litter size, differences in preweaning development timing, and differences in a highly heritable mouse quantitative jaw bone trait (mandible shape). These results suggest that, beyond the immediate cryoinjury damage to embryos, embryo freezing, while not highly detrimental, may not be wholly neutral either. We simply may not know everything we should about such fertility interventions.

Concerns for the genetic impact of the ART options presently in use have also been expressed that address a different level of human genetics than birth defects. These are concerns at a population genetics level that involve the statistical status of the human gene pool. The use of donor sperm or donor eggs in ART procedures has raised the question of the risk of unintended consanguinity (inbreeding) arising from marriages of young people born as a result of ART procedures and residing in a defined geographical region. While the odds of such a union are incredibly small, to limit the odds significantly, ethically acceptable standards have been set by the world-wide profession that limit the number of live-born children that can be derived from one sperm donor source to ten. The source is "retired" from use after the set number (ten) is documented. A similar circumstance (ten resultant children) exists in the USA for the use of female donor eggs. An impact on the gene pool of a different nature is a concern for the sperm microinjection and microinsertion ART procedures (SZI, ICSI, ROSNI). These procedures bypass blocks to reproduction that may be based on genetic mutations that render the male infertile or functionally sterile. The use of the ART procedure thus introduces a potential for passing on to one's progeny one or more defective genes that would not otherwise have been born into the gene pool of the next generations. In essence, mutant genes not meant by the design of nature to be passed on are being passed on through the creativity of human intervention. The chances of a major genetic population impact are unlikely at present. Consider instead a family in which the couple has three sons by such ART procedures with all three (infertile) sons receiving the mutant gene(s) requiring them in turn to resort to ART procedures to procreate. The original singular medical problem in our population would then be tripled.

A population genetics effect that has already been seen, however, is the marked increase in the number of twins, triplets, and higher-order births occurring in the USA and Canada. As a result the rate of birth of twins and triplets is increasing beyond what would normally be the case.

Concerns about the sex ratio of children born from ART procedures has also been a source of academic debate. While there is little evidence for concern regarding a shift in the ratio of males to females born from ART procedures, there are reports that the chance of obtaining a live-born male is significantly greater than for a female if all the IVF-produced embryos transferred for implantation contained no fewer than four cells. There is evidence in this regard from animal research that male and female embryos naturally develop at different rates in the reproductive tract and that male embryos develop faster in culture conditions *in vitro*. If such is the case with human embryos, it is possible that a mix of embryos of undefined sex, selected for more advanced cell numbers, may confer a positive bias (survival factor) for male embryos.

## The future "ART" of medically assisted reproduction

Through scientific discovery and technical innovation, we enlist the forces of the natural world to solve many of the uniquely human problems we face in the realm of human reproductive health. Discoveries from the preceding decades inspired and enriched our lives by teaching us about the mysteries of our reproductive system in a way that permits our developing reproductive technology applications to assist in bypassing its failures. Does ART imitate life? The answer is "yes in part". Since only 20–33% of all spontaneous conceptions produce a live-born outcome, the success rate for the basic, well-developed ART procedures of 14–30% per oocyte retrieval cycle does partly imitate life. The near future will undoubtedly see the live-born success rates increase to nearly 40%. It is theoretically possible that rates well beyond that may be attained. Animal model research has shown that live-born rates of up to 60% can be achieved for some species such as cows. In addition to the success rate improvements, it is highly likely that further medical developments for improving the matching needs of a well-nurtured embryo and an optimally receptive uterus will decrease the significant number of fertile couples for whom ART procedures currently fail to provide a child, while decreasing the risks involved for both the parents and child of an ART-formed family.

With the improvements noted above, we will undoubtedly see over the next decade the obsolescence and disappearance of a number of ART procedures from the current menu as they become recognized by the public and the profession as being too inefficient or risky to be continued as acceptable options. These consequences will follow from the conduct of badly needed clinical trials designed to distinguish the most effective ART applications and epidemiological studies analyzing long-term outcomes.

We will also see the development of new ART options not presently available. For example, recent scientific advances that provide an ability to induce very early male or female sex cells to develop functional status by *in vitro* maturation (IVM) technology will undoubtedly provide new ART that will continue to challenge social sensitivities and sensibilities as a result of the potential donor sources (fetal abortus or adult testes or ovaries from living or cadaver donors) of the germ cells and the extremes (postmortem conception, intergenerational conceptions, etc.) to which such ART options may be taken without proper oversight. The report in 1995 of animal research showing successful growth factor treatment for inducing the maturation of ovarian granulosa cell–oocyte complexes, grown as *in vitro* monolayer cultures of cells, into mature, egg-containing follicles suggests that effective ART options involving IVM technology will be very much a part of the future for medically assisted reproduction.

Present and future ART technologies will significantly contribute to aspects of reproductive medicine other than infertility therapies. Such procedures are an integral part of medical genetic research efforts attempting to use preimplantation genetic diagnosis (PGD) procedures to identify embryos produced by couples carrying inherited genetic diseases that cause catastrophic illness and/or death. Since only embryos free from the risk of known expressed genetic disease are used to establish a pregnancy by IVF-ET or a related technology, the couple achieves a healthy child without the risk of facing an elective abortion to terminate an undesired outcome.

ART procedures also provide critically important technologies for contraceptive development research involving approaches for preventing conception and implantation. ART procedures also produce embryos not used for transfers that serve as a source of embryonic stem cells for medical research. This ethically controversial research guides the development of such cells into limited-potential stem cells. Such cells are capable of replenishing themselves while retaining the ability to develop further into more advanced cells useful for somatic cell therapies designed to alleviate or cure disease. An example of this concept would be the ability to guide early embryonic stem cells to differentiate into bloodline (hemopoietic) stem cells capable of being grown in sufficient quantities for use in treating diseases in which the patient's own bone marrow stem cells have been destroyed. Should acceptable forms of germline gene therapy ever be identified, wherein genetic alterations in a person are passed on to all offspring, the ART procedures discussed earlier would be critical for their successful development.

### Suggestions for further reading

Berger, G. S., Goldstein, M. and Fuerst, M. *The Couples Guide to Fertility*, First Main Street book edition. New York, Doubleday, 1995. [Twenty-one resources for additional information listed on pp. 421–3.]

Glazer, E. S. *The Long-Awaited Stork*. New York: Lexington Books, 1990.

Rosenberg, H. S. and Epstein, Y. M. *Getting Pregnant When You Thought You Couldn't*. New York: Warner Books, Inc., 1993.

Silber, J. S. *How to Get Pregnant With the New Technology*. New York: Warner Books, Inc., 1991.

Tan, S. L., Jacobs, H. S. and Seibel, M. M. *Infertility – Your Questions Answered*. New York: Birch Lane Press, 1995.

*The Assisted Reproductive Technology Workbook*. Somerville, MA: Resolve, Inc., 1993.

Treiser, S. and Levinson, R. K. *A Woman Doctor's Guide to Infertility*. New York, Hyperion, 1994.

# "O brave new world": Rationality in reproduction†

ALBERT R. JONSEN

EDITORS' SUMMARY

Albert R. Jonsen, PhD, provides a historical and philosophical sketch
of reproductive technologies, picking up on theologian Professor Joseph
Fletcher's view that, to the extent it is under rational manipulation, reproduction becomes more human rather than less. Jonsen then argues that
the greatest rationality for reproductive technologies is one that meshes
the technical goals with the personal and social ones of normal gestation.
Technologies would then be ethical to the degree that they "measure up"
to assisting people with disabilities in reproductive powers, and would be
more problematic, even unethical, to the degree that they create new
problems for the social stability of gestating, nurturing, and supporting
child development. For the most part society has worked out these structures in normal family, heterosexual relationships. Newer procedures
would be measured against this standard.

Many years ago, the esteemed patriarch of bioethics, theologian Joseph Fletcher,
spoke loud and clear in favor of rationality in reproduction. By rationality, he
meant not merely limiting population growth, which he certainly favored, but
bringing to bear human analytic and creative intelligence on the random and
instinctive activities of sexual intercourse and procreation that we share with all

† An earlier version of this chapter appeared in *Cambridge Quarterly* (1995), 4(3), 263–7.

mammals. In his 1974 book *The Ethics of Genetic Control: Ending Reproductive Roulette*, he foresaw most of the issues that we are facing today. He reflected on artificial insemination, prenatal diagnosis, cloning, eugenics, ectogenesis, ovum transfers, and genetic engineering. All of these innovations passed his examination to the extent that each of them represents a way of exercising rational and responsible control over life and reproduction. The subtitle of his book proclaimed his faith. Professor Fletcher's dedication to rationality led him to make the astonishing statement, "Man is a maker and the more rationally contrived and deliberate anything is, the more human it is. Therefore, laboratory reproduction is radically human compared to conception by ordinary heterosexual intercourse."

Now here we are, 21 years later. Not long ago, reproductive technology made it into the media in a big way. Humans were cloned or, better, artificially "twinned." This feat was proclaimed a "miracle." We have become accustomed to seeing that word associated with medical technology. It is now trite. We seem to mean by it something that attracts attention; something that we could not do before and now can. But these are pale meanings of miracles. If we were to get serious about miracles, we would see them as inexplicable by all human powers. Miracles are the work of God, not humans. This is the opposite of Fletcher's rationality: standing in wondrous awe before an event that we did not, could not, produce. But this archaic meaning is lost: we are the miracle makers and here are the miracles.

Capable people explore the facts and issues that attend these modern reproductive miracles. I will not repeat the things they say. Rather, I intend to put them into a general context, historical and philosophical.

## Historical context

First, I will build a historical context. All of this started with a bizarre event in Philadelphia 110 years ago. In 1884, Dr William Pancoast, professor at Jefferson College of Medicine, impregnated a woman whose husband was infertile with sperm donated by "the best looking medical student" in his class. This first successful donor insemination was shrouded in utmost secrecy. Indeed, even the husband and wife were at first kept in the dark! This event was only reported 25 years later, a decade after Dr Pancoast's death. A controversy broke out. Many physicians and a multitude of the clergy and public figures condemned it as nothing more than mechanical adultery, equivalent to rape and clearly contrary to the law of God. The young man, however, who owed his existence to the "handsomest medical student" did some detective work and discovered his biological father, now a respected practitioner, to the delight of both. For them, there was no ethical problem.

Notice that the first artificial insemination was to remedy male infertility. During the 1930s and 1940s, two female physicians, Sophia Kleegman in the USA and Margaret Jackson in England, worked indefatigably to make this procedure common and accepted practice. As it became so, the ethical objections of the previous decades disappeared; some, including the Roman Catholic Church, continued to object, but in general the world found artificial insemination a benefit. In 1979, things changed. In England, Dr Patrick Steptoe and Professor Robert Edwards fertilized an ovum taken from Mrs Brown with sperm from Mr Brown in a Petri dish, cultured it for a few days and implanted it in Mrs Brown's uterine cavity whence it went on to become the bouncing baby Louise. Miracle of miracles! A baby made, as it was incorrectly said, in a test-tube. But, the baby, once made, was situated in the accustomed place and born in the accustomed way. The purpose, again, was to remedy a physiological problem: Mrs Brown's fallopian tubes were blocked by scarring. The technique of *in vitro* fertilization and the many modifications that have followed it, such as cryopreservation, embryo transfer, and donation of oocytes, aim primarily at female infertility, just as artificial insemination had aimed at male infertility.

When Louise Brown was born, the media celebrated the miracle. The temptation to recall Aldous Huxley's 1932 novel *Brave New World* was irresistible. Huxley had envisioned a future in which all babies were produced and monitored in hatcheries where the state scientists predestined, and preconditioned and "decanted" socialized human beings, as Alphas or Epsilons, as future sewer workers or world controllers. Baby Louise, tiny and inoffensive infant as she was, was the stimulus for the moral imagination. This moral imagination has not diminished as the technologies evolve. In December 1993, two modest scientists submitted a paper to their scientific society detailing their work in splitting human embryos. Within a day, the moral imagination created Huxleyan futures. The rather unexciting scientific procedure, improperly referred to by the scientists themselves and then by the media as "cloning," took on *Jurassic Park* proportions. The moral imagination of ethicists worked overtime and carbon-copy babies, babies to replace lost babies, identical twins and triplets produced at intervals all emerged from the fevered brow. One of our more sober ethicists suggested that this was, in effect, the theft of our inviolable right to our genetic heritage. And, of course, *Brave New World* was quoted again and again. Here is the *New York Times*' Gina Kolata: "One 'Brave New World' scenario made possible by embryo cloning is that parents might be able to save identical copies of embryos so that, if their child ever needed an organ transplant, the mother could give birth to the child's identical twin" (Oct 24, 1993).

It is interesting to return to the source of Huxley's phrase "brave new world." It comes from Shakespeare's *The Tempest*. Some of you, I am sure, have seen the remarkable film, *Prospero's Books*, which is *The Tempest*, done at its most cinematic. *The Tempest* is filled with strange creatures, sometimes beautiful, like Ariel, sometimes horrible, like Caliban. They are engineered by Prospero's magic art and dance and howl at his command. But these are not the creatures described by the phrase "brave new world." Rather, that phrase is uttered by Prospero's daughter, who having been raised among these strange creatures, encounters her first humans other than her father. She meets Ferdinand, whom she will eventually marry and the retinue of noble gentlemen cast by storm upon her remote island with the exclamation,

"O, wonder!
How many goodly creatures there are here!
How beauteous mankind is! O brave new world,
That has such people in't!"

*(Act V, scene i).*

She is astonished, not at the bizarre creatures that she has lived with all her life, but at quite normal, ordinarily proportioned, and modestly talented humans.

The real setting of the brave new world is the context within which we should set our philosophy of reproductive technology. I subtitled this chapter "Rationality in reproduction" not only to recall Professor Fletcher's belief that rationality could guide our use of these technologies, but to make a point about rationality itself.

## Philosophical context

Making points about rationality has been the occupation of philosophers for centuries. They have written much and come to no definitive conclusions. Neither will I. However, a few notes may be helpful. First, one sense of rationality coincides neatly with technology. Rationality can refer to the human ability to calculate ends and means; technology is the determination of ends and means with clarity sufficient to be built into a technique that can be replicated with ease and efficiency. Reproductive technologies are such a rationality. The gradual appreciation of the physiology of reproduction, together with the invention of micromanipulation, establishes a set of procedures toward a predetermined end, namely conception. While it is certainly true that the procedures do not always reach that end (the success rates of assisted reproduction clinics bear that out), the failures are due, not to the procedure, but to the complex environment within which they must work. It is quite conceivable that at some time in the future the technology and the environment will be made to mesh even more perfectly. Rationality, then, in this sense, is the ability to plan and implement procedures to accomplish ends with reliability and efficiency. This is, I think, the notion of rationality that entranced Fletcher and led him to utter the extraordinary words I quoted at the beginning of this chapter.

However, in our culture, rationality also means something rather different. In one of its etymological meanings, the latin word *ratio* means "reckoning." Rationality is the human ability to set any event or idea within a wider view, so as to reckon or measure the fit between them. Moral philosophers since Plato have attempted to discern how human actions "fit" into the vision of the human good. They have tried to measure acts and character by criteria of right and wrong, better and worse, noble and base. Certainly, their approaches are widely different, but their endeavors cast light on a fundamental human feature, the desire to see the contingencies and particulars of living within a vision, a framework, a structure of life.

Reproductive technologies stand at the juncture of these two diverse meanings of rationality: they are technological reasoning that must be placed into a broader context whereby they can be measured. It is probably not possible to produce a vision, framework or structure all at once and once for all. Rather, we will work at this fitfully and partially, much as the architects of Barcelona have worked at Gaudí's Sagrada Familia church over 60 years and, even when and if our vision is finished, it will, like Gaudís' masterpiece, probably still look unfinished.

## Starting point

Nevertheless, any endeavor must have a starting place. The ethics of reproductive technology cannot come into being as a pastiche of ethicists' somber comments and wild imaginings. Where should we start? I suggest that Miranda's exclamation, "O brave new world, that has such creatures in it," is as good a starting place as any. She saw well-proportioned, well-mannered, quite ordinary gentleman (there were no gentlewomen in that retinue). The original technologies aimed at sustaining the ordinary when the ordinary was disrupted by pathology: artificial insemination remedied male infertility, and *in vitro* fertilization remedied female infertility. In both cases, the technologies functioned as properly medical, that is, they were designed to restore a function that is part of, as philosopher Norman Daniels says, "normal species functioning."

As the technologies have advanced, it has become rational, in the first sense of the term, to go far beyond this original corrective purpose. While restrained within it, the technologies did little to change the broader context within which reproduction works. Even though artificial insemination and *in vitro* fertilization eliminated coitus as the means for conception, the gestation of the child was within a familiar unit of a married man and woman who intended to nourish and educate the child. Even though, in artificial insemination by donor, sperm from someone other than the husband is employed, the same process is followed, once the law learns not to consider the act as adultery.

I refer to the "familiar context" as the heterosexual couple intending to nurture and educate a child. I do not claim that this is the absolutely right-by-nature form for reproduction. However, that familiar form has a wide range of implications for social life. It allows a child to be identified; it creates a locus of responsibility for the child; it endows the child with a heritage and a history; it stabilizes inheritance of property. Reproduction within this context solves many social problems and we have come to rely on it over the centuries in our culture.

As soon as it becomes possible to break out of that familiar context, the range of previously solved social problems becomes again problematic. In the past, of course, breaking out was possible: children could be conceived out of wedlock, children could be orphaned or abandoned. These events did create social problems that were resolved by the invention of practices such as adoption, foster parenting, etc. However, these functioned within the familiar context of reproduction. The reproductive technologies, precisely because they are technological rationality, can revise the process of conception and gestation quite thoroughly. The possibility of five or six persons now having some part in the production of a child demonstrates that. This means that the familiar context is challenged at many points.

By recalling Miranda's exclamation about normal people, I propose that the ethics of reproduction fits most suitably within the familiar context of stable couples conceiving, nourishing and educating. The technologies themselves, which are ethically neutral, can be measured ethically to the extent that they serve that context. The primary aim of the technology should be to remedy infertility of males and females. As we begin to move away from that measure, we should do so with utmost caution. We may find it rational to do so, but the moves should be carefully evaluated for the effects on the overall context that serves as a measure. For example, the current state of *in vitro* fertilization allows for preimplantation diagnosis. This is a move away from remedying infertility; it constitutes a eugenic move. As this technique becomes possible, it should be considered only when there is the same known risk of serious genetic disease that would justify prenatal diagnosis.

I must make very clear that my reference to the stable heterosexual couple intending to nurture is not equivalent to the recent political and ideological references to "family values." First, I do not understand what "family values" is supposed to mean, and I can imagine a variety of genuine values fostered by a variety of family structures. Second, the nuclear family, itself something of an artifact of recent times and familiar cultures, is clearly changing as a social phenomenon and it is unclear whether earlier forms are better or worse than the modified forms with which we have become familiar. My meaning is not intended to consecrate one form of family or even one form of reproduction. Rather, I am seeking a base point from which to evaluate changes and modifications.

For example, it might be surprising to hear me say that I do not find any serious ethical objection to the practice of utilizing the techniques of assisted reproduction for single women. This is clearly a departure from the standard of the stable heterosexual couple. However, I make that move away from the standard because I know of no empirical data that suggest that children raised by single women, or even by homosexual couples, fare worse in any way, if social and economic variables are controlled for. This is a departure from the norm that does not appear to create the sorts of problem and dislocation that other forms of reproductive technology create.

Surrogacy, however, at least commercial surrogacy, does create problems relative to the identity and responsibility for offspring that we have not yet learned how to manage. It is for that reason that some jurisdictions in the USA and abroad have even made commercial surrogacy illegal.

It is also very important to understand that I am not saying, as some critics of reproductive technology say, that the use of these technologies will destroy the family or go counter to the best interests of the children. This may or may not be. I am saying that the way to think rationally about the ethical problems of assisted reproduction is to take as the base point the social institution that has served as the solution to a wide variety of problems that arise whenever a new instance of the human species appears: the basic questions "whose baby is this?" "What tribe does it belong to?" "In what religion should it be marked?" "Who will take care of it?" have for many centuries and across many cultures been answered by the family structure of stable heterosexual couples (although in most cultures those couples were surrounded by extended but integral family unlike our nuclear family culture).

My contention is that, insofar as the rationality of technology makes it possible to bypass this structure, the problems that the structure has traditionally answered must be resolved before we can accept the practice as ethical. The rationality of ethics, as distinguished from the rationality of technology, attempts to assure that newness finds a fit in a pattern of solutions; this will come about either by devising new solutions or by limiting the scope of the newness.

## Conclusion

Certainly many of the particular questions about assisted reproduction and reproductive technology will be examined with intelligence and compassion in the future. All of them will be rational, for that is the mode of discourse common to the academic and professional in our culture and, because rationality can go in different directions, some of them may differ. Out of this sort of discourse, carried out over time and in many cultures, a view, a vision, a framework will come into being.

However, one last word of caution: reproduction is not entirely rational. Children are conceived in passion; the having and raising of children is surrounded by emotion. Rational rules for reproductive technologies must take this human fact into account. Technology is not all rational either: it generates fame, money, power. It is a product that is itself productive and generates all the human emotions that such productivity can feed: pride and possessiveness, generosity and greed. Rational rules about reproductive technology may never be able to encompass these human facts. Still, as persons concerned about the welfare and rights of persons and the good order and good life of our society, we must endeavor to devise reasonable plans for this extraordinary technology.

## Suggestions for further reading

Alpern, K. D. (ed.) *The Ethics of Reproductive Technology*. New York: Oxford University Press, 1992.

Fletcher, J. *The Ethics of Genetic Control: Ending Reproductive Roulette*. Garden City, NY: Anchor Press, 1974.

Glover, J. *Ethics of New Reproductive Technologies*. Dekalb: Northern Illinois Press, 1989.

Meyers, D. T., Kipnis, K. and Murphey, C. F. *Kindred Matters: Rethinking the Philosophy of the Family*. Ithaca, NY: Cornell University Press, 1993.

Ramsay, D. *Fabricated Man*. New Haven: Yale University Press, 1993.

Steinbock, B. *Life Before Birth: The Moral and Legal Status of Embroyos and Fetuses*. New York: Oxford University Press, 1994.

# Reproduction, abortion, and rights

ROSAMOND RHODES

EDITORS' SUMMARY

Rosamond Rhodes, PhD, tackles some of the most difficult ethical issues of our time. With the rise of reproductive technology greater and greater power is given to human beings over forms of human life, from gametes to fetuses. Rhodes argues first that all these issues revolve around the right to life, and will be judged largely on our view of abortion. After briefly describing arguments for and against abortion, and finding them all wanting on either side, she proposes an "assentist" view of rights and obligations. This view essentially argues that there are no intrinsic rights of nonperson human beings, and that all their rights are derived from freely assented obligations assumed by parents. This permits the morality of such technologies as assisted reproduction, surrogate motherhood, *in vitro* fertilization, and, perhaps, fetal reduction in the womb of a woman who is trying to get pregnant.

This is a good chapter for furthering understanding of how our science and technology require us to rethink our views of the value of human life.

Fetal tissue research, birth control, assisted reproduction, pregnancy reduction, surrogacy – ultimately our judgment about the morality of any of these reproduction-related issues depends upon what we think about abortion. The abortion debate dominated the American political scene throughout the 1980s and continues to be an important and controversial issue in the 1990s. Almost everyone has strong feelings about whether or not abortion is morally permissible and, if faced with the need to make a personal decision, about whether or not abortion would be the choice.

Unfortunately, in spite of the voluminous literature on the subject, the arguments for accepting abortion have remained unclear and, hence, insufficiently examined by both pro-life and pro-choice advocates in the general public. And most recently, people for whom abortion is not a central concern have treated the abortion issue as an impossible subject to discuss. This new attitude of militantly avoiding discussion may stem from utter frustration with the choked debate, enlightened tolerance, or even a reading of Ronald Dworkin's *Life's Dominions* in which he tries to explain the impasse as an unresolvable religious argument.

Regardless of the inclination to avoid the problem, it is important to continue trying to understand the arguments for and against permitting abortion because so many other pressing reproductive issues turn on the ethical acceptability of abortion. Before explaining the relation of abortion positions to the ethics of related reproduction issues, this essay lays out the most popular arguments for allowing abortion, points out objections that have been raised, and goes on to present a new, and perhaps more persuasive, argument for accepting abortion.

## Abortion

Opponents of legalized abortion argue that abortions are not morally permissible. This claim follows from the principle that no innocent human life may be taken and from the position that fetuses are human. Proponents of legalized abortion claim that abortion is permissible. This claim still needs to be defended by a coherent, sustainable, convincing argument. Advocates of the right to abortion typically argue that fetuses are not persons and, hence, have no right to life, while delivered infants are human and have at least this right. As yet these advocates have failed to show how or why the fetus is transformed into a franchised "person" by moving from inside the womb to outside or by reaching a certain stage of development. Supporters of abortion rights also are seen to lack a moral principle to justify killing human fetuses. If we should view abortion as a morally permissible act and not merely as an act that is sometimes convenient, then a persuasive moral argument should be developed.

Many attempts to solve the problem of abortion use claims about the right to life. There are, however, other frameworks for thinking about abortion. Some people approach the issue by considering nature, some focus on consequences, some direct their attention to principles, and some highlight the importance of nurturing. Although these positions contribute to the discussion, the right to life remains at the heart of the issue. So that we can appreciate their contribution to both the problem and the solution, let us examine these several approaches in turn.

## Preliminary considerations: Nature, consequences, principles

### *Nature*

Some people object to abortion on the ground that it is unnatural. However, what they mean by "unnatural" or precisely what about abortion concerns them is not clear. The objectors might be concerned that abortion is an intrusion into God's domain and thereby a violation of the natural order. Or they might see it as immoral because it perverts human nature. Or abortion might be objectionable because it violates some sense of natural law. Unless religious tenets are involved, however, these objections do not seem compelling.

If the issue is interfering with life, we must notice that intervening to preserve life is often the right thing to do. So, intervention to terminate life should not be wrong solely because it involves tinkering with life. If aggressive intervention with advanced technology to preserve life is not considered a violation of the natural order, it seems arbitrary to conclude that intervention to terminate life is immoral on the grounds that such action is not the place of human beings.

Spontaneous abortion is part of a natural process in which a woman's body prematurely expels a fetus. Often the naturally rejected fetuses are defective in some way or, owing to the maternal condition, would not have developed into normal human beings. Nevertheless, spontaneous abortion is a part of the natural biological system of reproduction, and is neither morally good nor morally bad, because only voluntary actions have ethical status. In contrast, induced abortion is a purposeful, deliberate act. But since spontaneous abortion is part of a process that is consistent with nature, it is hard to see why induced abortion is a perversion of human nature. If abortion is called perverse because it is considered murder, then the objection is that abortion is killing in violation of a right to life, and the argument actually has to do with rights and not nature.

### *Consequences*

A common reason for approving or permitting abortion is the severe anguish that would be suffered by the mother or family if pregnancy continues to term. Terminating a pregnancy could cause pain to the one killed, but allowing the fetus to live could cause many more to suffer. Avoiding such pain may seem to justify the destruction of a potential human being. Yet, because the verdicts of such consequentialist formulations can sometimes offend even the most callused sensitivities, this simple consequentialist line of argument captures no high ground in the abortion debate.[1]

## Principles

The most crucial principle in the abortion debate is "preserve human life." Other principles may also be involved: "avoid pain," "preserve families," "preserve freedom" (even for mother), "develop talents," "distribute goods justly." When a woman is considering whether or not to terminate her pregnancy several of these principles are implicit in the consideration, and the decision is difficult precisely because following these valued principles would require incompatible actions. Whatever the choice, some principle will be sacrificed. The conflict between the acts required by the several relevant principles might sometimes be resolved by deciding to abort a fetus. Yet, for a person who greatly values human life, the other considerations would hardly shift the balance away from preserving life. But someone who takes other principles to be more important could favor abortion. The subjectivity and variety of solutions to particular cases create an obvious problem for an approach that calls for weighing principles. Thus, for a more generalizable answer to the abortion question we must look elsewhere.

## Abortion and the right to life

Many people have suggested that the right to life lies at the heart of a solution to the abortion issue. However, there is a similar flaw in typical versions of right-to-life arguments from both the pro-life and the pro-choice positions. Both positions accept that someone's right to life ought to be respected and that whoever has a right to life should not be killed. They differ on whether the fetus has a right to life. On the one hand, the pro-choice advocate argues or asserts that fetuses are neither rational agents nor social beings and so have no right to life. On the other hand, the anti-abortionist argues or asserts that life is present from the moment of conception and that fetuses are human because they look like babies, have a human genetic code, and so have a right to life.

Both positions draw invalid inferences of the same sort. The validity problem, which undermines both opponents' arguments, involves the inference of a moral conclusion (an "ought") from a factual, nonmoral premise (an "is"). For example, from the factual statement that it *is* 3:00 p.m. or that it *is* Monday no one should logically draw a moral conclusion that I *ought* to be writing a letter to my friend. That moral conclusion could logically follow only from a moral statement such as "I promised my friend I would mail my letter to him by Monday and I *ought* to keep my promises." The pro-choicers maintain that psychological characteristics (e.g. on-going consciousness of self, emotions, ability to feel pain, conceiving goals and plans) make a moral difference, whereas anti-abortionists consider biological characteristics to make a moral difference. Proponents of both positions are tripped by the pitfall of sliding from *is* to *ought* by converting a psychological quality or a biological quality into a moral category. Neither the biologically based argument against abortion nor the psychologically based argument for allowing abortion offers an acceptable right-to-life answer. Yet, this critique does not show that a right-to-life argument cannot be constructed, but only that certain forms of rights arguments are not successful in the abortion debate.

## Nurture

In recent literature, nurturing, rather than the prohibition on killing, has been put forward as the appropriate moral concept for addressing the abortion question. Advocates of this position complain that conservatives embrace a traditional but distorted view of the role of women which holds that sexually active women are morally required to bear and nurture children. Nevertheless, they charge that liberals tend to ignore the genuine obligation of parents to nurture their offspring.

It seems that the overlooked concept of nurture is significant in making sense of the ethics of abortion, but if we want to gain insights into the morality of the issue it is also obvious that we cannot ignore the issue of killing the fetus. In what follows I shall provide a new kind of right-to-life argument, one that pays attention to the moral weight of both the right-to-life and the duty of nurture. This position reflects and elaborates the fundamental intuition that the duty to provide nurture takes hold when a woman accepts a pregnancy. This approach, which derives rights from agreement or assent, can generate a right-to-life solution to the abortion question and avoid the question-begging moves of other right-to-life arguments. This account is also useful in distinguishing between cases, and its conclusions match many common intuitions about when abortion is permissible and when it is not.

## An assentist right-to-life argument

Assuming a correlation between rights and duties and that when one person has a right another person has a duty related to that right, any discussion of rights is also a discussion of duties. To appreciate the correlativity of rights and duties, think of freedom of speech. If someone has the right to speak, others have the duty not to interfere with that right. If someone has the duty not to interfere with another's speaking, that other has the right to speak. From this widely shared theoretical perspective, every right of A against B implies that B has an obligation or duty to A. In the context of the abortion debate, an argument claiming that a fetus, A, has a right to life should also tell us who has the concomitant duty to respect this right to life. Identifying the one who is duty bound then requires accounting for how that duty originated.

Moral theory sets out two accounts of how rights originate – natural rights theory and what I shall call assentist theory (in political philosophy what is usually called contractarian theory). Natural rights theorists maintain that certain things have rights as an intrinsic feature of their nature, that those rights are obvious, and that the naturally occurring rights create obligations in others. Unfortunately, the debate over abortion is as heated as it is because some people clearly "see" that fetuses have the right to life and others just as clearly "see" that fetuses have no rights at all. Unfortunately, the natural rights perspective provides no way to determine whether the emperor (fetus) has no clothes or whether some viewers suffer from moral blindness.

The assentist focuses primarily on obligations rather than rights. The assentist maintains that there are no naturally occurring moral obligations and so no naturally occurring moral rights. For the assentist, all moral obligations arise only from undertaking particular commitments. In this analysis, the right to life could arise only from someone's right-creating act of assuming an obligation. This position seems most fruitful for unraveling the abortion question.[2]

To explain the assentist position further, consider that if the right to life does not come simply from being conceived by a woman or from possessing some particular anatomical (e.g. blond hair, blue eyes) or psychological (e.g. desiring, calculating) characteristics, then it is reasonable to consider whether the right to life arises in the way that other rights do. Our right to speak freely comes from our government's accepting the obligations spelled out in the First Amendment to the United States Constitution. The rights of a weekend guest arise when a host takes on the duties of hosting him. The right of a patient with kidney failure to use one of your kidneys arises only when you accept the obligation. In each of these situations the right of A against B comes from B's accepting the obligation to A.

Applying this scheme of analysis suggests that when the parent assumes the duty of caring for and nurturing the child, the child gets the right to be cared for and nurtured.[3] An infant's right to life comes from parents accepting responsibility for the care and nurturing of their offspring (*care* until the child can be independent and able to assume obligations as a moral agent; *nurturing* to develop and train the child to be a responsible moral agent). Using this analytic scheme the origin of obligation can be pinpointed in time for each parent. Parents who are trying to conceive are bound to avoid situations that they believe might endanger the health of their baby from the time they begin trying. For example, since it is known that the health of infants may be adversely affected by a mother's smoking or drinking alcohol during the first days after conception (before the fact of pregnancy could be known), she would be bound not to smoke or drink once she started trying to become pregnant. When an unplanned pregnancy occurs, the parents have no obligation to the fetus until they decide to have the child. Because the fetus, at that point, has no right to life, there is no violation of the fetus' right when a mother decides to abort the pregnancy. When a pregnant woman decides to bear her child and then place the child for adoption, she accepts the duties of acting to preserve her baby's health and arranging for a placement that will provide the baby with adequate care and nurturing. Adoptive parents, while not engendering a right to life, have the obligation to care for and nurture the baby only because they have accepted such duties along with the baby, not because the baby has some inherent right.

An assentist approach permits the abortion of any fetus before the parents invest the fetus with rights. Still, it may seem that abortion could be avoided whenever someone, other than the parents, is willing to take on the obligation for the care and nurturing of the infant or, in some hypothetical world, of the fetus. However, such is not the case. If all rights arise from human relationships, someone has a right only because of a special relation. In bearing the obligation for all of their acts, parents have the responsibility to make decisions about the offspring they conceive. That obligation gives only them the right to decide whether or not to bring their fetus to term, and no benevolent institution has the right to wrest that decision from them.[4]

## Potentiality and the abortion question

This analysis of parents creating their child's right to life by assuming obligations implicitly assumes a classical notion of potentiality. Only persons (i.e. moral agents) have rights. So, only the potential to become a person allows a fetus to be invested with the right to life. Ironically, a common-sense argument against abortion, not yet addressed in this chapter, also employs the concept of potentiality.

Briefly, this pro-life argument claims that the potential for personhood or humanhood gives the fetus the right to life: i.e. whatever has the potential for becoming a human being actually has the right to life. However, use of the concepts of potentiality and actuality are empty here because the alleged connection is essential rather than contingently developmental. The claim depends on a bald assumption that the right to life is an essential quality of potential human beings. But again this analysis is unacceptable because no justification for the crucial premise is offered to those who have neither discovered its *a priori* truth nor accepted a religious warrant for the connection.

Actual human beings and human fetuses are potential possessors of a large array of moral rights and moral obligations, but they do not have any rights until the rights are actualized. Human beings are able to acquire moral rights and assume obligations just because they actually have that potential. As a potentiality, the right to life could then be actualized or not by a parental act of undertaking or refusing the obligation to care for and nurture the fetus. Until and unless that change occurs, killing the fetus could not violate its right to life since it still has none. An adult, acting as the source of change, must be the originating source of change for actualizing the right to life in a potential bearer of that right.

## Implications

The analysis of potentiality and actuality within an assentist response to the abortion question refines the abortion answer that I have offered. According to the assentist, morality arises only by freely choosing our response to the situations that present themselves. To the extent that any unintended pregnancy is "chance," "luck," or even an "act of God," no one is morally responsible. There is nothing moral or immoral about these events.

Potential parents have the obligation to make a decision about their potential offspring. The moral choices are then either to reject parental duties and accept the responsibility for abortion or to accept and fulfill parental duties. Understanding the acceptance of parental responsibility as the source of the right to life provides definitive answers to questions related to reproduction. The assentist perspective provides a framework for addressing the morality of fetal tissue research, birth control, assisted reproduction, and surrogacy.

## Fetal tissue research

Those who find abortion ethically unacceptable because it violates the sanctity of life are inclined also to reject, on the same grounds, the use of fetal tissue in medical research. Others who object to abortion on the same or different grounds might nevertheless accept fetal tissue research. Once an abortion is performed, using tissue from the dead fetus could be justified by trying to improve or prolong other lives. Provided that the pregnancy is not undertaken or that the abortion is not performed to supply fetal tissue, those who value life strongly could accede to using the tissue to preserve life as the better choice if the tissue would otherwise be discarded.

Of course the assentist, who accepts abortion before the right to life has been bestowed, should accept fetal tissue research. Nevertheless, an additional issue arises. Since the fetus is hers, the woman who decides to terminate her pregnancy does not give up her right to determine what will become of the aborted fetus. She could choose to have it buried, cremated, preserved as a model for students of embryology, or used for research or therapy. She could abandon it as one does a broken umbrella, giving up all rights to its future use to whomsoever may want it for whatever purpose. It is not obvious that a woman who chooses to terminate a pregnancy either considers or understands the possibilities of what will become of the aborted fetus in the same way that most people can appreciate the implications of tossing away a damaged umbrella. Since the right to determine the disposition of the fetal remains could be significant to her, and since the person who chooses abortion may have strong preferences about what should or should not be done with the aborted fetus, it is important to explain the alternatives to the woman and to allow her to make the choice. Using fetal tissue for research is morally acceptable from the assentist position so long as the woman who has chosen to terminate her pregnancy also either specifically agrees to the use of the fetal tissue for research or explicitly gives up her right to determine what will be done with the tissue.

## Birth control

If no one has a natural duty to reproduce, no woman who uses birth control to avoid reproduction could be failing a natural obligation on that account. And since eggs, sperm, and fetuses do not have inherent rights, using any means to avoid fertilization or to avoid implantation or to destroy a fetus, could not violate their rights. All sufficient means to prevent reproduction (e.g. condoms, pills, intra-uterine devices, implants, injections, etc.) would be equally acceptable when their use did not violate the rights of others as when, for instance, someone had made a personal commitment to another (e.g. a spouse) to try to have a child. So, except in those cases where a special undertaking commits a person to try to have a child, the use of birth control could not be a violation of rights.

Yet, since actions often signify the assumption of responsibility, given the general availability of contraception in our society, the decision to employ or to avoid the use of birth control could have special moral significance on my analysis. Clearly the regular use of birth control expresses a refusal to undertake parental responsibilities. Not using birth control during intercourse while intending to use a morning-after pill or intending to undergo an abortive procedure in case of a pregnancy could also count as a refusal of parental responsibility. However, the sexually active person who chooses to use no birth control and yet has no clear intention about having or not having a child is morally irresponsible. And the woman who becomes pregnant and does not decide to terminate her pregnancy until many months have passed is an even more troubling case. There may be a point in a pregnancy at which a woman who does not seek an abortion should be assumed to have implicitly undertaken a rights-vesting commitment to the fetus. On this view, the ambivalent woman who uses no birth control and first chooses to have the child and then changes her mind and opts for an abortion does violate the fetus' right to life. While such cases present serious ethical qualms for the assentist, the legal position of legislating only actions would permit the abortions and leave these internal matters of intention to be decided by the court of personal conscience.

## Assisted reproduction

Employing any of the technology-assisted means of achieving a pregnancy may be ethically unacceptable to people who believe that interfering with reproduction is meddling in God's domain, but assisted reproduction is perfectly acceptable on an assentist account. Undertaking a procedure to increase the likelihood of a pregnancy is a clear act of commitment to the resulting fetus. Because only those who choose to have a child would even consider employing the technology, those who pursue a pregnancy with the aid of special technology should be understood to have vested (or at least conditionally vested) the fetus(es) with the right to life.

While the assentist position sanctions assisted reproduction, following an assentist line could be especially problematic when multiple fetuses are created. Several of the most commonly employed techniques (e.g. hormone therapy, *in vitro* fertilization) can result in a woman being impregnated with more than one fetus at a time. Since this consequence of the use of the technology is well known, it is unclear whether the pregnant woman should be thought of as having created the right to life in each fetus, or whether she should be considered to have created only a contingent right to life in each fetus and to have reserved the right to terminate those in excess of three, or two, or even one.

If the concept of a contingent right to life is coherent, it could also be used to shed light on the ethics of terminating chosen pregnancies when the fetus is found to be seriously defective or deformed or found to have a serious genetic abnormality. The assentist view could be modified so that a woman who chooses to have a child could be seen as investing her fetus with a contingent right to life only if the fetus is capable of achieving moral capacities and independence. When a prenatal diagnosis of a serious abnormality is made, the woman who had conditionally granted her fetus a right to life could either choose to terminate her pregnancy because the fetus did not meet the minimal conditions of reasonable parental expectations, or choose to take on the special and life-long obligations of providing care and nurturance for a seriously handicapped child.

## Surrogacy

In technology-assisted reproduction when people who want a baby are not able to produce the requisite eggs or sperm, a human donor contributes the gametes. The donor, who may or may not be paid for the donation, time, and inconvenience, agrees to the use of his sperm or her eggs and gives up parental rights to the resulting child. The gamete recipient accepts the responsibility for the offspring. From an assentist-rights perspective such practices are unproblematic because no rights are violated in these transfers.

In a surrogacy arrangement a couple employs a woman to be the egg source, to be impregnated by the future father's sperm, and to carry a fetus through gestation. If the prospective parents accept the obligations to the resulting child and if the surrogate mother genuinely accepts that she will give up her parental rights to the child, the agreement could be ethical because no rights are violated and because the obligations are satisfied.

Serious problems have arisen with surrogacy arrangements, some of which have reached the courts and the media. The question raised by publicized cases is whether someone's agreement to be a surrogate mother can be counted as genuine consent. Philosophers have argued that some acts cannot be genuinely agreed to (e.g. killing or hurting oneself, selling oneself into slavery). When a person appears to agree to one of these acts, others must not count the apparent consent as genuine moral commitment. Opponents of surrogacy claim that consent to a surrogacy contract is so much a violation of self-interest and the payments offered for the service are so coercive to a needy woman as to make genuine agreement impossible.

This objection would be especially serious for an assentist who sees agreement as the source of obligation. But, while surrogacy may seem too great an emotional burden for some women (and therefore extra precautions should be taken in the psychological screening of prospective surrogates), good evidence suggests that surrogacy can be morally acceptable. Many women have been able to give up children for adoption, thereby both foregoing parental rights to the baby they have carried and accepting the obligation to allow their child to be raised as the child of others. Adoption, therefore, should count as *prima facie* evidence for continuing to regard surrogacy as ethically acceptable.

## Conclusion

Abortion and related moral issues of reproduction will remain charged and divisive issues well into the foreseeable future, largely because the positions are derived from different comprehensive ethical doctrines. While tolerance of the beliefs of others is taken as the distinctive virtue of liberal society, murder and repression fall beyond the pale of tolerance. So long as pro-life advocates continue to see abortion as murder and so long as pro-choice advocates continue to see limitations on access to abortion as repression, there is little hope that either side can come to tolerate the practice advocated by their opponents.

This chapter has offered an alternative structure for understanding the morality of abortion. The assentist approach allows the problem to be addressed from a perspective outside of the firmly drawn battle lines. An assentist analysis of questions related to reproduction also helps to identify significant moral issues that might not be seen from other points of view. In the interest of political tolerance and philosophical charity, the assentist line of argument deserves the attention of proponents on either side of the impasse as well as anyone searching for a way of grasping the ethical core of the abortion issue.

### NOTES

1. More sophisticated consequentialist accounts might avoid this line of criticism.
2. Hobbes, Rawls, Nozick, Thomson and existentialists such as Sartre could be said to hold such assentist positions.
3. It is often claimed that the child has the corresponding obligation to care for aged parents and that parents have a right to such care, but it is inconceivable that infants so bind themselves as has already been pointed out by Immanuel Kant (De obligatio activa et passiva. In *Lectures on Ethics* (Harper Torchbooks, New York, 1963), p.21.)
4. To further illustrate this point, imagine that a strong young man wants to take on the obligation of carrying my heavy package for me. It seems that I have the right to refuse the assistance and he has no right to impose his assistance on me.

   Imagine a wealthy dowager who believes that every able child should have ballet lessons because they can be shown to be the most beneficial form of exercise. If she were willing to take on the obligation of providing those classes, would she have the right to have my male child participate or would I have the right to refuse? It seems clear that the dowager would not have the right, while I would.

### Suggestions for further reading

Brody, B. A., *Abortion and the Sanctity of Human Life: A Philosophical View*. Cambridge, MA: MIT Press, 1975.
Dworkin, R., *Life's Dominions*. New York: Alfred A. Knopf, 1993.

Feinberg, J. Abortion. In *Matters of Life and Death*, ed. T. L. Regan, pp. 185–202. New York: Random House, 1980.

Feinberg, J. (ed.) *The Problem of Abortion*, 2d edn. Belmont, CA: Wadsworth, 1984.

Gomberg, P. Abortion and the morality of nurturance, *Canadian Journal of Philosophy* (1991), 21(4), 513–24.

Hull, R. T. *Ethical Issues in the New Reproductive Technologies*. Belmont, CA: Wadsworth, 1989.

Knight, J. W. and Callahan, J. C. *Preventing Birth: Controversy*. Salt Lake City, UT: University of Utah Press, 1989.

Marquis, D. Why abortion is immoral. *Journal of Philosophy* (1989), 86, 183–202.

Noonan, J. T., Jr. An almost absolute value in history. In *The Morality of Abortion: Legal and Historical Perspectives*, ed. John T. Noonan Jr, pp. 51–9. Cambridge, MA: Harvard University Press, 1970.

Singer, P. *Practical Ethics*. Cambridge: Cambridge University Press, 1979.

Thomson, J. J. A defense of abortion. *Philosophy and Public Affairs* (1971), 1(1), 47–66.

Tooley, M. *Abortion and Infanticide*. Oxford: Oxford University Press, 1983.

Warren, M. A. On the moral and legal status of abortion. *The Monist* (1973) 57(1), 48–52.

# The critically ill neonate

JAMES M. ADAMS

EDITORS' SUMMARY

James M. Adams, MD, demonstrates how for one century improvements
in the commitment and care for newborns have led step by step to ethical
issues. The first was whether such improvements should be pursued at
all, since defective newborns may just have been nature's mistakes. As
each problem of maintaining temperature, avoiding and treating infections, providing rest and sleep, and helping underdeveloped lungs grow
was addressed, the quality of life of the "graduates" from the newborn
intensive care unit was scrutinized as was the impact of returning many
of these children to their families. Underlying the scientific advancements
have been cultural commitments to the value of the life of a child that
have also grown correspondingly more explicit.

The field of neonatology appeared quietly in the late 1950s with the development
of specially staffed and equipped nurseries for the intensive care of sick and mal-
formed infants. Since these early beginnings, neonatal intensive care has become
the most expensive medical care in the world; one heavily driven by technology
and increasingly dependent upon it. During the subsequent three decades, a
tremendous outpouring of research and resulting improvements in patient care
have led to the survival of many infants whose future was previously deemed
hopeless. These advances have brought neonatal care to the frontiers of human
viability – but they have also produced increasingly complex ethical dilemmas.

Both the science and economics of health care are changing rapidly today, as
is the social background in which medical care must be delivered. To understand

how this rapidly changing health care environment may affect the critically ill neonate it is necessary to examine the historical phases of development of this medical specialty.

## The infant welfare movement

During the first half of the twentieth century perinatal and infant mortality were high. By the end of this period striking improvements had been made and the field of pediatrics had emerged as the driving force in newborn care. Ironically, much of the new focus on perinatal health actually had its roots in the military expansion in Europe in the 1870s. The emergence of Germany as a dominant power raised concerns in other European nations that their falling birth rates and high perinatal mortality would limit the availability of soldiers for national defense. This resulted in the Infant Welfare Reform Movement, which had as its goal a reduction in mortality in the first year of life and placed emphasis on hygiene, nutrition and the outcome of pregnancy.

At the turn of the century the birth rate in the USA was high. In crowded industrial cities, swelled by a rising tide of immigrants, living conditions were poor and many infants perished. Labor and birth were viewed as natural processes, and the infant was largely a by-product. Few communities attempted to maintain any meaningful birth statistics. There was little child advocacy and, in fact, children were frequently exploited.

Physical trauma or fetal asphyxia (loss of oxygen) often accompanied the birth process, with subsequent long-term consequences for infant survivors. Smaller premature babies and those with major malformations usually succumbed.

This is not to imply, however, that vigorous attempts were not made to improve perinatal health and preserve the lives of sick infants. Even before 1900, Etienne Tarnier and Pierre Budin in Paris had organized a hospital nursery, developed infant incubators and popularized tubal feedings. In 1901 an early model for prenatal care was introduced with the opening of beds for sick mothers at Edinburgh's Royal Infirmary. These early pioneers established temperature, nutrition and recognition of certain diseases as the basic components of newborn care.

The most serious obstacle to improving perinatal health in the USA was the absence of any national birth registration system or comprehensive statistics. By 1915, however, the vigorous work of individual reformers and health advocate organizations had resulted in the passage of model birth registration laws and the beginnings of a national data base. At that time the American infant mortality rate was almost 100 per 1000 live births. Half of these deaths occurred in the first month of life.

During the next three decades pediatricians assumed an ever-increasing role in newborn care and nursery development. During this time it was recognized that drugs given during labor may depress the neonate. The impact of poverty and its related social ills became clear. Of particular importance was the recognition that premature deaths were a major determinant of infant mortality rates.

Even at the turn of the century, the question of whether advanced care of the newborn was good for society generated considerable debate. Charles Darwin's hypotheses suggested that fetal and neonatal death might be in harmony with natural selection. These ideas were used by some physicians to encourage therapeutic passivity. The public largely accepted this status quo. Nurseries for the care of premature and sick babies continued to appear, however, and in 1922 J. H. Hess initiated follow-up studies of premature nursery graduates in an attempt to answer the question of whether these infants were worth saving. Further debate on this issue was largely silenced by the turmoil of the Great Depression and the Second World War. The question returned with renewed focus in the postwar years, however, when retrolental fibroplasia resulted in an epidemic of blindness in oxygen-treated premature infants.

### The era of the neonatal intensive care unit

In 1958, the first of a series of studies appeared that demonstrated that control of the thermal environment with incubators significantly reduced the death rate among sick neonates. These benchmark studies reaffirmed temperature control as a cornerstone of newborn care. They also forged a partnership of basic research, technology and clinical practice that not only changed the outcome of patient care but also initiated an era of explosive growth of neonatal intensive care units (NICUs) in the industrialized world.

The main goal of this growth was a reduction in mortality, particularly among premature infants. The result was two decades of increasing invasiveness and complexity of care. At the end of this period, a striking reduction in deaths had indeed been achieved, but a number of new dilemmas had also appeared.

The foundation for the rapid growth of this new medical specialty was the accumulation of a large body of perinatal outcome data in Canada and Europe in the years following the Second World War. These data showed that neonatal deaths accounted for almost three-fourths of infant mortality and that two-thirds of neonatal deaths were related to low birth weight. As a result, attempts were made to organize perinatal care into primary, secondary and tertiary tiers (subspecialty perinatal intensive care) in defined geographic regions.

During this period care became increasingly dependent on technology. The use of warming devices, cardiopulmonary monitors, intravascular infusion pumps and blood gas analysis became routine. Respiratory care became both more aggressive and more invasive. Umbilical artery catheters, tracheal intubation and mechanical ventilatory support became widespread even for the smallest premature infants of 23–25 weeks' gestation. Daily care dictated numerous laboratory tests, radiographs, and the need for highly skilled caregivers and support personnel increased. The NICU became a round-the-clock beehive of constant light, noise and activity with many infants subjected to frequent handling and little quiet sleep.

By 1980, a significant reduction in neonatal mortality risk had indeed been achieved. Of particular note, significantly improved survival of babies weighing less than 1500 grams occurred when such infants were born in the most technically advanced subspecialty perinatal centers rather than community institutions. These differences in outcome persisted even when controlled for inter-hospital variables such as race and multiple birth. The driving force for these impressive gains was a large body of research that provided much new understanding of newborn physiology, particularly that of cardiopulmonary disease.

These gains, however, were not without cost. Secondary infection was common in small premature bodies spending two to three months in the hospital. Bronchopulmonary dysplasia, a chronic lung disorder, was increasingly recognized following mechanical ventilation for hyaline membrane disease. The problem of blindness related to retinopathy of prematurity stubbornly persisted despite close monitoring and strict guidelines for the use of oxygen. These and other problems were increasingly concentrated in the very smallest prematures.

In addition, medical devices had become ever more complex, with an associated increase in risk and a heavy burden upon care providers to understand and maintain proficiency in their use. Medical equipment became a primary determinant of cost and even the most objective of neonatologists began to feel that intensive care could not be successful without massive investment in ever-changing technology. The contributions of new technology during this period were indeed considerable, particularly the introduction of techniques such as ultrasonography. Some of the focus on medical devices, however, came at the expense of education and the role of clinical reasoning.

Of particular concern was the neurologic outcome of the surviving infant of very low birth weight. By the end of the 1970s survival of babies weighing 750 grams or less had reached 50% in some centers. Accompanying these improved results were fears that improved survival would produce growing numbers of handicapped individuals needing long-term support from society. Of greatest concern about the quality of life was the observation that intraventricular brain hemorrhage was common in small premature babies and was a common cause of permanent brain damage.

## A kinder, gentler care

By the mid 1980s results had advanced to the point that reproductive health care could no longer focus on survival alone. The major role of the NICU had become the care of the very low birth weight infant. Two important factors once again emerged to change the focus of care of the critically ill neonate.

The first of these was the availability of exogenous surfactant (obtained from cows) for the treatment of hyaline membrane disease, due to premature lungs. After nearly 30 years of research into the biology of surfactant and its relationship to prenatal events and postnatal lung function, surfactant replacement therapy became available for clinical use. This therapy promptly cut the death rate associated with hyaline membrane disease in half and directly prompted a further fall in infant mortality.

The second of these factors was the emergence of team care in the NICU. Physicians, nurses, respiratory therapists and other support personnel initiated new mechanisms to combine activities in an integrated fashion and promote continuity. Physicians relinquished many traditional roles to nurses. Attempts were made to focus team education on specific goals of care.

As a result of these changes, the issue of quality of life again came under intense scrutiny. This new focus has concentrated specifically on the impact of the NICU environment on outcome, as well as the role of the family in the care of the compromised baby.

In recent years many nurseries caring for small premature infants have modified the intensive care environment in an attempt to reduce specific morbidity (disease processes). The most vigorous of these efforts has been directed toward reducing the occurrence of germinal matrix brain hemorrhage, a major cause of neurologic disease in very low birth weight babies. Attempts are made to minimize handling, avoid painful stimuli and maintain stability of blood pressure in these small premature babies who have poorly developed regulation of cerebral circulation. In addition, noise and light are minimized and attempts are made to avoid the undesirable physiologic consequences of the stress response on control of breathing, growth and sleep states.

Substantial undisturbed sleep time is allowed in a quiet, darkened area. Preliminary research suggests this type of modified environment, combined with provision of early intervention services to enhance cognitive and behavioral development, may promote significant reductions in long-term brain dysfunction in low birth weight survivors. Along with these environmental modifications there has been increased reliance on noninvasive monitoring and a reduction in painful procedures.

Of critical importance in this evolution of the modern NICU is the expanding role of the family in the day-to-day care of the small or sick neonate. Family-centered care has been enhanced by use of unlimited visitation, sibling visitation and encouragement of early direct contact between parents and their sick infant. Though the role of the family in decision-making remains a complex issue, there is growing acceptance of parents as major partners in this process.

## The critically ill neonate – today and tomorrow

Infant mortality has now fallen from 100 to fewer than 10 deaths per 1000 live births in most industrialized nations and is less than half that in some, largely resulting from a steady fall in low birth weight mortality in recent years. Care in the NICU is now dominated by the small premature infant and the baby with major malformations. Cost of care is related directly to duration of hospitalization, which in turn is a function of birth weight and chronicity of disease.

The explosive rise in health care costs and the new financial realities of a changing world economy have only intensified ethical dilemmas that have existed for decades. For more than a century, debate has centered around the basic question of how many potentially normal survivors a society should sacrifice to avoid a damaged one.

Survival among premature infants weighing less than 1500 grams now exceeds 80%. Since the 1950s the proportion of normal survivors has more than tripled, with handicap rates in most centers being below 20%. Survival of babies of 25 weeks' gestation has reached 70% and these infants are candidates for aggresive support. It is those of 23–24 weeks who now truly represent the frontiers of viability.

Although several perinatal centers have demonstrated decreased handicap rates in these smallest of premature survivors, epidemiologic studies from many industrialized nations suggest a rise in the prevalence of cerebral palsy, largely related to the increasing contribution of very immature infants.

It is paradoxical that, while society's concern for the quality of life has grown, judicial and statutory protection for children and the handicapped has become increasingly stringent. Attempts to implement the United States Federal "Baby Doe"[1] regulations led to the adoption of similar criteria for withholding or withdrawing life support in several states. The Handicapped American Act and position statements of the American Academy of Pediatrics have further reinforced the rights of handicapped infants, including those of premature babies. The explosive growth of malpractice litigation in the USA has clearly compromised the decision-making process. These factors have seriously restricted the role of parents in decisions regarding provision of medical care for their infants. At the same time, it is recognized that the interests of a sick child may at times conflict with that of his or her family.

These issues must be approached with some urgency. Neonatologists will continue to grapple with complex ethical decisions; particularly the issue of when to withhold or remove life support. These decisions are unique, since they involve a patient unable to give his or her own consent and require mediation exclusively by third parties. Nursery practices must also come under renewed scrutiny. In particular, the contribution of tests and expensive technology to improved outcome must be verified by objective study.

The small or chronically compromised newborn is politically vulnerable in an era of shrinking health care budgets. In a patient population that can neither vote nor voice its own needs, the rights of these patients will continue to require a certain level of legislative and judicial protection. At the same time, the decision-making process must be freed from the fear of malpractice litigation and must become increasingly responsive to input from families and society.

Future advances in perinatal outcome are unlikely to come from the NICU. Because the patient is completely dependent on the environment, nursery care after the occurrence of disease can never be made adequately cost-effective. Further improvements in outcome must be made at the preventive level. To date, only limited reductions in the rate of premature births have been achieved in most nations. Some gains have been realized, however, and it is clear that this must become one of the highest priorities for future research and clinical focus in the next era of health care evolution.

NOTE

1. The "Baby Doe" regulations were instituted by the US Congress to protect handicapped newborns from being denied treatment on the basis of their handicap.

## Suggestions for further reading

Als, H., Lawhon, G., Duffy, F. H., McAnulty, G. B., *et al.*
Individualized developmental care for the very low-birth-weight preterm infant. *Journal of the American Medical Association* (1994), **272**, 853–8.

Beard, R. W., Bentall, A. P., Brundell, J. M., *et al.* Recommendations for the improvement of infant care during the perinatal period in the United Kingdom. *Report of the BPA/ROCG Liasion Committee*, March 1978.

Behrman, R. E. Preventing low birth weight: a pediatric perspective. *Journal of Pediatrics* (1985), **107**, 842–4.

Bhushan, V., Paneth, N. and Kiely, J. L. Impact of improved survival of very low birth weight infants on recent secular trends in the prevalence of cerebral palsy. *Pediatrics* (1993), **91**, 1094–100.

Desmond, M. M. A review of newborn medicine in America: European past and guiding ideology. *American Review of Perinatology* (1991), **8**, 308–22.

Harrison, H. The principles for family-centered neonatal care. *Pediatrics* (1993), **92**, 643–50.

Horbar, J. D., Wright, E. C., Onstad, L., *et al*. Decreasing mortality associated with the introduction of surfactant therapy: An observational study of neonates weighing 601 to 1300 grams at birth. *Pediatrics* (1993), **92**, 191–6.

Paneth, N. P., Hilly, J. L., Wallenstein, S., *et al*. Newborn intensive care and neonatal mortality in low birth weight infants. *New England Journal of Medicine* (1982), **307**, 149–55.

Silverman, W. A., Fertig, J. W. and Berger, A. P. The influence of the thermal environment upon the survival of newly born premature infants. *Pediatrics* (1958), **22**, 876–86.

*Special Report: The Robert Wood Johnson Foundation*, Princeton, NJ, **2**, 1978.

The Investigators of the Vermont–Oxford Trials Network Database Project. The Vermont–Oxford Trials Network: Very low birth weight outcomes for 1990. *Pediatrics* (1993), **91**, 540–5.

Wegman, M. E. Annual summary of vital statistics – 1993. *Pediatrics* (1994), **94**, 792–802.

# Medical technology and the child

AMNON GOLDWORTH

EDITORS' SUMMARY

Amnon Goldworth, PhD, discusses the case of Baby L as a good example of the problems arising from new technologies in the Newborn Intensive Care Unit. While we are able to prolong the lives of such babies, when in the past they would have died, we are concerned about the best interests of the child. Is it enough that they live, albeit with severe mental and developmental deficits? Should our technology be curtailed in such cases, since so much care and nurturing will be required by the parents that they will have virtually no time for their other duties? How can we make judgments about the quality of another's life? Can parents push physicians to provide care the latter consider medically futile? These are just a few of the questions Professor Goldworth poses.

## Technology and some attendant problems

Technology is the application of knowledge in order to gratify human wants and desires. Its success is marked by an expansion of human choices. But there are problems generated by new opportunities for action. One problem derives from the fact that an action taken may fail to achieve the expected result, not because the technology fails, but because what it produces is not what is wanted. Another problem is based on the fact that the action taken produces undesirable as well as desirable results. For instance, we are effectively transported from here to there by an automobile, but only at a cost in environmental degradation. The third problem emerges from the fact that new technology generates novel conditions whose consequences are only partially understood or not understood at all. This precludes the ready application of moral principles, where these are called for, or the appeal to moral precedents, where these are needed, because new conditions call for new thinking. These problems are dramatically evident in the case of children at the early stages of life.

## Benefits and costs in neonatal medicine

The last three decades have seen dramatic changes in the care of neonates that have led to a reduction in neonatal mortality. Some infants now survive who are born as young as 22 weeks of age and who weigh as little as 450 grams. This is due to improvements in the delivery of care found in neonatal intensive care units. But such improvements are themselves the product of scientific discovery and attendant technological applications such as warming devices, mechanical ventilation, blood gas analysis, tracheal intubation and the use of surfactant. However, the dramatic improvement in the rate of survival of infants who are at risk has had negative effects. The average daily monetary cost of treatment of such infants in the USA is $2000. In addition, there are problems (medical iatrogenesis) that are generated by the application of medical technology itself. Feeding tubes are conduits for infection; needed oxygen can cause blindness; mechanical ventilation can generate chronic lung disease. In addition, there are suspected negative effects of intensive care on the neurological and psychological development of the surviving child. Such undesirable and mixed effects make it difficult to decide whether, when and how technologically innovative treatments are to be applied. We are also confronted with unprecedented situations whose moral worth is difficult to ascertain. Should we make every effort to sustain the life of an infant born without a cerebral cortex (an anencephalic infant) and therefore incapable of thinking or feeling? What should the nature and extent of the treatment be for an infant who has suffered a neurologically devastating bleed? These situations generate controversy, as the following case illustrates.

## Baby L

Baby L was born at 36 weeks and weighed less than one kilogram. There were complications during the last trimester of the pregnancy caused by distention of the pelvis due to obstruction of the ureter (fetal hydronephrosis) and deficiency of amniotic fluid (oligohydramnios), and her condition at birth was poor (Apgar: 1 at 1 minute, 4 at 5 minutes, 5 at 10 minutes). She was stabilized and weaned from mechanical ventilation. However, she was not responsive neurologically except to pain. She also had stomach surgery (a gastrostomy and a Nissen fundoplication) and surgery (a tracheostomy) to facilitate the passage of air and the evacuation of secretions. During her intial 14 month stay in hospital, she had episodes of choking (aspiration) and uncontrolled seizures. In the ensuing months, she was repeatedly hospitalized for pneumonia and systemic infection. After 23 months, she required mechanical ventilation and cardiovascular support for worsening pneumonia and sepsis. During this time, her mother rejected the opinion of her daughter's health care providers that further intervention would be futile and inhumane because it would prolong the child's pain and suffering. She insisted that everything be done to ensure her daughter's survival and ultimately took the matter to court. The court decided to transfer the child to another physician who was willing to treat her. As last reported, Baby L was still alive two years later, but was seriously disabled and possessed the mental status of a 3 month old.[1] Although the chiefs of service, primary care physicians, nurses, hospital lawyer and chairpersons of the institutional ethics committee were in agreement that further medical intervention was unwarranted, consensus was absent in later discussion of the Baby L case.[2] This was due to problems stemming from the uses of the notion of best interest and the notion of futility. I now explore these two notions briefly.

## Best interest

Baby L's health care team and some of the later commentators believed that her best interests would not be served by further life-preserving efforts. However, several of the later commentators did not agree. One believed that where a request to save the life of a child comes from a knowledgeable and committed parent and there is some benefit to be obtained from further treatment, it should be provided.[3] Several others questioned medical judgment that could threaten the interests of the incompetent patient. They believed that the important question was whether continued existence for such a patient was better than no life at all.[4]

## Futility

Baby L's health care providers also believed that continuing treatment would be futile. But this judgment is fraught with ambiguity. How we value an action is dependent on the value of its goal and the probability of achieving it. Both of these elements can generate disagreement. The health care providers concentrated on the neurologic condition of the child. But several of the later commentators argued that the goal was to reverse the life-threatening respiratory condition brought about by the pneumonia.[5] This reversal was not futile, since it could be achieved.

We can be objectively certain that a treatment is futile only when there is agreement by all interested parties concerning the goal of the treatment and the probability of success is zero. Should the probability be more than zero, then we may be left with a disagreement about the value of the treatment based on a difference in assessing the value of the goal of treatment.

Where there is a difference of opinion about the goal of treatment then what is perceived to be successful treatment will also be a point of contention. Most medical treatments produce some change in the existing condition of the patient. Where such a change may be viewed as unimportant by the physician because it fails to alter the long-term effect, it may be intrinsically important to the patient or surrogate. For health care providers, further treatment of Baby L was judged to be a net loss, notwithstanding the non-futile outcome obtained by treating the pneumonia and sepsis. For Baby L's mother, further treatment was judged to be a net benefit, notwithstanding its inability to alter the child's chronic condition. It appears then that futility, which is intended to serve as an objective criterion, is, in most instances, a subjective one. It is objective only in those very rare instances when the probability of successful treatment is zero.

## The need for public policy

In considering the best interests of Baby L, the health care staff and commentators were ignoring the best interests of others. This is proper given their duty to care for this particular child. But, given a world of limited resources, this exclusivity is bound to generate unmet needs. How we provide for some when we cannot provide for all requires, from the moral point of view, that we ration medical care under the constraints imposed by distributive justice. This is determined by the benefits and burdens distributed by society that it believes satisfy conditions of fairness or equity. Determining that a given treatment is futile can be viewed as a means of rationing medical resources. But, it cannot be said to satisfy conditions of fairness or equity given the uncertainty of its application in most contexts.

What fairness or equity requires in the Baby L case, I will leave to the reader to decide. But, if we are to avoid contentiousness and the idiosyncratic perspectives of physicians, parents and lawyers, then the issue of the rational allocation of scarce resources must be addressed by the whole community as a matter of social need and resolved in terms of public policy.

## Baby L's future

Baby L's case illustrates the power of the new technology. At an earlier time, she would probably have not lived for many hours. Whether her survival in present circumstances can be viewed as a benefit to her is, as we have already seen, very much an open question.

We concentrated on this child's past life. During this time, there was a significant amount of maternal nurture. If we consider her future, we can see that this child, with an arrested development of three months, will need constant nurture if she is to continue to live. If she lives, we know that she will continue to lack most of the capacities and attributes that constitute normal growth and development. She will not have any autonomy or responsibility and she will lack the ability to communicate except in the most elementary ways. For these reasons, she will not possess moral agency. She will also lack the ability to form significant human relationships. She will be unable to comprehend herself because she lacks a concept of self. She will be unable to comprehend the world as more than a buzzing confusion because she may be devoid of any capacity to organize experiences. There is a deep sadness to such an existence which is not felt by her, but rather by us. And this, too, is a price we pay for advances in technology.

NOTES

1. Paris, J. J., Crane, P. K. and Reardon, F. Physicians' refusal of requested treatment: The case of Baby L. New England Journal of Medicine (1990), **322** (14), 1012–13.
2. Fleischman, A. R., Perelman, R. H. and Fost, N. C., *et. al.* Physicians' refusal of requested treatment. *Journal of Perinatology* (1990), X (4), 407–15. It should be understood that all further notes refer to this article.
3. Fleischman, A. R. *ibid.*, p. 407.
4. Perelman, R. H., Fost, N. C., *ibid.*, p. 413.
5. Fleischman, A. R., *ibid.*, p. 407. Perelman, R. H., Fost, N. C. *ibid.*, p. 413.

## Suggestions for further reading

Belkin, L. *First Do No Harm*. New York, London: Simon & Schuster, 1993.

Frohock, F. J. *Special Care: Medical Decision at the Beginnings of Life*. Chicago, London: University of Chicago Press, 1986.

Jonsen, A. R. and Garland, M. J. *Ethics of Newborn Intensive Care*. San Francisco: Health Policy Program, School of Medicine, University of California, and Berkeley: Institute of Governmental Studies, University of California, 1976.

Lyon, J. *Playing God in the Nursery*. New York, London: W.W. Norton and Co., 1985.

President's Commission for the Study of Ethical Problems in Medicine and Biomedical and Behavioral Research, *Deciding to Forego Life and Sustaining Treatment: Ethical, Medical and Legal Issues in Treatment Decisions*. Washington DC: Government Printing Office, 1983.

# On caring for children

MARY B. MAHOWALD

EDITORS' SUMMARY

Mary B. Mahowald, PhD, delves into the changes that have occurred during the past 50 years in our views of the moral status of children, and how those changes alter the way in which we ought to treat their ability to consent to treatment, to understand death, and determine their own medical treatment when that involves technological intervention. In addition to providing telling cases of the need for dialogue and even negotiation with children, she also argues that a new kind of bioethics is needed, one based on caring. Her view of the pediatricians' role toward that of children is like that of a mother who wants neither to give birth too soon nor to hang on too long. This chapter supports fundamental changes in our views of children occasioned by technological changes in medicine.

Children come in even more different shapes and sizes than adults. They also come in as broad a range of abilities, competence, cultures and races. However, one element of comparison by which children clearly outstrip adults is their potential; one aspect by which most adults outstrip most children is their experience. All of these similarities and differences are relevant to ethical decisions in health care.

While newborns are the least experienced of persons, they also have the greatest potential. They are, after all, at the very beginning of their lives. Yet, as we saw in the chapter by Adams (this volume), infants have not always been regarded as full-fledged persons. Unlike their older counterparts, newborns are not consistently treated as if they have the fundamental right to life that is generally granted to persons as such. Parental decisions regarding treatment or nontreatment of their newborns have been determinative particularly in situations where infants are severely anomalous or premature. Physicians have supported parental decisions in this regard through "therapeutic passivity." Although the "Baby Doe" controversies of the 1980s led to legal clarification of the obligation to treat disabled infants (see p. 97, note 1), it did not radically alter the tendency to view infants as less-than-fully persons.

Nor is this tendency only applicable to newborns. Further along but still at the beginning of their lives, older children are sometimes treated as less-than-fully persons. At the turn of the century, as Adams has noted, there was little child advocacy, and children were frequently exploited both at home and in the workplace. Although the situation has improved since then, children are still often regarded as a class of individuals whose very personhood is dependent on others. This dependency, which is greatest in the newborn, constitutes a vulnerability common to those in need of health care, regardless of their age.

Encounters between children and health care professionals arise in healthy, normal situations, as well as disease or disability states. Even in the womb, potential children are often monitored, and their emergence into the world from their mothers' bodies is typically observed and assisted by health professionals. Sometimes as we have seen, this means high technology care in the neonatal intensive care unit. As development continues, both physically and psychologically, children often become involved in the health care system, e.g. through well-child care, routine procedures such as vaccinations, and treatment for occasional but normal needs such as infections or accidental injuries.

Health care workers who treat children throughout the span of childhood face a comparable challenge to one faced by parents: to recognize and respond to the developmental level of each child. While laws are relatively clear about the onset of adulthood with its concomitant rights and responsibilities, parents and children alike often see a discordance between this legal clarity and their experience. Some children, after all, are more mature than some adults, and some children are much less mature than other children of the same age. For many a pediatric practitioner, the daunting task of caring is to adjust to the uniqueness of each child through the course of his or her development.

This chapter is intended to illuminate the task of caring for children by focusing on its central concepts and their implications. With regard to caring, I contrast the meaning of care in the health care system with the meanings of care recently developed by psychologist Carol Gilligan and philosopher Nel Noddings. The pitfalls of a care ethic as experienced by parents and elaborated by feminists are also considered. As a means by which paternalism may be avoided and the developing autonomy of children respected, I discuss possibilities for joining an ethic of care to an ethic of justice.

With regard to children, I offer examples of how they differ both generally (e.g. by age group) and individually (even within the same age group), and argue that the developing moral agency of children deserves more regard than is legally required from health caregivers. I also consider children's concepts of disease and death, and their implications for decisions by and about them. Finally, I recount and comment on specific cases that illustrate the challenge of respecting the developmental level of children and adolescents in pediatric practice.[1]

## Care: Its meaning and pitfalls

The term "care" has long been used in medical practice, often interchangeably with "treatment." Recently, however, "care" has been identified with an ethical orientation applicable to other areas of life. The care orientation has been contrasted with the justice orientation of principle-based, traditional ethics. Justice, according to most philosophers, requires a certain "blindness," i.e. the ability to close our eyes to the ties that may irrationally bind us to some individuals more than others, so that we can abstractly consider issues whose resolutions are universally applicable. Gilligan maintains that a care-based approach gives greater weight to relationships and attachments to persons than to individual rights and impartial judgments that ignore different circumstances. Her studies have shown that women in general are more likely than men to base ethical decisions on care rather than justice.

Not surprisingly, Noddings points to maternal nurturance as the paradigm for a caring relationship. Maternal nurturance, however, is triggered naturally or spontaneously through the experience of motherhood. Ethical caring, in contrast, is undertaken deliberately, as in the physician–patient relationship. Noddings describes two elements of ethical caring: engrossment and motivational displacement. Engrossment refers to an ongoing preoccupation or concern; it corresponds to the dictionary's definition of care as a troubled or burdened state of mind, the burden being the interests of the other. Motivational displacement refers to an identification between the other's interests and one's own. Although this identification may be analogous to empathy, motivational displacement means not simply that I have a real sense of the other's needs or feelings, but that the other's needs or feelings really are mine as well. In health care it is not that I am disappointed that a patient suffers pain or loss of mobility, or that a family member suffers because of a loved one's pain, but I as caregiver suffer the pain or loss myself. In advocating for children or parents, therefore, I am advocating for myself.

Considering how the mother–child relationship is started sheds light on how a care-based approach is applicable across the whole spectrum of pediatric practice. When the fetus is ready to move from womb to world, ready to survive apart from a pregnant woman, the mother literally pushes the baby out of her body. The umbilical cord is cut, and now she really is a mother, with a unique, ongoing, and irreversible relationship to her child. The child is still dependent, but not uniquely, on her, and throughout the child's life, the mother continues to foster the child's independence and autonomy. Her parental challenge is to determine just how much independence the child can handle without undue risk.

Another aspect of maternal nurturance that sheds light on a care approach to pediatric ethics is the fact that mothers often nurture a number of children at the same time, each with differing needs, desires, and possibilities. Only a foolish parent would define her obligation to each child as the same, and only a foolish parent would deny that she herself has limitations. Just as the care of infants is generally more time consuming (although not necessarily more demanding) than the care of teenagers, so optimal care of each one of multiple patients, even those of the same age, demands different expenditures of time, energy, and expertise. Like good parents, good clinicians seek to overcome their own (inevitable) limitations by asking for help; the consult model of medical practice is thus supported by a care model of moral reasoning.

A care ethic emphasizes context rather than abstractions. So, too, of course does clinical reasoning and clinical ethics. Just as physicians tailor the principles of therapeutics to fit the health needs of particular patients, so too must they "let the context be their guide" in applying ethical principles of respect for autonomy, beneficence (acting in the best interests of others), and justice to children. The context to be considered in ethical decision-making, however, is broader than one comprising medical burdens and benefits alone; it involves relationships with parents, parental beliefs and wishes, and the beliefs and wishes of other caregivers as well as the children.

Most parents are well aware that children, even rather young ones, can manipulate them – so too in the relationship between children and health professionals. From an ethical point of view, the possible exploitation of caregivers raises concerns about justice. But consideration of justice returns us to principle-based reasoning, to the necessity of avoiding exploitation through a critical perspective on care. Gilligan describes justice and care as *different* perspectives, claiming that most people use both models of reasoning. Marilyn Friedman proposes an integration of the two models, arguing that caring for another requires insistence that the one cared for treats the carer justly. To allow oneself to be manipulated is not caring behavior. Caring means helping the other to be a better person, and this sometimes entails "tough love." The implications of this view for practitioners as well as parents of either sex are obvious. Just caring for children involves recognition and encouragement of their developing autonomy, along with responsibility for the consequences of its expression.

## Children: Developing moral agents

Although the law maintains a rather sharp distinction between children and adults, some children are clearly as mature as some adults. The principle of respect for autonomy applies to individuals of any age who are capable of exercising autonomy. This principle is grounded in the assumption that autonomy or self-determination is a good in its own right, and that human beings can only act morally to the extent that they are free to do so. Infants are not moral agents because they are incapable of comprehending the moral or immoral dimensions of specific actions; older children and adolescents act as moral agents to the extent that they recognize and choose moral or immoral actions as such. As with adults, it is often difficult if not impossible to determine whether an individual child is acting as a moral agent in a specific case.

Consider, however, the following situations suggesting moral agency in children:

1. A group of preschoolers are playing hide-and-seek. A "hider" complains that the "seeker" has cheated because she peeked while counting to 10.
2. A 10-year-old is a potential blood marrow donor for his sibling who has leukemia. On listening to an explanation of the procedure with its harms and risks to him and its possible benefits to his brother, he says he wants to do this for his brother.
3. A 14-year-old who is mentally retarded scolds her younger sisters for fighting with each other.
4. A 7-year-old secretively takes cash from his mother's purse, intending to buy candy at the neighborhood store.
5. A 9-year-old is aware that she is dying, but does not broach the subject with her parents. She tells the nurse she does not want to "make them cry."
6. A 16-year-old mother gives consent to lifesaving treatment for her seriously ill newborn.

In each of these examples, the behavior of the child or adolescent suggests a capacity to act as a moral agent. Whether the individual is in fact expressing that capacity remains an open question because ability does not necessarily imply exercise of the ability in a particular situation. In each case, however, the individual apparently makes a judgment that involves moral values.

The first and third situations present the most problematic scenarios in support of the child's capacity for moral agency: for case 1, the child's very young age seems to preclude the possibility; for case 3, mental retardation seems to preclude it. However, leaving aside the question of a child's moral consciousness, the pre-schooler's perception that cheating is something to complain about coincides with the general perception of adults in that regard. The validity of the child's perspective seems evident here because an ethical principle called fairness has been betrayed. But fairness only makes sense in the context of social awareness; the child's sense of the wrongness of cheating stems from that awareness. By complaining she apparently wishes to end the seeker's cheating behavior, or incite another (for example, the teacher) to end it.

Adolescents are generally credited with capacity for making more responsible decisions than younger children. The feature that further complicates case 3, however, is that the 14-year-old under consideration is retarded. Retardation spans a spectrum from profound to mild. Scolding a sibling for fighting suggests that the scolder is not severely retarded. Not only does the scolding indicate an assumption of some degree of competence, it also suggests a recognition of the importance of relationships and moral cognition that fighting is wrong, at least in some circumstances. Mental retardation means a slower pace of learning or development but does not imply that understanding of relationships, values, or principles is impossible.

The second situation describes a borderline case of competence for a decision to undergo a low risk procedure for the sake of another. The moral content of the decision is obvious, but whether a particular 10-year-old child understands the risk and intent of the procedure is questionable. In the case of a 5-year-old potential donor, most practitioners would follow the parents' wishes whether the child "agreed" with them or not. They would not oppose the dissent of the 15-year-old, even with parental permission. But the case of the 10-year-old would need to be settled on an individual basis. Whatever the age, assessment of competence should be distinguished from assessment of the moral content of the decision, because one does not imply the other. Moreover, if we are to respect the autonomy of competent potential donors on a consistent basis, this respect applies to refusal as well as consent to bone marrow donation. If a refusal is challenged for the sake of the prospective recipient, this should be acknowledged as a violation of autonomy, whether the competent donor is a child or an adult. The violation of autonomy may be morally defended in either case on grounds of obligatory beneficence towards the recipient. Legally, no adult would be forced to undergo bone marrow donation for the sake of another.

The 7-year-old who secretively takes cash from his mother's purse indicates moral awareness through his secretiveness. Recognition of need for money, of where to find it, and of how to get it connotes basic understanding of the cause–effect relationship. Developmental studies clearly establish a sense of the self as independent or autonomous even before age 3. Nonetheless, for children as well as adults, consistency of behavior is not determinative of whether an individual's action or decision is autonomous. No correlation has been demonstrated between rational processes concerning morality and the actual behaviors selected by individuals. In the absence of empirical evidence to the contrary, then, we cannot legitimately disenfranchise children from the moral enterprise solely on the grounds that their behaviors are inconsistent in themselves, or with the moral principles they claim as theirs.

The dying, 9-year-old girl in case 5 apparently has some meaningful concept of death. In fact, her decision not to broach the subject with her parents suggests that she well understands its seriousness and sadness. Her parents are probably aware of the gravity of the situation, and she may realize that. Even if she does not, however, her decision not to discuss the matter with them illustrates a crucial requirement of moral agency, namely recognition that one's own behavior influences others, whether negatively or positively. It also reflects the sensitivity to relationships on which a care model of reasoning is based.

Regarding case 6, most would agree that a normal 16-year-old is capable of making moral (and immoral) decisions. Indeed the law has predominantly acknowledged this through the "mature minor precedent," even while setting 18 years as the age of legal majority. Basically, this precedent of the United States judicial system is a long-standing practice by which parental consent for treatment of minors over 14 years of age has not been required so long as the adolescent is judged competent, and there is some impediment (which may simply be the adolescent's request that parents not be informed) to obtaining parental consent. The probability of competence for a 16-year-old, the status of emancipated minor, and the assumption of a special relationship between (even a teenaged) mother and her child constitute a convincing case for claiming that the young woman described in case 6 is legally and morally capable of responsible decision-making for her newborn. As for the content of a decision to provide life-saving treatment to a seriously ill infant, the moral dimensions of the decision are obvious and momentous.

The range of ages, mental ability, and circumstances of the situations described above illustrate variables that influence not only the capacity for moral decision-making, but the content of these decisions. This variability is further illustrated in children's concepts of illness and death.

## Children's concepts of illness and death

Children's concepts of illness have been correlated with Swiss psychologist Jean Piaget's stages of cognitive development: prelogical thinking, which is characteristic of children between 2 and 6 years old; concrete logical thinking, which is manifest in those between 7 and 10; and formal logical thinking, which is typical in children who are 11 years and older. Prelogical thinking involves a rudimentary sense of the cause–effect relationship and is immediately triggered by familiar spatial or temporal cues. In their studies of children's concepts of illness, psychologists Roger Bibace and Mary E. Walsh identify two types of prelogical explanations: phenomenism and contagion. The phenomenist explanation interprets the cause of illness as an external concrete phenomenon that is spatially or temporarily remote. Asked what causes colds, the child at this stage might answer: "The sun," or "God." When the cause of illness is seen as proximate but not touching the child, the concept is one of contagion. Still at a prelogical stage, the child may think that colds are caught "from the outside," or "from someone who gets near you."

Children who have reached the stage of concrete logical reasoning are generally able to distinguish between what is internal and external to themselves. According to Bibace and Walsh, their explanations of illness then include concepts of contamination or internalization. The child may think people "catch" colds from other persons (contamination), or from harmful bacteria that have entered the body by swallowing or inhaling (internalization). At the stage of formal logical thinking, children have a better sense of the distinction between themselves and the world, allowing them to point to the source of illness within the body even while describing an external agent as its ultimate cause. Physiological explanations show an understanding of the nonfunctioning or malfunctioning of an internal organ or process. Psychophysiological explanations extend the recognition of internal physiological causes of illness to recognition of psychological causes as well. A child who is 11 years old, for example, is likely to understand that a heart attack may be caused by underlying tension.

Bibace and Walsh's account of children's developmental concepts of illness support the view that children may be moral agents at an early age. It also suggests possibilities for fostering moral agency by attending to the ways in which children understand illness. For example, physiological explanations of the virus that causes chicken pox are probably futile for a 7-year-old, whereas a concrete description of how more pox will appear, are likely to itch, and so on, will be quite meaningful. Because young children tend to focus on immediacy, preparation of a 5-year-old for surgery might concentrate on observable events such as the bright lights in the operating room and the doctors' gowns and masks, whereas preparation of a 12-year-old might include a description of anatomical details involving the surgery itself.

Preparing a child for death may be more difficult than preparing one for surgery, but the preparation may again be facilitated by recognition of developmental stages of children's concepts. According to psychologist Susan Carey, experts agree on three periods in that development. The first and earliest (which Carey attributes to youngsters 5 years old and under) is a concept of death as sleep or separation. While the emotion this concept evokes is a sad one, death is not recognized as final or inevitable because people ordinarily wake from sleep and may return after their separations.

In the second stage of the child's emerging concept (early elementary years), the child views death as a permanent cessation of existence. The cause of death, however, is generally thought to be external. There is as yet no sense that death may be caused solely from within the body, or as a result of internal processes initiated by external events. In the third stage (about 10 years and older), death is seen as an inevitable biological process. The notion of death's finality is sometimes tempered by belief in an afterlife (as it is in some adults), but there is still the recognition that the dead person will not return "in this life."

Despite Carey's claim of universal agreement about the stages of children's concepts of death, some authors maintain that most children understand the essential components of a concept of death (irreversibility, nonfunctionality, and universality) as early as 7 years of age. Regardless of whether the experts agree, however, some children are obviously more precocious than others in their conceptual development. Through the ages, for some it is the experience of their own terminal illness, or being closely associated with dying persons and deaths, that explains this precocity. It is important, therefore, to pay attention to individual exposure to death as well as generic variation in children's concepts of death.

With regard to both serious illness and impending death, philosopher Gareth Matthews argues that disclosure to children of their prognoses is morally more defensible than withholding such information. He claims that the usual arguments for nondisclosure are based on two worries: first, that the bad news will be successfully communicated to the child; and, second, that the child will be unable to cope with the news. Regarding the first worry, Matthews maintains that "one will simply not succeed in getting across the mature message, that is, the full meaning of impending death." Concerning the second worry, children may be helped to cope with the news of death by their inability to fully comprehend it. Even for young children who conceive of death as sleep from which one will eventually awaken, the news of death is probably not so awful as it is for adults. According to Matthews, adults' own difficulties in dealing with the deaths of children probably underlie their inclination to withhold information from children.

Marilyn Bluebond-Langer, a social scientist, has amply demonstrated the capacity of older children to understand and deal with their impending deaths. She describes five stages through which children dying of cancer progress. First, they think: "I am seriously ill"; second, "I am seriously ill but will get better"; third, "I am always ill but will get better"; fourth, "I am always ill and will not get better"; and fifth, "I am dying." While categorizing the stages, Bluebond-Langer found that chronological age was irrelevant to their sequence. The experience of the disease itself, with all the unpredictable elements of that experience, prompted the different stages. This is a strong argument for recognition of the limitation of generalizations based solely on chronological age. While drawing on the advice of experts in cognitive development, it is thus important to relate their finding not only to individual variations in cognitive ability but also to children's experience as individuals. Beyond parenting, the field of pediatrics presents an important challenge in this regard.

### Implications for pediatric practice

Children often meet the requirements of informed consent, despite lack of acknowledgment to that effect by the legal system. The usual requirements for informed consent are competence for responsible decision-making, understanding of pertinent information, and voluntariness. All three factors are found in some children at least some of the time, and not found in all adults all of the time. In general, moral agency should be respected in whomever it occurs, and to the extent that it occurs. Depending on the child and the circumstances, this may mean their participation in decision-making, assent or dissent to decisions made by others, or their fully informed consent. Truly informed and free consent may thus be viewed as an ideal to be approximated by children as well as adults.

As I have already suggested, participation in decision-making is possible even at a very young age. The participation may be as minimal as the child's developmental awareness allows. The following case is an example of failure to respect a young child's developmental awareness by allowing her to participate in a tragic situation and its implications. As this particular case unfolded, a seriously ill mother's right to see her child was also denied.

> A family of four experienced a devastating automobile accident in which the 7-year-old boy and his father were killed. The 5-year-old girl was transported to the nearest hospital tertiary care center, where she was placed in the pediatric intensive care unit after undergoing emergency surgery. Her mother, also in critical condition, was transported to the same hospital, treated, and placed in the surgical intensive care unit. Although the child was expected to recover, the woman was thought to be near death.
>
> When told of the death of her husband and son, and of her daughter's and her own condition, the woman asked to see her daughter. Concurrently, the little girl asked several times to see her parents. She was told that they were in another part of the hospital

and not able to visit her yet. The pediatric surgeon insisted that a visit between mother and child not be arranged, maintaining that the child would be too upset by the sight of her mother's injuries and illness. "As the child's physician," she claimed, "I am obligated to prevent anything that might compromise my patient's welfare."

Whether the surgeon had a legal right to refuse the parent's request is doubtful, but that is not my point here. The woman's autonomy was overridden, and the child's desire was also overruled. Nurses challenged the physician's refusal, but by the time a meeting was convened to resolve the disagreement, the woman had lapsed into unconsciousness, making her request unfulfillable. Although the child's desire to see her mother might still have been granted, the surgeon's refusal was honored.

If seeing her mother would have compromised the child's chance of recovery, the physician had a valid but inadequate reason for opposing both patients' wishes. Obviously, the degree of compromise and its risk needed to be assessed. If these were minimal, respect for both patients' wishes should have overridden the principle of beneficence; if the compromise and risk were grave, beneficence might have overridden the reduced capacity for autonomy of the child, and possibly also that of her mother.

Apparently, the pediatric surgeon in this case was unaware of studies regarding children's concepts of illness and disease and of their capacity for dealing with such tragic events. Had she recognized this limitation, she might have consulted someone with expertise in these matters. In fact, a child psychologist who heads a program for counseling children and their parents on how to deal with the deaths and dying of loved ones was available at the pediatric surgeon's own institution. Had he been asked, he would have visited both child and mother and offered advice. Unfortunately, he was not invited to do so.

Participation of young children in health care decisions is mainly a matter of sensitive disclosure. There are times, of course, when the child has *no* capacity for understanding. Even at an early age, however, most children have at least rudimentary understanding of death, life, health, and disease, concepts basic to communication with them about their care and that of their family members. Such communication is an essential component of respect for their developing autonomy. If the child cannot understand these concepts, then mere disclosure is unlikely to be harmful; if the child *may* understand them, disclosure is obligatory in proportion to the probability of understanding and in proportion to the risk of harm from the disclosure. Generally, the thought of what might happen is worse than the reality. But if disclosure would precipitate resistance to a procedure crucial to restoration of a child's health, foregoing it is justified. The assumption in such circumstances is that the child's limited understanding has reduced her capacity for autonomous decision-making.

While disclosure is a component of respect for autonomy, it is not equivalent to the obligation of obtaining assent, or respecting dissent, once disclosure is provided. The pediatrician who explains to a 7-year-old child what a lumbar puncture involves does not require the child's assent to the procedure so long as it is crucial to proper diagnosis and treatment of a serious infection such as bacterial meningitis. But the same child may decide whether to present his or her left or right arm for a blood sample, and may also decline to participate in a nonrisk research project that the treating physician proposes. Assent and dissent are modes of participation, but not equivalent to fully informed and free consent.

Fully informed consent of children is best approximated in cases such as the following:

> A 12-year-old boy had congenital strabismus (eye muscle imbalance) for which he had been operated on at 2 years of age. Although the surgery achieved partial correction, the imbalance became more obvious as the boy got older, and he was self-conscious about it. At a routine visit the ophthalmologist told the boy that new surgical techniques could be utilized to permanently improve the appearance of his eyes. It would not, however, improve his vision. The boy said he did not want surgery.
>
> Several weeks later, the boy changed his mind and told his parents he might like to have the operation, but would first like to ask the doctor some questions about it. Responding to a message left by the boy's mother at his office, the physician called the boy and said: "Fire away. Ask me anything you want because you should know what it would be like before you decide to go through with this kind of operation. You don't have to have it unless you want it."

Pressure from parents might, of course, reduce the autonomy of children capable of informed consent to elective procedures such as the preceding. Such pressures may come from caregivers as well. Psychological and social constraints affect children as well as adults. Thus the voluntariness essential to consent is present in various degrees in different situations. The caregiver or parent who wants to maximize respect for a child's autonomy attempts as far as possible to reduce those pressures.

Admittedly, had the 12-year-old declined life-saving rather than elective surgery, the argument for respecting his autonomy would not have been compelling. But what about situations in which treatment is medically indicated, yet not immediately necessary in order to preserve life? Consider, for example, the following case.

A 15-year-old girl had severe scoliosis (curvature of the spine),
requiring surgery in order to prevent respiratory compromise. The
girl's mother had consented to the surgery, but the girl insisted she did
not want it. Despite her objections, the girl was taken to the operating
room on the appointed day. As the orthopedic surgeon arrived, he
heard her scream "Don't do this to me. I told them I didn't want it."
The surgeon declined to perform the surgery, and sent the patient back
to the floor, where he later apologized to her for not having learned of
her wishes beforehand. At that point he initiated the first of several
discussions in which he "made a deal" with the patient. He explained
that she might die within the next few years if she didn't have the
surgery, and that she could choose the time for it so long as this
occurred within the next half year. The physician's effort to give the
adolescent more control over her medical course was effective. Two
months later, she voluntarily underwent the surgery.

In neither of the preceding cases were the physicians legally obliged to obtain
consent for surgery from their pediatric patients. If they were to respect the autonomy of children, however, both needed to elicit the consent of their pediatric
patients. To their credit, both the ophthalmologist and the orthopedist distinguished between legal and moral aspects of decision-making, recognizing that
the two may be at odds in children as well as adults. They contributed to the
moral development of their patients by maintaining a more discerning and critical
attitude toward differences in individuals, regardless of their age.

Possibly the best metaphor for the role of the pediatric practitioner is that of
motherhood. Ordinarily, a woman becomes a mother by pushing her fetus from
womb to world. The optimal time for her crucial push, while it varies from individual to individual, is determined by the natural course of development. Premature pushes are perilous for children and sometimes for their mothers. Balancing
the need for protection from harm while nurturing autonomy defines the task of
parenthood. Following birth, the challenge of both parents is to foster development towards greater independence until eventually the child is an adult, i.e.
mature enough to care for others. On either side of the relationship there may be
"growing pains." So too for pediatric practice. But caring for children, whether
by parents or by health professionals, has its rewards as well as challenges.

NOTE

1. Much of the material used here on children's moral agency, their concepts of illness
   and death, and implications for pediatric practice appeared in Chapter 11 of my
   *Women and Children in Health Care: An Unequal Majority* (New York: Oxford
   University Press, 1993), pp. 186–201.

## Suggestions for further reading

Bibace, R. and Walsh, M. E. (eds.) *Children's Conceptions of Health, Illness, and Bodily Functions*. San Francisco: Jossey-Bass, Inc., Publishers, 1981.

Bluebond-Langer, M. *The Private World of Dying Children*. Princeton, NJ: Princeton University Press, 1978.

Carey, S. *Conceptual Change in Childhood*. Cambridge, MA: MIT Press, 1988.

Friedman, M. Beyond caring: The demoralization of gender. In *An Epic of Care*, ed. M. J. Larrabee, pp. 258–73. New York: Routledge, Chapman & Hall, 1993.

Gilligan, C. *In a Different Voice: Psychological Theory and Women's Development*. Cambridge, MA: Harvard University Press, 1982.

Kopelman, L. M. and Moskop, J. C. (eds.) *Children and Health Care: Moral and Social Issues*. Boston, MA: Kluwer Academic Publishers, 1989.

Mahowald, M. B. *Women and Children in Health Care: An Unequal Majority*. New York: Oxford University Press, 1993.

Matthews, G. B. Children's conceptions of illness and death. In *Children and Health Care: Moral and Social Issues*, ed. L. M. Kopelman and J. C. Moskop, pp. 140–00. Boston, MA: Kluwer Academic Publishing, 1989.

Noddings, N. *Caring: A Feminine Approach to Ethics and Moral Education*. Berkeley, CA: University of California Press, 1984.

Piaget, J. *The Moral Development of the Child*. New York: Free Press, 1969.

Schell, R. and Hall, E. *Developmental Psychology Today*, 4th edn. New York: Random House, 1983.

Sherwin, S. *No Longer Patient: Feminist Ethics and Health Care*. Philadelphia, PA: Temple University Press, 1992.

# Clinical transplantation

## ROBERT A. SELLS

### EDITORS' SUMMARY

Robert A. Sells, MB, FRCS, a transplant surgeon, discusses the advances in medical technology, surgical skills, and immunosuppressive drugs that make today's transplant surgery so successful. By far, most organs come from cadavers, although, due to scarcity, live donor transplantation is increasing. Arguments about the ethical issues caused by these advances are many and complex – from the definition of death to monetary payments for organs from live donors. Dr Sells is concerned especially about a market economy for organs, since it may decrease the level of altruism and the very nature of the medical intervention itself. There is also the ever-present danger of obtaining organs from vulnerable and poor persons through potentially coercive economic incentives. Transplantation technology requires rethinking the goals of medical intervention and the requirements we place on one another in society.

In the early days surgery was a simple mechanistic science. It developed, stepwise, as new discoveries revealed how organs worked, how they went wrong, and how they could be surgically interfered with to alter the progression of a disease process. Anesthesia and antisepsis opened up the human body to new opportunities: it became possible to unblock or bypass obstructed tubes such as coronary arteries or bowel, to cut out tumors, and to remodel the body's fabric. The successful rearrangement of body parts depends on rapid and effective healing of tissue kept free from infection by an effective immune system.

This reductionist definition of surgery suggests an image of the surgeon as an artisan or engineer, the human body as a machine whose components can be mended when they go wrong, and the operating room as a garage floor. That image incenses most surgeons, but it is a popular public perception. So the replacement of a terminally diseased body part, such as a kidney, liver or heart, seems to fit the mechanistic image. But the technical difficulties are formidable: preserving the function of the excised donor organ during its journey from donor to recipient, preparing the patient for, and sustaining him or her during a lengthy, arduous and, perhaps, bloody operation, and joining the organ's blood vessels seamlessly to those of the recipient. And each of these hurdles requires a technical solution, the successful conjunction of each being crucial to the success of the whole transplant operation.

## The cadaveric organ donor

The principal source of transplanted organs is the human corpse. The conditions that must be fulfilled to remove live organs from cadavers in such a way that they will work immediately in the recipient are exacting. The notion of a patient being dead whilst his or her vital organs (other than the brain) remain alive has been universally accepted by doctors, at least in Western countries, although the issue is still debated by politicians, governments, philosophers and religious authorities. Before the advent of the artificial ventilator, the traditional and unambiguous sign of death was permanent cardiac standstill. However, it is now common for a patient to be resuscitated by prompt ventilation at the time of respiratory arrest due to brain trauma: the brain may be lethally injured (as in brain trauma, or following hemorrhage within the rigid skull, squashing the delicate brain tissue) but the heart and other organs will continue to function with appropriate intensive physiological support.

The decision to abandon resuscitation and thence consider organ donation can be taken only after death of the brain is known to have occurred. The clinical techniques for the determination of brain death were developed in 1969, and have been published and accepted almost universally within the profession. The absence of brainstem reflexes, including the crucial sign of the patient's inability to breathe spontaneously, are the necessary and sufficient signals of brain death. When two experienced practitioners, neither of whom are associated with the care of a potential organ recipient, confirm the absence of these reflexes, then the death certificate may be signed. The technique has certainly stood the test of time: no case has been described in the medical literature of a patient who recovered any brain function following diagnosis of death by these clinical criteria, when left alone and not used as an organ donor.

There are some rare conditions that approximate to, but do not constitute, brain death as defined above: the "persistent vegetative state" may result from a temporary episode of oxygen lack that destroys higher brain function. Resuscitation of such individuals results in an intact brainstem, with spontaneous breathing and blood pressure control, but no sensation, expression, or coordinated movement. Such a patient is totally dependent on nursing care; given a warm environment and adequate feeding, such tragic cases may survive for many years. Anencephaly (a congenital absence of the higher brain) is technically similar to the persistent vegetative state. In the past these seriously incapacitated neonates have been allowed to die, their capacity for independent life being nil.

Can one use cases of the persistent vegetative state and anencephaly as cadaver donors? A consequentialist line of thinking might persuade a doctor that he or she ought to try to make the death of that individual as useful as possible. But in the case of the persistent vegetative state, organ donation would require a wilful, elective, putting to death of the patient, and as such would be an illegal as well as an immoral act. Organs have been removed from anencephalic neonatal children, but there is professional and ethical unease concerning the presence of normal brainstem anatomy and function during organ explantation, and our worry is exacerbated by our limited understanding of neurophysiology in the neonate. Both groups are now excluded as organ donors.

## Rejection and immunosuppression

After the organ has been successfully transplanted, a new biological hurdle has to be overcome before long-term success may be expected. This is the immunological problem of rejection: immunity to infection is provided by specialized cells (lymphocytes) that recognize foreign constituents (antigens) on invading organisms, mount a chemical attack that destroys the membrane of the invading bacterium (or fungus or parasite) and neutralizes foreign viruses. Lymphocytes patrol the body, circulating in the vascular spaces; each is derived from a clone of progenitor cells whose DNA is programmed to react to and destroy one of a multitude of antigens to which the human organism may be exposed. The mechanism by which such a huge spectrum of immune reactivity is generated depends on a process called "accelerated somatic mutation;" this takes place in the lymphoid tissue during the embryonic and neonatal phase of life so that the young human is equipped to deal with bacterial and viral infection from an early age.

Organs from other people, or other species, are immediately recognized as foreign by the host lymphocytes, as the recipient blood circulates through the transplant. Cells programmed by their clonal ancestors "recognize" the new alien tissue, proliferate in response to messenger molecules released by that encounter, and the new transplant becomes a battlefield of inflammatory events. Damage to the graft culminates in the infiltration of killer cells and antibodies, which respectively destroy the functioning graft tissue, and clog the vessels that supply its life-line of arterial blood.

The trigger for rejection therefore is antigenic disparity between donor and recipient. Where no such disparity exists (as between identical twins, who have the same DNA) no rejection will take place. The pioneering achievement of a successful kidney transplant was done in Boston by Drs Murray and Harrison on 23 December 1954. The donor was an identical twin, whose kidney was transplanted into a patient suffering end-stage renal failure due to chronic nephritis. No long-term dialysis treatment was available, and an early death due to fluid accumulation in the lungs, toxic poisoning of the blood, and heart failure, was inevitable. However, this life-saving gift gave the sick twin eight healthy years before he succumbed to a heart attack.

But for organ transplants to be successful between *non-identical* individuals, a therapy had to be devised that prevented the recognition and attack on foreign antigenic (allograft) tissue; at the time of the Boston twin transplant (which required no immunosuppression), the only realistic way this could be done was by destroying most of the white cells in the body by X-irradiation, a technique too hazardous to be employed routinely. The pharmacological challenge was to produce a drug regimen that specifically inhibited those cells which initiate and execute graft rejection, without serious injury to those other bone marrow components that resist infection, prevent anemia, and control blood clotting.

The first drug to meet these exacting demands, azathioprine (Imuran), was first developed by Dr Roy Calne experimentally in England, and was successfully used in patients in Boston by Dr Murray's team in the early 1960s. Since then, our insights into the detailed biological mechanisms of graft rejection have become much clearer, and more refined drug regimens have emerged that greatly alter the delicate "balance of survival" in favor of the patient. Nonidentical kidney transplants using cadaver organs are now routine, many patients surviving five years and more with a well-functioning kidney transplant. The quality of life is near normal, and although the operation and maintenance immunosuppression is expensive, it is very much cheaper than chronic dialysis. Transplanted patients usually return to work, lose their dependence on state support, and of course pay their taxes.

## The increase in demand for organs and its ethical consequences

Thus equipped with a sure means of diagnosing brain death, surgical methods of explanting and implanting the new organs, and suppressing rejection without necessarily injuring the host, clinical transplants boomed: in 1993, nearly 10 000 patients were treated in the USA, with liver, heart, lung, pancreas, as well as renal transplants. The results of these operations commonly equal or exceed the survival and quality of life obtained with many cancer procedures: a renal transplant recipient may expect a normal life with a functioning transplant for up to five years in 70% of cases, similar survival figures being obtained routinely in recipients of heart and liver. Transplantation of the pancreas for diabetes complicated by renal failure offers 80% graft survival at two years, and the expectation of sight in a blind patient receiving a corneal graft is approximately 85%.

The scarcity of cadaveric organs has led to an increase in the number of kidneys transplanted from living donors. In many countries, where chronic dialysis cannot be provided outside the private sector, and where cadaver donors are infrequently available due to cultural or legal restrictions, the only chance of preventing death from end-stage renal disease may be a living related, or a paid unrelated, kidney donor. In a few centers in the USA, Germany and Japan, a shortage of cadaveric livers has led to segmental liver transplants being donated from parents to their children. Segmental pancreas transplants have also been removed from living related donors, the recipients having been diabetic siblings or offspring. Given the increase in success of living donor transplants as judged by graft and patient survival, and given the scarcity of cadaveric organs, live donor transplants have come to be regarded as a pragmatic necessity. However, there is a widely held view in the transplant profession that the only justification for the use of living donors in renal transplantation is a continuing shortage of cadaveric organs.

But, in some developing countries, the poor and socially disadvantaged have come to be regarded as an expandable resource from whom organs may be bought. This situation has produced a lively debate within the medical, ethical, and legal professions concerning the coercive influences that may be brought to bear on healthy people by the plight of chronically sick patients who could be cured by a transplant. Must living organ donation always be a voluntary act? Or is financial gain or other coercion ever justifiable as a means of bridging the gap between the supply and demand for kidneys? The donation of organs by mentally incompetent adults and children has also raised serious ethical concerns about possible exploitation of such individuals if consent to donate is provided by surrogates or guardians. The ethical consensus is that these people should not be used as living related donors of nonregenerative organs. Nine Canadian provinces have enacted legislation that prohibits minors and mentally incompetent adults from making live donations.

The predominant ethical issue in transplantation over the last few years has been the attempt in some Western countries to develop a market in organs. In a free market economy it is very tempting to consider a transplant organ as a commodity. Some economists certainly espouse this view, and, in doing so, come into a head-on collision course with the traditional medical view that an organ transplant is a priceless gift to the recipient and organ donation should always be an altruistic act. The medical view goes on to state strongly that organs should not be bought or sold, either directly or within the framework of a futures market; in this, the doctors in the West are supported by the World Health Organization, which has ruled against physicians being involved in transplantation if they have any reason to believe that the organs have been the subject of commercial transactions. The philosophical view has been expressed: that self-harm per se is not enough to forbid donation (living related donors are after all used in the West) and risk taking per se is not irrational, nor forbidden (there is no law against rock climbing or deep-sea diving); so the mere existence of risk cannot permit a general prohibition that curtails the freedom of all to sell their organs.

The consensus medical response to this libertarian approach can be summarized as follows:

1. A full understanding of the risk is essential for properly informed consent to any form of surgery; since paid organ donors will always be relatively poor, and may be under-privileged and under-educated, the donor's full understanding of these risks cannot be guaranteed. Laws are therefore required to protect these individuals from exploitation by payment for their organs, despite the definite limitation on the freedom of everyone that such legislation imposes.

2. Doctors believe that operations should be performed for therapeutic indications, and not for the acquisition of money no matter how well that money could be spent.

3. A rewarded donor from an impoverished background cannot be presumed to give proper voluntary consent, since the coercive influence of a substantial payment is likely to distort the balance of risk versus benefit in a person's mind, and in conditions of severe impoverishment could amount to a compulsion to donate on behalf of improvement in lifestyle of the donor's family.

4. In a free market for organs, profit is the first objective and medical standards will fall, no matter how well regulated the market is.

5. Legal paid donation would undermine altruistic donation.

6. Traffick in organs divides society because the donor will always be relatively poor and the recipient relatively rich.

7. Commodification of the body is objectionable.

In transplantation, the technical advances have been developed ahead of the ethical rules governing their application, just as humanity's ability to manufacture things runs a step ahead of its ability to use them with taste or discretion. Transplant surgeons have never had a "watch dog" committee overseeing their activity: ethical comment has followed the publication of clinical results. Transplant surgeons generally have welcomed the comments of moral philosophers and lawyers, without whose active interest the passing of safe laws to allow cadaveric and living transplantation would not have been possible.

Perhaps the most difficult aspect of this debate is the knowledge that the ethical principles may be easily agreed by the exponents of transplantation in the West, but their consistent application through legislation and social and cultural acceptance will be a difficult goal to reach world wide. It is not so much the differences in ethical approach that divide the developing countries from the West in this issue, but the flagrant abuse of ethical principles in transplant surgery that appears to be so difficult to prohibit in those countries where the market in organs has flourished. We may only hope, perhaps piously, that appropriate changes in the law will enable cadaver organs to be removed without risk of criminal charges against the surgeon, and that, as the utility of cadaver transplants is perceived, so the pressure for a continuing market in organs will recede, in those countries where it is still practiced.

## Suggestions for further reading

Gillon, R. (ed.) *Principles of Health Care Ethics*. Chichester, John Wiley and Sons, 1993.

Kjellstrand, C. M. and Dossetor, J. B. (eds.) *Ethical Problems in Dialysis and Transplantation*. Dordrecht, The Netherlands; Kluwer Academic Publishers, 1992.

Morris, P. J. (ed.) *Kidney Transplantation Principles and Practice*, 3rd edn. Philadelphia; W. B. Saunders and Company, 1988.

Moore, F. D. *Transplant: The Give and Take of Tissue Transplantation*. New York; Simon & Schuster, 1972.

# Transplantation and ethics

RAANAN GILLON

EDITORS' SUMMARY

Raanan Gillon, BA, MB, BS, FRCP, proposes that a useful method of analyzing ethical issues in transplantation is the four-principle approach, with the additional consideration of the scope of each principle. This approach helps us to identify our conflicting moral obligations in specific circumstances. Thus, should autonomy be extended to a teenager who wants to donate her kidney to a younger brother? Further analysis from the standpoint of benefits and harms leads to questions about the good to be pursued in such an instance, or even more broadly in adult care or medical research. Even if the good is a potential improvement for the patient and a recognition of altruism for the donor, are we justified in deliberately inflicting harm? If so, how much? Finally, allocation of scarce organs is clearly an issue of distributive justice. Dr Gillon applies the four principles analytic framework to the issue of commerce in organs. While for particular buyers and sellers this may be beneficial, autonomy respecting and just, he points out that many believe its anticipated societal harms outweigh these advantages.

## An approach to moral analysis

As a standard basis for moral assessment in biomedical ethics that is compatible with a very wide variety of moral theories, I recommend and use the "four principles plus scope" approach advocated by the Americans Beauchamp and Childress in their famous and standard text book, *Principles of Biomedical Ethics* and also in my own shorter and smaller introduction, *Philosophical Medical Ethics*. According to this approach, biomedical moral problems can be analyzed in terms of four basic moral principles or values that almost everyone, on reflection, accepts as being *prima facie* valid – along with reflection about the scope of application of each of these principles (i.e. reflection concerning to whom or to what we owe the moral obligations entailed by accepting the principles). The four *prima facie* principles, in no particular order of precedence, are:

- respect for people's autonomy or self-determination (insofar as such respect is consistent with equal respect for the autonomy of all potentially affected);
- nonmaleficence or the obligation not to harm;
- beneficence, or the obligation to benefit;
- and justice – roughly the obligation to treat competing claims fairly.

"*Prima facie*" was used by the English philosopher W. D. Ross to mean that the principle is morally binding in a particular situation unless it conflicts with one of the others. Then choices have to be made between conflicting moral obligations.

The four principles approach does not tell us how to choose between conflicting moral obligations when conflicts arise in particular contexts; different moral cultures, even different individual moral stances, will choose differently, a source of great dissatisfaction to those who believe that in any moral dilemma there is only one morally acceptable answer. But even such critics should agree that at least the four principles plus scope approach helps us to see what our conflicting moral obligations are in particular circumstances; critics may simply add their own explanation of how and why the principles should be prioritized one way rather than another when they conflict in given circumstances. The rest of us must balance them, harmonize them, prioritize them or simply choose between them, as best we can, and often in ways heavily influenced by the norms of our cultures, religions, professions and chosen "life stances."

## Four principles and their scope of application

It seems that almost everyone, regardless of background, philosophy, religion, politics or culture, can accept these four principles. If one does not, it is useful to ask oneself which of these principles one rejects and why – and also which additional moral principles or values one accepts that one believes cannot be accommodated within some combination of one or more of the four principles. With the exception of certain religious obligations not expected to apply to those who do not share that religion, I have so far found none.

## Respect for autonomy

The first principle is respect for people's autonomy, insofar as such respect is compatible with equal respect for the autonomy of all potentially affected. Autonomy is essentially deliberated self-rule or self-determination – making thought-out choices for oneself. This ability to make thought-out choices for ourselves is probably the main moral characteristic that distinguishes us from other animals. Each of us has only to reflect on how morally indignant we become (or would become) if people try to impose upon us choices for ourselves with which we disagree to realize how important is the principle of respect for autonomy.

### Scope: Children and donation of organs for transplantation

As with each of the principles, considerable problems arise in deciding who or what falls within the scope of this principle – who counts as an autonomous – or an adequately autonomous – agent whose autonomy ought to be respected. Every parent will acknowledge that babies cannot make thought-out choices and are thus not autonomous at all. But what about young children who are beginning to think for themselves, but do not yet have the knowledge or experience to be regarded as adequately autonomous to have their decisions respected when we regard their decisions as wrong? Similarly in medical ethics, when are children adequately autonomous to have their own decisions respected? For example, in the context of transplantation, when is a child mature (autonomous) enough to have his or her decision respected to donate a kidney or some bone marrow or even some blood (for a blood transfusion is a sort of transplant) to a brother or sister?

### Emotional pressure, autonomy and organ donation

Still pursuing the question of the scope of autonomy, is a donor who is emotionally very involved with the potential recipient adequately autonomous to make a free decision to take the risks involved and donate his or her organ to the loved one – or is the donor's autonomy likely to be excessively impaired by strong emotions? And what about the possibility that the donor does not really want to give up a kidney or other organ but is the victim of "emotional blackmail"? In some cultures family obligation overrides almost everything else, and if a family member – especially a powerful family member – requires a transplant, woe betide any potential donor who refuses to "volunteer." Can a decision to "volunteer" in such circumstances be adequately autonomous for doctors to respect and accept? How can adequate safeguards be established to protect donors against having their organs removed if they do not truly wish to be donors?

## Nonmaleficence and beneficence and organ donation

The second *prima facie* moral principle that we almost all accept is that of non-maleficence, or the obligation not to harm each other. This obligation is accepted even when a corresponding obligation to benefit (beneficence) in the particular circumstances is not accepted. (For instance, even if I reject a moral obligation to give money to a beggar, I have no doubt that I am morally obliged not to kick him in the face: or, even if I reject an obligation of beneficence requiring me to donate one of my kidneys to a total stranger who requires a kidney transplant, still I will acknowledge that I must not go and harm that stranger.) That obligation not to harm others would at first sight seem to prohibit doctors operating on perfectly healthy people to remove one of their kidneys, for undoubtedly any operation produces some harm (the cutting itself, the small but inescapable risk of a general anesthetic, the postoperative pain, risks of infection, the small risk of complications of surgery including the tiny risk of dying).

Normally in medicine the risk or inevitability of harm is justified by the anticipated benefit to the patient of taking that risk. Which brings us to the next principle, beneficence, or the *prima facie* obligation to try to benefit. Whenever one tries to benefit another there is always a risk of harm, so that beneficence always has to be considered along with the principle of nonmaleficence, with the objective being to provide net benefit with as little harm as possible. In ordinary medical practice an operation and its attendant risk of harm is only contemplated if that operation can reasonably be expected to result in net medical benefit for the patient, despite the anticipated risks of the operation. Thus, if a patient has a cancer in a kidney the risk of harm from the operation is normally greatly outweighed by the benefit of removing the cancer, which may otherwise spread and kill the patient. Conversely, if the cancer has already spread and removal of the kidney is therefore unlikely to benefit the patient, then the operation is riskier to the patient than its potential benefits and is therefore not justified.

When it comes to removing a kidney from a volunteer in order to transplant it into a patient who needs that kidney, the net medical benefit argument does not work so far as the donor's operation is concerned – for he or she takes the risks without any corresponding medical benefit. The benefit goes to someone else – the recipient. This moral issue can be perceived from different perspectives.

### Different perspectives on harms and benefits
The most important perspective is surely that of the potential donor. He or she is contemplating a small but significant risk of harm, even the remote risk of death, not to mention considerable discomfort and inconvenience, with no corresponding benefit to himself or herself but with enormous potential corresponding benefit to

the recipient. There would be widespread agreement that there is no moral *obligation* to sacrifice one's kidney (or even one's bone marrow or blood) to benefit others – but there would be equally widespread agreement that any such action would be morally admirable, a fine example of altruism or good action above and beyond the call of duty. (There would not be universal agreement about this – for some people and for some cultures the scope of one's obligation of beneficence might well be seen as requiring such self-sacrifice in the interests of, for example one's child, sibling, spouse or friend.)

Even if the donor has thought carefully about the matter and is clear about his or her desire to donate a kidney – and even if there is no problem about coercion whether by moral pressure or threats or any other sort of coercion – there is still the perspective of the medical profession to consider. As indicated above there is a traditional and fundamental moral norm in medicine, going back at least 2500 years to the time of Hippocrates, according to which doctors only impose harm or risk of harm on their patient if they believe this is likely to produce net medical benefit for that patient. Yet here a distinct though improbable risk of harm, plus pain, disturbance, and inconvenience is borne by the donor patient without any prospect of medical benefit for himself or herself. Should the profession accept such altruism and operate? And even if the medical profession might be prepared to give up its ancient concern to make sure that it only inflicts harm or risk of harm if this is for the particular patient's benefit, what about the perspective of society more broadly? Do our societies wish to allow doctors, with their often extraordinary powers over their patients' lives, to inflict medical harm or risk of harm on one person simply in order to benefit another or even many others?

This is a problem that relates not just to transplant donors but also more widely in the context of medical research. Many societies and their medical professions have arrived at a compromise on this issue, both for medical research and for transplantation. Briefly put, it is that doctors may inflict small degrees of harm or risk of harm on subjects in the medical interests of others, so long as the risk of harm really is very small, (certainly not greater than the degree of risk ordinarily encountered in everyday life) and so long as the subjects, or their proper proxies in the case of children, are thoroughly briefed about the pros and cons of their proposed involvement.

In such circumstances doctors may accept volunteers for medical research that is not in the latter's personal medical interest – and doctors may operate on donors of organs and tissues. In this way, societies and their medical professions have attempted to balance, on the one hand, the traditional and morally important reluctance of the medical profession to impose medical risks on patients except for the anticipated benefit of those particular patients, and, on the other hand, the widespread and also morally important desire to allow people to volunteer themselves, including parts of themselves, for the medical benefit of others.

Furthermore, given additional safeguards, children too can "be volunteered" by their proper proxies – normally their parents – both for medical research that will not benefit them but is hoped to benefit others, and as organ donors. Again the criterion of very small risk of harm is fundamental, and argument continues about what sort and degree of "minimal harms" for which parents and doctors may properly be allowed to volunteer children. Few would doubt that parents should be allowed to volunteer their children as blood donors in certain circumstances; but what about bone marrow donation, kidney donation, partial liver donation?

The criterion that may be most helpful in making such decisions is comparison of the anticipated medical risks to the ordinary risks of everyday life to which parents are already permitted by their societies to subject their children. For example, the very ordinary activity of driving one's child to a hospital to see a sick relative imposes a risk on that child of dying or of other harms resulting from a possible car crash. Yet to prohibit parents from imposing such risks would create intolerable restrictions on everyday life. It would also greatly undermine the notion that some degree of risk-taking in the interests of others is a moral good. But if parents are to be allowed to impose such risks on their children in everyday life, for these sorts of reasons, why should they be prohibited from doing so in the context of beneficial, very low risk medical research or donation of tissues and organs?

### Death, brain death, and organ donation

One thing is rarely in doubt: no one should be allowed intentionally to sacrifice his or her life (let alone the life of another) in order to provide organs for others, even if by sacrificing one life several might be saved, and even if the potential donor autonomously wishes to volunteer. (It is worth reflecting on the contrast here with war ethics, where societies allow, indeed sometimes require, individual soldiers to sacrifice themselves and military leaders knowingly to send soldiers to their deaths in order to try to protect their nations against aggressors.) While such sacrifice is morally rejected in the context of organ transplantation, there has been a major debate about whether the use of "brain dead" organ donors does or does not in fact involve killing those donors. There is widespread social and professional agreement that it does not, on the grounds that brain death really is death. However, opponents have argued that to use as organ donors so called "beating heart cadavers" (for example, victims of severe head injuries whose vital functions are for a while kept going by artificial respiration, nutrition and hydration but whose

brains, including their brainstems, are permanently destroyed) is in fact to kill
dying patients, by removing their vital organs, and *not* to remove organs from
already dead bodies.

In briefly defending the contemporary widespread acceptance of brain death
criteria of death, suffice it to say here that death of human beings is a dual concept
that involves both biological death (the permanent cessation of the integrated functioning of a living biological organism, the human being) and personal death (the
permanent cessation of existence of a human person). Modern scientific understanding of the functioning of the brain makes it clear that functioning of the
brainstem – a small region at the base of the brain – is a necessary condition for
both biological human life (because the brainstem controls and integrates the
body's biological functions including, especially, breathing) and personal human
life (because a functioning brainstem is necessary for the capacity for consciousness, and, it is widely agreed, to be a person one must have the capacity to be
conscious).

Thus, if the brainstem of a human being has permanently ceased to function,
that human being is dead both as a human biological organism and as a human
person. Therefore criteria for brainstem death are proper criteria for human death.
The fact that many of the biological functions of such brain-dead humans can be
kept going for a few days by artificial means is of enormous benefit to potential
recipients of transplanted kidneys, hearts, livers, lungs and other organs, because
those organs can be kept oxygenated and undamaged until they are actually transplanted – but the unconscious, warm, human being whose heart and lungs are
thus "kept going" on the mechanical ventilator is as unequivocally dead as the
cold corpse in the hospital morgue. It is, however, understandably more difficult
for many – especially if they know little or no biology – to understand that this is
so, and, as above, why it is so.

### Brain death as an issue of scope
This debate about brain death, incidentally, is not essentially about the substance
of our moral obligations. Rather it is one about the *scope* of those obligations.
Thus, for example, we all agree that we have a moral obligation not to remove
each others' beating hearts, an obligation not to remove people's beating hearts
even to benefit other people, and even if those hearts are not going to be of much
use to their owners for much longer because those owners are dying. But what
do we mean by "each other" – what do we mean by "people"? These are scope
questions. In particular, does a brain-dead human being with a beating heart, who
is kept going on a ventilator, fall within the scope of these agreed moral obligations
to "each other" to other *people*? For the reasons given above the answer is widely
accepted to be "no" just as it is accepted to be "no" for the cold corpse in the

morgue. By this account, while dead human bodies certainly fall within the scope of our obligations of respect for the dead, they do not fall within the scope of our socially accepted but very different obligations to each other – to other *people* when alive.

## Justice and transplantation

The fourth of the four *prima facie* principles on which we can all agree is the principle of justice or fairness. Essentially this is the obligation to treat conflicting claims fairly or justly, especially for health care ethics, in the context of scarce resources, of respect for people's rights, and of respect for morally acceptable laws.

Of course there is considerable disagreement about what constitutes being fair, or just. Most agree that it is something to do with treating people as equals, but that it is not merely to do with treating them equally. (It would not be just or fair, for example, to give equal amounts of the organs available for transplantation to every one in the population.) In the context of justice people have to be equal in some morally relevant way (for example, in relation to the available organs they have to have equal *need* for those organs to be regarded as equals). And just as equals should be treated equally, justice requires that unequals should *not* be treated equally. For example, those who do not need medication should not get medication.

Similarly, so far as rights-based justice is concerned, people with equal rights should be treated as equals, whereas those who do not have those rights should be treated unequally. For example, justice requires that those who do not have a right to a state benefit should be treated differently from those who do have such a right. Similarly with legal justice – those who offend against a given (morally acceptable) law to the same degree should be treated equally (including equal concern for any mitigating circumstances); and so on.

To this extent most people will agree that we are all bound by the *prima facie* obligation of justice. However, when it comes to filling in the content of the principle or principles of justice there is far less agreement. Essentially, as I perhaps idiosyncratically would see it, they disagree either about how much weight to give to the moral concerns that so often conflict when we wish to be fair or just, or else about the scope of application of those moral concerns. Thus in relation to scope, if, as argued above, brain dead humans do not fall within the scope of our normal socially accepted moral obligations to each other, then brain dead humans may justifiably be regarded as "unequals" so far as our obligations of justice are concerned. In relation to conflict between moral values, one theory of justice may give priority to the moral obligation of beneficence to those in need (and therefore, for example, argue that justice requires high taxes in order to provide a national health service to meet all those needs). Another theory of justice may give priority to the moral obligation to respect people's autonomy, including their right to use democratic processes to increase their autonomy by reducing their tax burden at the expense of their obligations of beneficence to those in medical need. Yet both approaches will appeal to justice. Other theories of justice will try to incorporate both of these conflicting moral obligations, and prioritize them in various different circumstances.

In the context of transplantation, various issues of justice may arise. As with all forms of medical treatment there is the question of how much of our resources to put into this type of treatment (and indeed into research and development of this type of treatment) as compared with other treatments. For many, the debate, while problematic, is no more so for transplantation than for more standard types of medical treatment such as provision of antibiotics for infections or appendectomies for appendicitis. Some, however, find the whole idea of using parts of others' bodies sufficiently repulsive, sufficiently "unnatural" to oppose transplantation altogether, or at least to give it very low priority in the queue for scarce resources. Much could be said in opposition to this view, but suffice to say here that "unnatural" is a highly ambiguous concept. One needs to be very clear which of its many possible meanings is intended, and then to argue one's case sticking to that one meaning.

In any case, transplantation medicine confronts at least as many of the host of problems in the realm of justice as do all other types of medical care. Distributive justice asks how are resources to be fairly allocated to the provision of this sort of medical care as against other sorts of medical care, both at a "macro" level (e.g. when United States state governments decide how much transplantation medicine to provide via Medicare or Medicaid versus how much pregnancy care or treatment of children, or care of the elderly); at a "micro" level (e.g. when transplant surgeons have to decide how to allocate their limited operating time and resources among the too many patients who would benefit from a transplant) and at an intermediate or "meso" level between the two.

Rights-based justice requires transplantation medicine, just like other types of medicine, to respect patients' rights. For example every patient has a right to reject proposed surgery, whether the patient is a potential recipient of a transplant or a potential living donor.

And legal justice requires transplantation medicine, just like other types of medicine, to respect morally acceptable laws. (I can only assert rather than argue that I believe that it is the way a law comes into existence that determines its moral acceptability, rather than its content. On this basis, laws may be morally acceptable even though one personally disapproves of them, if they have been passed as a result of democratic and thus essentially autonomy-respecting processes, within a legislative system that respects the moral requirements encompassed by the four *prima facie* principles.)

*Selling and buying organs for transplantation – moral considerations*
In this context it may be worth considering briefly the marked disagreement that shows signs of developing internationally over the question of whether or not it should be legally acceptable to sell and buy organs for transplantation. Many countries have already outlawed such commerce. Yet the moral analysis is by no means clear-cut.

Certainly from the perspectives of the potential donor/seller of the organ, and of the recipient of the organ, a properly controlled system of organ sale should be morally acceptable in many cases. Why? Because in many cases their autonomy will be respected, they will obtain net benefit over harm, and the transaction will be just in terms of distributive justice, in terms of rights-based justice and (in countries that do not outlaw the practice) no law will be broken.

The complexity of the case is demonstrated when counterarguments to this thumbnail sketch of a justification are considered. Thus, are the donor/sellers adequately autonomous to make such decisions? It may be claimed that since the sellers of organs are likely to be poor and socially exploited they are in fact being coerced into their decision by their poverty and by their social exploitation. This is a morally dangerous argument for it implies that the decisions of poor people, if based on a desire to make money, may properly be ignored whereas the decisions of rich people who do not need the money so badly may be respected – a highly paternalistic attitude to poor people, which further diminishes respect for their autonomy. The stronger version of this counterargument states that whenever a person does something for money, the need or desire for money is coercive and thus the decision cannot be autonomous.

Two counter-counterarguments(!) seem relevant. First, all of us who work for money could thus be properly regarded as inadequately autonomous to have our decisions respected if they were influenced by need or desire for money (thus allowing others paternalistically to impose decisions upon us for our own good). Second, the argument betrays a mistaken understanding of what respect for autonomy is; it is not respect only for those people and or those decisions that are "fully" autonomous. Given our human frailty, not one of us is fully autonomous, and few if any of our decisions are fully autonomous. Rather, respect for autonomy requires that *adequately* autonomous people are respected along with their *adequately* autonomous decisions.

Of course our decisions are usually constrained by mundane reality – including the need to make a living, please people, compromise, decide on the basis of less than complete information, and so on. But only if we are below a certain threshold of ability to make adequately autonomous decisions; and/or if a particular decision is subject to a level of coercion that makes that decision fall below a certain threshold of autonomy, can others properly step in and ignore our decisions and impose their own views on how we would better lead our lives.

Undoubtedly, people and cultures differ about where to draw the line of adequacy, below which they permit themselves to ignore a person's decision for himself or herself, in favor of decisions that *they* believe will be better for the person. But it is difficult to understand how the fact of payment in itself could justify such paternalism, even if the recipient is poor and already exploited by the social system within which he or she lives. Indeed, one can envisage cases where a poor person who is paid a substantial sum for a kidney might as a result actually alleviate the burden of social exploitation under which he or she labors and in fact be empowered by such payment. For example, in a notorious United Kingdom case, a Turkish peasant sold his kidney in order to purchase hospital care for his sick daughter. Was he not to some extent empowered to overcome the negative effects of poverty and possible social exploitation?

Nevertheless one can well imagine circumstances in which donors/sellers are indeed unable to make adequately autonomous decisions, not so much because of their poverty but because of the intensity of personal pressures that are brought to bear on them, for example by family members or even by criminal gangsters using them, as it were, as organ prostitutes. However, this argument would fail if it were possible to regulate the operation of commercial transplantation procedures so that potential donors/sellers suspected of having been thus coerced or exploited would be detected and excluded from the programs.

As for net benefit over harm, while both the recipient of the organ and the seller might well benefit considerably, given the right sort of circumstances, (a) they might not, and (b) they are not the only ones potentially affected by a social policy that allows commercialization of transplantation.

Opponents argue that overall social effects would produce net harm rather than net benefit. Among the detrimental effects they predict are discouragement of altruistic donation – for once people learn that it is possible to sell their organs are they not far more likely to sell rather than to give? A further predicted negative effect is the concentration of donors/sellers within the realms of the poor – for are not the poor far more likely to want to sell their organs than are the better off? Not only is this likely to increase socially harmful resentment and a sense of exploitation among the poor themselves – it is also likely to increase the dangers to recipients of "commercial" organs. This is because the poor and therefore their organs are more likely to be unhealthy. Opponents cite transmission by such commercial transplantation of diseases, including acquired immunodeficiency syndrome (AIDS), as examples to support this objection. Furthermore, the selling of organs is also predicted to cause social harm in encouraging a view that commodifies human beings and their biological parts rather than respecting them as integrally whole people.

Opponents of commercial transplantation also challenge the extent and probability of benefit to both recipients and donors/sellers. While donors/sellers *might* be actually empowered by the money they obtain for their organs, far more likely, opponents argue, they will actually be cheated by being given only a small proportion of the money paid by the recipient for the organ – and not enough to produce any proportionately compensating benefit. And, where poor donors/sellers of organs are not eligible for free health care, opponents argue that the risks of such operations to the donors/sellers themselves are far greater than proponents of the sale of organs generally acknowledge.

Thus harm–benefit assessments are complex and contentious, and particularly difficult when they are predictive and prospective rather than retrospective, and where they concern populations rather than individuals or small groups. Yet they are an essential part of moral assessment in the context of beneficence and nonmaleficence, and we must try to do them as best we can in coming to a decision about the likely effects of a proposal such as allowing the selling of organs by living donors/sellers.

Finally, so far as justice is concerned, while it may be conceded that distributive justice may be served insofar as the buying and selling of organs increases the possibility of meeting the needs of potentially fatally ill patients, egalitarian views of justice are likely to oppose such commercialization on the grounds that it is unjust in increasing the gulf between the rich and the poor, for only the rich are likely to be able to afford such medical care. (Though note that there would be nothing to stop a national health service or indeed philanthropists from paying donors of organs and then themselves donating the bought organs to recipients – and note too that this sort of egalitarian justice counterargument is equally effective against any sort of private medical care.)

In the context of rights-based justice there do not seem to be arguments peculiar to the buying and selling of organs that do not apply more generally in the transplantation debate. So far as legal justice is concerned, undoubtedly the existence of morally acceptable laws against such commercialization, as in the United Kingdom, should act as a powerful moral disincentive to participating in any such operations. But if we accept that laws should only be created and/or preserved if they are morally justifiable, we should look behind the mere existence of such laws and at the moral justifications for their creation or preservation when we decide whether or not to favor their creation where they do not already exist, or their abolition where they do.

## Conclusion

The considerations sketched above should help us to make up our minds. Our decisions, however, are bound to turn in the end on how much weight or importance we give to one moral concern rather than another when they conflict, as well as to our decisions about the scope of application of these obligations. The four principles plus scope approach does not help us do that. But it does help us to be clearer about what our conflicting moral obligations actually are. Then we must choose.

### Suggestions for further reading

Beauchamp, T. L. and Childress, J. F. *Principles of Biomedical Ethics*, 4th edn. New York: Oxford University Press, 1994.

Brecher, B. Organs for transplant: Donation or payment? In *Principles of Healthcare Ethics*, ed. R. Gillon and A. Lloyd, pp. 993–1002. New York, Chichester: Wiley, 1994.

Evans, M. Against brainstem death. In *Principles of Healthcare Ethics*, eds. R. Gillon and A. Lloyd, pp. 1041–51. New York, Chichester: Wiley, 1994.

Gillon, R. *Philosophical Medical Ethics*. Chichester, NY: Wiley, 1986.

Lamb, D. What is death? In *Principles of Healthcare Ethics*, eds. R. Gillon and A. Lloyd, pp. 1027–40. New York, Chichester: Wiley, 1994.

Radcliffe-Richards, J. From him that hath not. In *Organ Replacement Therapy: Ethics, Justice and Commerce*, eds. W. Land and J. Dossitor, pp. 190–6. Berlin: Springer-Verlag, 1991.

Sells, R. Transplants. (eds.) In *Principles of Healthcare Ethics*, eds. R. Gillon and A. Lloyd, pp. 1003–26. New York, Chichester: Wiley, 1994.

Titmuss, R. M. *The Gift Relationship – From Human Blood to Social Policy*. London: George Allen and Unwin, 1970.

# Legalizing payment for transplantable cadaveric organs†

JAMES F. BLUMSTEIN

EDITORS' SUMMARY

James F. Blumstein, LLB, presents a very strongly argued case for financial incentives to relieve a shortage of organs. He argues that financial
concerns help to shape medical decisions, and should not be feared. At
odds with the official altruistic view of donation of organs, his commercial
view is highly controversial. There is no obstacle to influencing behavior
by using monetary incentives, he argues, and that this is the way other
forms of care in the health care system are provided. Such incentives
would relieve the shortage of organs and benefit many patients who today
must wait despairingly on the lists for an available organ.

A newcomer to the field of organ transplantation has many of the experiences of
an anthropologist being exposed to a new culture, a new way of thinking. There
are holy totems and sacred cows that permeate the thinking and have profoundly
influenced the development of public policy in the organ transplant arena. These
traditional values – e.g. an almost sacrosanct altruism and hostility to financial
incentives – have underpinned policy formulation in this field. It is important to
reexamine and reevaluate these entrenched ways of thinking as organ transplantation has evolved into mainstream, therapeutic medical care.

---

† Earlier versions of this chapter appeared in *Transplantation Proceedings* (1992), **24**, 2190–7
and *Health Matrix* (1993), **3**, 1–30.

There is widespread agreement that a shortfall exists in the number of organs made available for transplantation and that lack of availability inhibits the further utilization of therapeutically promising organ transplantation techniques. Nearly 30 000 people are on waiting lists for organ transplants in the USA. These waiting lists have gotten longer over time, with about 200 people being added to the lists each month. Despite this quite fundamental supply-side problem, leaders in the organ transplant community have been strikingly hostile to markets and to the use of financial incentives to increase the availability of transplantable organs. There has been a strong, visceral, adverse reaction to the introduction of commerce in the field of transplantable organs. This *Weltanschauung* is reminiscent of an earlier era when the existence of a role for markets in health care was hotly contested. Shibboleth and shamanism have thrived at the expense of rigorous analysis.

In short, ideology, as much as technology, has driven organ transplantation policy. An intellectual orthodoxy, traceable to the United States Department of Health and Human Services 1986 *Report of the Task Force on Organ Transplantation, Organ Transplantation Issues and Recommendations* (OTTF Report), has permeated the field. The OTTF Report has had remarkable influence on the development of thinking and the evolution of policy. In view of the organ shortage problem, another viewpoint and frame of reference need expression.

## Organ transplantation policy values: The ethical foundation of altruism

What values have underpinned organ transplantation policy in the USA? Even a superficial exposure to this field reveals an intense commitment to altruism. This is deemed a moral imperative. For example, the OTTF Report (p. 28) stated that a core value shaping organ transplantation policy was the goal of "[p]romoting a sense of community through acts of generosity." Despite the widespread recognition of the shortage of transplantable organs, there persists an insistence on the exclusion or elimination of financial incentives from all facets of organ supply, acquisition or distribution. Yet, financial incentives could enhance the availability of organs for transplantation.

## Organ transplantation values within the broader health policy context: The use of incentives

A fundamental element of health policy over the last 15 years has been the recognition that competition and markets have an important role to play in the health policy arena. The values underpinning organ transplant policy are in distinct tension with those values. They also seem strangely at odds with a policy of encouraging an increased supply of transplantable organs. What are the market-oriented values that strikingly contrast with the values underpinning organ transplant policy?

Not so long ago, the use of incentives in medical care was rejected on the related grounds of ethics and effectiveness. Ethically, the objection stemmed from the ideological commitment in some quarters that unrestricted access to medical care on the basis of medical need was the appropriate normative benchmark. This was a component of the rhetorical espousal of medical care as a right. If one believes that access to medical care should be costless to users, the imposition of financial disincentives is directly in conflict with that principle. Obviously, if one starts from the premise that an individual's utilization of medical services should bear no economic consequences for the beneficiary of the treatment, the use of financial disincentives will have unacceptable distributive effects.

In terms of effectiveness, financial incentives were questioned because of the prevailing medical view that money did not affect how patients were treated. It was assumed that there was a correct course of treatment, and that was a professionally determined decision. Science, not economic incentives, drove medical care diagnosis and treatment decisions. Therefore, financial incentives could not be effective because they did not influence medical care decision-making. And, implicitly, to the extent that scientific and professional judgments might indeed be influenced by financial considerations, that was an inappropriate deviation from the clinically correct scientific pathway.

The ethical issue is now seen as more richly textured than it once was. The rhetoric of rights and equality has been deemphasized, replaced in responsible circles by concern about the role of government in providing for core services to individuals unable to afford medical care.[1] Further, there is now a broader understanding that establishing a relationship between utilization of medical resources and expenditures by patients as consumers is not always inappropriate. Medical care is not monolithic; in some areas, it may be troublesome to use financial incentives, but, with respect to other types of care, use of incentives to shape behavior might be acceptable.

With respect to effectiveness, it is now commonplace to consider the effects of incentives on conduct in the health arena. This analysis is no longer an oddity but is quite mainstream. There is a recognition that economically unrestrained decision-making in medical care, as in other areas, has consequences in terms of resource utilization. Thus, the ideological commitment to unrestricted access has run into a hardboiled economic reality: elimination of the financial dimensions from medical care decision-making increases the overall use of resources. Financial incentives influence much more than distributive values.

The existence of clinical uncertainty, reflected in the widely divergent and unexplained procedure rates among providers in seemingly similar circumstances, further suggests an appropriate role for incentives. The evidence about divergent procedure rates undermines the assumption of a monolithically correct mode of treatment. In this world of uncertainty, it is hardly inappropriate to consider financial issues in decision-making. The assumption can no longer be indulged that financial incentives undermine scientifically clear-cut clinical pathways.

Principles of competition and incentives have influenced the rest of the health policy arena in the past decade. Those principles are strikingly at odds with the strongly held, fundamental principles of altruism and communitarianism that are so widespread in the organ transplantation arena. The logical next question is what would organ transplantation policy look like if it were more compatible with mainstream trends in health policy?

The answer is that it would allow for the introduction of incentives; that is, it would permit commerce in organ transplantation. Eliminating the exclusive reliance on altruism would recognize and acknowledge the priority of overcoming organ supply shortages and yet retain fidelity to principles of autonomy and individual choice of donors or their families.

## Does organ transplantation warrant a different policy approach from that of maintream medical care?

I have described mainstream public policy in the overall health care arena, explained how organ transplantation policy does not coincide with the mainstream, and stated what organ transplantation policy might look like if it did coincide. Does the field of organ transplantation warrant such a different policy approach from that of mainstream medical care?

### *The presumption for incentives in a market system*

I would start with the assumption that organ transplantation policies should allow for financial incentives in the absence of convincing arguments to the contrary. Evidence from elsewhere in the health arena shows that incentives affect behavior. Evidence from abroad shows that financial incentives dramatically increase levels of transplantable organ supply. The issue is increasingly being raised and discussed favorably in professional meetings and forums. And a recent survey performed under the auspices of the United Network for Organ Sharing (UNOS) and published in the *Lancet* demonstrates that a majority of the respondents believed that some form of compensation should be offered in the USA to donors of transplantable organs; only 2% of those surveyed commented that use of financial incentives would be immoral or unethical. I would require that those seeking to maintain the ban on incentives in organ transplantation persuasively make the case against commerce empirically or ethically. "In a nation whose institutions have relied on market mechanisms for making basic economic choices," governmental action that prohibits the use of incentives, which constitute a fundamental component of the market system, "bears a burden of persuasion."[2]

## The advantages of a market approach

What are the advantages of the market approach that would allow individuals (or parents on behalf of their children) to enter into forward contracts while alive and in good health for the use of their organs for transplantation after their death? I want to set forth two rationales simply and succinctly and then examine, at greater length, criticisms leveled at the use of markets.

First, there is the libertarian argument in support of the use of incentives and markets. This position emphasizes respect for the autonomy of the donor (and the ability of the donor to choose), deemphasizes paternalism, and strengthens the hand of the individual rather than the family. Payment to "donate" (a misnomer in this context) allows a person to determine his or her own organ's fate, respects the right of the buyer to contract, and recognizes the ability of the medically needy donee beneficiary to benefit from the transaction.

In addition to the libertarian argument, there is also the utilitarian argument; that is, would or could incentives increase organ supply? Permitting contracts for the sale of organs and making provision for a registry of potential donors would provide pressure to pursue transplants aggressively. A source of potential suppliers could be expected to come forward, and, once a contract had been entered into, the purchaser and the ultimate beneficiary would be forceful advocates for the effective use of transplantable organs to save lives.

## Criticisms of markets

Criticisms leveled at the use of markets for increasing the supply of transplantable organs can be either empirical or ethical in character. It is worth reiterating that the burden of persuasion should lie with those advocating that market transactions remain illegal.

### Empirical criticisms

Many critics of the use of markets and incentives focus on an empirical claim: use of financial incentives – allowing the market to function in this area – will not result in an increase in the supply of transplantable organs. The same claim was once made about financial incentives in medicine generally. But empirical evidence now firmly refutes this. Incentives work.

Objectors raise various concerns about the state of mind of potential donors – fears, uncertainty and ignorance. These deal not with the feasibility of a system of incentives but with the price that would be needed to induce supply. It is artificial to think in terms of absolutes – people will or will not contract for ana- tomical "gifts." It is better to think of what inducements are needed to encourage sufficient supply so as to satisfy the demand. The issue is one of degree, not of absolutes.[2]

Some critics claim that the introduction of commercial incentives could result in reduced altruistic donating and a net reduction of organ supply. Although experiences from other countries suggest that financial incentives increase organ supply, the empirical issue is a serious and legitimate one. Since such transactions have been illegal in the USA since 1984, no reliable American data exist. Where disagreements are empirical, not ethical, in character, the appropriate scientific approach is to run a controlled experiment. Define a region, repeal the federal ban on the purchase and sale of organs for transplant in that area, and, with proper controls and monitoring, see what happens.[3]

This experimental approach is not acceptable to many market critics. They worry that the altruistic system will be undermined and that the damage will be irreversible. The thought seems to be that the use of incentives is like an incurable infectious disease – once it is unleashed, it will deal a fateful and fatal blow to the altruistic underpinnings of the existing system of organ donation, a blow from which the existing system would not recover.

This is a hard position to counter because of the absence of firm data. The evidence from India and Egypt indicates that inducements do work. Evidence from other areas of health care suggest the same thing. It is not reasonable to maintain an empirically based criticism of financial incentives and simultaneously deny society the opportunity at least to have an experiment, even if only in a region and not in the nation as a whole.

In appraising the empirical criticism, the analyst must remember that the burden of persuasion is on those seeking to outlaw market-oriented behavior in our democratic society. In the absence of firm evidence, or an experiment, we should legalize commerce in organs for transplant, as other commerce in medical care is now permitted.

## Ethical criticisms

I now consider the ethical criticisms of commerce in transplantable organs. I deal with these in specific contexts, taking on the harder claims first.

### The effect of markets on the distribution of organs

Much of the ethical concern regarding commerce in transplantable organs focuses on the issue of distribution: who gets the organ available for transplantation and what is the effect of a market in transplantable organs on the distribution of organs? This pair of questions focuses on the demand side of the market.

The ethical thesis is that organs are different from other commodities or services that are distributed by the market system. Organs, argue critics of markets, should be allocated by medical criteria, not by financial considerations. This claim needs to be taken seriously. Medical resources generally are not allocated solely on the basis of medical need; the question is whether organ transplantation is different from other forms of medical care generally and from comparable life-saving therapies.

In my view, the original banning of market transactions in transplantable organs stemmed from an understandable yet ultimately unsophisticated linkage of issues surrounding the demand and the supply sides of the market. A market in transplantable organs can function on the supply side and, if desired on ethical grounds, society can leave the demand side to a nonmarket form of distribution.

The concern by ethical critics of commerce in transplantable organs is with the effect of wealth inequality on the distribution of available organs. There is a special claim that, while wealth inequality is acceptable as a general matter, it is unacceptable as a basis for deciding which persons are to be recipients of organ transplants.

The problem, however, can be resolved by public subsidy for those whose inadequate level of wealth bars access. The kidney program is an example of a publicly financed program for a specific illness and a specific set of procedures. To establish a principled basis for this type of categorical public support for kidney transplantation, however, advocates must be prepared to justify a kidney transplant program in comparison to other transplant therapies, such as heart or liver transplants, which are more likely to deal with life-threatening situations and which are not funded as generously by the federal government. Also, those who make claims of special consideration for transplant programs must be prepared to demonstrate that the justification for special status for organ transplantation does not apply as persuasively to nontransplant treatments of other life-threatening illnesses (e.g. public financing of drugs such as AZT for AIDS patients). Does society have an obligation to provide a public subsidy to make available and distribute this type of life-enhancing or life-prolonging drug? And if so, must the drug be made available without a fee, so that commerce is completely eliminated in the allocation of the scarce resource? If wealth can make a difference with regard to AZT, for example, why cannot financial considerations enter into organ allocation decisions?

Special consideration for organ transplants must also distinguish not only other life-saving but also other quality-of-life-enhancing procedures. Dialysis, after all, is an alternative to kidney transplantation, albeit less desirable therapeutically. Thus, since an alternative treatment regimen exists, kidney transplantation is not necessarily a life-saving procedure. Other parts of the kidney transplantation process require a fee. Whereas kidneys cannot be paid for by the patient, and kidney donors cannot receive compensation for their beneficence, organ procurement organizations can be paid for their organ procurement efforts by hospitals. Drugs are paid for, hospital stays are paid for, physicians are compensated. Money matters in every dimension of organ transplantation. It is not so clear why ethical critics become so fastidious, so squeamish, about paying for the life-giving organ itself.

There is an irony at work here. Advocates for funding by third-party payers claim that organ transplantation is in fact ordinary and necessary medical care – mainstream, nonexperimental medicine indistinguishable from other life-saving and life-enhancing treatments. The claim for insurance coverage requires that organ transplantation be viewed as just another effective therapy, like many others covered and paid for under traditional medical insurance. The claim for third-party coverage rests on the mainstream status, the lack of specialness of organ transplants. Yet that very specialness forms the ethical foundation for the underlying hostility to commerce in transplantable organs. There is a clear tension between these two positions.

Further, for these ethical objections to commerce in organs to make sense, there must be a willingness on the part of the objectors to exalt these distributive values above overall lifesaving and quality-of-life-enhancing objectives. This is true because, for the ethical discussion, we must assume that, empirically, commerce in transplantable organs will result in an increased supply of such organs and, consequently, more saved and quality-of-life-enhanced lives from transplantation procedures.

Some have made the forthright argument that it is better not to save lives in order to maintain distributional equity. I find that argument troubling. If one assumes that a price induces more supply, and that a wealthy person's life is thereby saved, how is the poor person harmed as compared with the status quo? One must take the position that it is better to deprive the wealthy person of a transplant, which hypothetically would not otherwise be available, in order to preserve some sense of egalitarian justice. This is a difficult outcome to impose on a person in the name of fairness, since the economically disadvantaged person is not benefited in any tangible way by prohibiting the wealthier person from using his or her resources to pay for an otherwise unavailable transplantable organ.

This is a genuinely troubling ethical dilemma that is worthy of further intellectual investigation, debate and discussion. As a pragmatic matter, however, the issue can be finessed in the name of incremental reform. For the time being, at least, market-oriented reforms can be concentrated on the introduction of supply-side commercial incentives, leaving intact a nonmarket-driven system for the distribution of transplantable organs on the demand side.

### The effect of markets on live donors

A second difficult question is whether a system of commerce in transplantable organs should permit payment for use of organs provided by (i.e. sold by) live donors. This ethically serious and troubling issue also can be finessed. Market-oriented reforms can focus, at least at the initial stage, on the use of markets exclusively for the sale of cadaveric organs, preferably by a forward contract.

The ethical concern with live donors is coercion, but the coercion claim may not be as much of a problem as some would argue. While there is an increase in choice, it is coercion only if we equate coercion with hard choices.

We do allow people to choose risk for a price. Life with one kidney is risky. But the question is whether this is a socially acceptable level of risk that should be subject to private choice and decision-making, or whether the risk should be banned by collective action through paternalistic governmental regulation. The risk of living with one kidney is quite moderate, equivalent to driving back and forth to work 16 miles (25 kilometers) a day. Society tolerates that level of risk in other areas, why not in the transplant area as well? Clearly the recipient/purchaser, whose health is improved, benefits. Arguably, so does the "donor"/seller, who can use the proceeds for other advantageous purposes.

One must hasten to add that, if any system of commerce is established involving live-"donor" organs, safeguards are necessary to assure voluntarism and to bar other uses of body parts via coerced, not induced, sale. Still, despite such safeguards, the paternalistic concern regarding live-"donor" organ sales is real and pervasive.

Also, there is worry about what some view as organ imperialism – the sale of organs by poor or Third World persons to provide organs for wealthy people. Rationally, the analyst may note that the sellers deem themselves better off, do not consider the risk to be excessive, and deem organ donation to be an avenue of opportunity. The skeptic may even call paternalistic objections to this activity elitist or illustrative of a certain "feel-good" morality. Yet a typical reaction is one, perhaps, of revulsion. And although this may be irrational, nonrational, or a form of symbolic hypocrisy, the objections exist and persist.

### The effect of a market for cadaveric organ sale on the "donor's" family and on society

It is now appropriate to address the issue of cadaveric organ sale. There are two concerns here – the potential effect of a market in cadaveric organs on the family of the "donor" and the potentially dehumanizing effect of organ sales on society.

The family issue is a legitimate concern. By selling his or her organs during life by forward contracting for transplantation at the time of death, an individual takes charge of the disposition of his or her organs at death. In the world of estate planning, this is not unusual, rather it is the norm. By allowing sales and by enforcing these forward contracts for organ transplantation, society validates the autonomy of the individual. At the same time, society takes away the ability of the family to veto the decedent's decision to allow use of his or her organs for transplantation purposes. This undoubtedly detracts from the "silver lining" phenomenon through which family members, in the exercise of altruism, feel good about giving the organs of a loved one to save the life of another person.

The psychological satisfaction of the family in this circumstance can be considerable, but the autonomy of the patient, if it is to be adequately respected, must outweigh the family's concern. This is, of course, the normal pattern with respect to inheritance, and it is the clear determination of the existing legal regime under the United States Uniform Anatomical Gift Act, which already vests legal authority in the individual to donate his or her organs for transplantation irrespective of the wishes of the family. The family veto recognized in the transplant community is an extralegal custom not validated by existing law. Indeed, the Act was expressly drafted to overcome the family veto, giving primacy to the autonomy of the individual donor. Given the existing legal framework, which does not recognize the family veto, the supposed loss of the family's psychological well-being is a weak claim. Ultimately, the autonomy of the donor and the welfare of the beneficiaries who receive the transplantable organs must outweigh the claims of the family. The establishment of a market for cadaveric organs will take nothing from families to which they are currently legally entitled.[4]

The potentially adverse effect on society of a market in cadaveric organs stems from a concern about the commodification of human body parts. This is an abstract, hazy issue. For example, Dr Leon R. Kass, who objects to the use of markets in transplantable organs, recognizes that his objections "appeal . . . largely to certain hard-to-articulate intuitions and sensibilities that . . . belong intimately to the human experience of our own humanity."

One set of objections to commodification of human body parts focuses on the value of communitarianism. Establishment of a market in cadaveric organs could pose a threat to the value of altruism. I just do not see this as a transplant issue. It deals with other, broader, philosophical issues about how society should be organized and about how people should be motivated to live their lives. Advocates of communitarianism generally are suspicious of what they regard as the atomization of society that stems from reliance on markets for making economic allocation decisions. They are hostile to market transactions, and they worry that commodification of human body parts places yet another set of decisions into an economic context with which they are none too pleased to begin with. Again, I view this type of concern to be quite unrelated to organ transplantation issues per se but rather related to broader humanitarian concerns about how market economies function in general.

Significantly, in the context of a market for cadaveric organs, there is no issue of coercion as there could be in the context of live-"donor" organ sales. Similarly, there is no real issue of organ imperialism, there is no concern regarding irreversibility, and there is no problem of exploitation of the poor. According to Cohen, "[I]n the cadaver market the vendors are neither rich nor poor, merely dead."

The loss of altruism should not in itself be viewed as a problem, except if it results in reduced supply of transplantable organs. Supply is an empirical not an ethical concern. The issue is not whether or not altruism is a good thing. The question is whether market transactions should be made illegal. Advocates of markets in cadaveric organs have no desire to make altruism a felony. Altruism can coexist peacefully with a flourishing market for cadaver organs. The claim of market proponents is based upon principles of freedom, autonomy, and choice. Indeed, when one carefully examines the argument for preserving altruism by outlawing market transactions, one wonders whether the real fear is that legalization of market transactions will in fact work; that is, given a choice, people would choose to participate in a market and would abandon altruism. Unpacked, the argument to outlaw market exchanges to preserve altruism is in reality an argument to coerce altruism. This surely is a strange way of promoting the supposed good feeling of communitarian solidarity that comes from voluntary donations of the organs of a recently deceased loved one for the benefit of another human being.

To make the case against market transactions in transplantable organs, advocates must establish the unique features of organs and organ transplantation. Since, in a market economy, market-based transactions are the norm, those seeking to curtail the operation of a market must show that there are special reasons justifying the restriction. I will now argue that organs and their transplantation are not unique in this policy-relevant sense.

There are numerous other life-saving or life-enhancing therapies for which sales are not prohibited. There is no ban on the sale of alternatives to organ transplantation, such as kidney dialysis. There is no ban on the sale of substitutes for failed body parts, such as artificial organs and other artificial body parts. Thus, we are left with a gnawing concern because the transplantable organ derives from a dead human body. This is not an objection to the use of the organs of a cadaver for transplantation purposes, since a donated organ is acceptable. It is just a question of how we induce donors or families to donate those organs. Does this, as Leon Kass has suggested, really "come perilously close to selling our souls"? The

UNOS-sponsored survey, which showed support for compensation for the use of transplantable organs, would certainly call that predictive judgment into question.

Analysts must balance the hazy, abstract concern about the effect on society of the use of financial incentives to motivate individuals to supply their organs for transplantation at the time of death against other, fundamental values. When life and death (for recipients) are in the balance and when libertarian values of individual autonomy are involved (for the "donor"), there is an insufficient justification for a ban on contracting while an individual is alive for the use of that individual's organs for transplantation at the time of death.

## Conclusion

There are important advantages in allowing commerce in forward contracts for transplantable cadaveric organs.

- There is a shift in the locus of decision-making – away from the bedside of the dying family member to an earlier time when an individual can make a determination about organ sale or donation while he or she is healthy and can act coolly and rationally.
- Recognition of market transactions promotes and validates the autonomy of the individual donor/seller.
- Legalization of market exchanges for cadaveric organs creates a legal and public-relations counterforce at the time of a "donor's" death so that the owner/purchaser of the organ and the potential and identified recipient of the transplantable organ can counteract the extralegal influence of the reluctant family of the potential donor and possibly of the attending physician as well.
- Use of financial incentives is likely to induce a greater supply of needed transplantable organs than the current, exclusively altruistic system.

These advantages take on added significance because of the acute need for transplantable organs and the dearth of available organ supply under the current system. This is where the success of organ transplantation makes a difference. The squeamishness about markets could be indulged when the stakes were not so high. The life-saving ability of organ transplantation means that organ supply shortages are costing lives. The claim by organ transplantation experts to mainstream status within the medical community, along with third-party payment for what is now considered to be ordinary and necessary medical treatment, suggests that it is now time to emphasize the similarities between organ transplantation and other forms of life-saving and quality-of-life-enhancing medical procedures rather than emphasizing the differences. Values regarding organ transplantation fit within the mainstream. They are not unique. We have allowed this "ghettoization" of organ transplantation policy within the health policy arena to go on for too long.

At this point I would not argue for a complete, full-scale market approach. I do not now call for creation of a market for live-"donor" organs. Nor do I now call for experimenting with a market on the demand side for the distribution of transplantable organs. But I do call for a controlled supply-side experiment with the sale of cadaveric organs. Permitting the sale of cadaveric organs in advance through forward contracts, with the concomitant establishment of a computerized donor registry, would represent a reasonable, modest, incremental experiment that is well worth trying. This is especially true given the lives at stake. This would allow for a constructive blend of altruism and self-interest, and nurture the hope that this combination would help to reduce the existing shortage of transplantable organs. Upon analysis, my reluctant conclusion is that opposition to experimentation with the sale of organs is based upon prejudice, not reason.

NOTES

1  See *President's Commission for the Study of Ethical Problems in Medicine and Biomedical and Behavioral Research, Securing Access to Health Care: A Report on the Ethical Implications of Differences in the Availability of Health Services* (1983), p. 18.
2  Quality, which is a concern, arguably could be monitored where patients, by contract, have agreed to the use of their organs for transplant at their death. The purchaser would have an incentive to keep the seller's organ healthy, and in most scenarios the seller would have the same incentive. This makes the organ market distinguishable from the blood sale market, where quality is a serious concern.
3  For a proposal to perform a pilot study of the effect on organ supply of a $1000 death benefit for organ donation.
4  For individuals who wish to provide this form of psychic satisfaction to members of their family, the option of delegating this choice to family members would continue to exist under the Uniform Anatomical Gift Act. The choice for altruism as the basis for an anatomical gift at death would still be available.

## Suggestions for further reading

Blair, R. D. and Kaserman, D. L. The economics and ethics of alternative cadaveric organ procurement policies. *Yale Journal on Regulation* (1991), **8**, 403–52.

Blumstein, J. F. Federal organ transplantation policy: A time for reassessment? *University of California at Davis Law Review* (1989), **22**, 451–97.

Blumstein, J. F. and Sloan, F. A. Health planning and regulation through certificate of need: an overview. *Utah Law Review* (1978), p. 3.

Cohen, L. R. Increasing the supply of transplant organs: The virtues of a futures market. *George Washington Law Review* (1989), **58**, 1–51.

Guttmann, R. D. The meaning of "The economics and ethics of alternative cadaveric organ procurement policies." *Yale Journal on Regulation* (1991), **8**, 453–62.

Hansmann, H. The economics and ethics of markets for human organs. In *Organ Transplantation Policy: Issues and Prospects*, ed. J. F. Blumstein and F. A. Sloan, pp. 57–85. Durham, NC: Duke University Press, 1989.

Kass, L. R. Organs for sale? Propriety, property, and the price of progress. *The Public Interest* (1992), **107**, 65–86.

Kittur, D. S., Hogan, M. M., Thukral, V. K., McGaw, L. J. and Alexander, J. W. Incentives for organ donation. *Lancet* (1991), **338**, 1441–3.

Peters, T. G. Life or death: The issue of payment in cadaveric organ donation. *Journal of the American Medical Association* (1991), **265**, 1302–5.

Radin, M. J. Market-inalienability. *Harvard Law Review* (1987), **100**, 1849–937.

Schwindt, R. and Vining, A. R. Proposal for a future delivery market for transplant organs. *Journal of Health Politics, Policy and Law* (1986), **11**, 483–500.

US Department of Health and Human Services. *Report of the Task Force on Organ Transplantation, Organ Transplantation Issues and Recommendations.* Washington: Government Printing Office, 1986.

Vining, A. R. and Schwindt, R. Have a heart: Increasing the supply of transplant organs for infants and children. *Journal of Policy Analysis and Management* (1988), **7**, 706–10.

# Scientific advances in aging

JOHN E. MORLEY

EDITORS' SUMMARY

John E. Morley, MB, BCh, articulates the advances that have occurred recently in addressing major problems that the elderly have in their quality of life. After noting how longevity is highly susceptible to environmental factors, particularly diet and support services, Dr Morley notes how what he calls "high touch" help for the elderly is far more important for their well-being than "high technology" medicine. Some examples are to watch for diet and vitamin intake, care for posture and balance through exercise, offer support to caregivers, and adequately treat depression. Unless many of the recent discoveries are better assimilated by primary care physicians, the elderly remain at high risk for many of their health problems.

## The hormonal fountain of youth

Since Ponce de León left Puerto Rico to search for the fountain of youth in America, only to discover Florida in 1513 (some might suggest that based on the migration of modern American elderly to Florida his quest was successful), the attempt to prolong life through the use of a variety of substances has become scientifically fashionable. This search led to the recognition that as we grow older a number of the secretions of the ductless glands are reduced. The classical hormonal decrease that occurs with aging is the dramatic fall of estrogens in women that occurs at the time of menopause. More recently, it has been recognized that secretion of the male hormone, testosterone, also declines with age, though over a longer time. Other hormones that clearly decline with age include growth hormone, vitamin D, and dehydroepiandrosterone.

Spurred on by the findings that replacement of estrogen hormones decreases bone loss and death from myocardial infarction in women of northern European origin and the recognition that some of the effects of hormone deficiency mimic the appearance of aging, attempts have been made to slow the aging process with hormone therapy. Growth hormone is produced by the pituitary gland and is responsible for producing growth in children. Its role in adults is less certain, but may play a part in increasing muscle strength. Examples of persons with excessive production of growth hormone have suggested that prolonged secretion of growth hormone may lead to weakness, glaucoma, and interference with the nerves to the limbs (carpal tunnel syndrome). Goliath, the Philistine of biblical fame, might have been a pituitary giant, initially extremely strong due to his excessive growth hormone production, but who eventually grew weaker and was slain by little David and his sling. The world's tallest man once lived in Alton, Illinois. He too had excessive growth hormone production and died in his early twenties.

Despite these dramatic examples of the potential dangers of growth hormone, a study was undertaken to examine the effects on the aging process of growth hormone administration in healthy older males. After six months of treatment, the results, while not dramatic, suggested that growth hormone treatment might reduce body fat. Unfortunately, after a year of treatment many of the subjects had developed side-effects, especially tingling in their fingers from carpal tunnel syndrome. They gave up therapy. Thus, it appears that growth hormone must exit as a potential fountain of youth. Nevertheless, my colleagues and I successfully used growth hormone therapy for a short period of time to successfully reverse severe malnutrition in older persons.

Older men with fractured hips have been demonstrated to have very low testosterone levels. Testosterone is well known to improve muscle strength. We, therefore, undertook a study to examine the effect of testosterone treatment in older men with low testosterone levels. After three months, testosterone improved muscle strength and increased the number of red cells. This increase in red cells is potentially dangerous as it could lead to strokes. Ongoing studies are attempting to determine the effects of longer therapy with testosterone on muscle strength.

As we grow older, our skin cannot manufacture the precursor of vitamin D from ultraviolet light as well as it did when we were young. Also the kidney is less effective at converting vitamin D to its most active form. Vitamin D plays an important role in maintaining bone strength. A recent study of residents in a French nursing home has clearly established that daily vitamin D and calcium ingestion reduce the risk of hip fracture.

The role of hormone replacement in older persons remains tantalizing but unclear. Continued studies are appropriate. It is important that these studies focus on who is most likely to benefit from hormone therapy, as it is unlikely that these hormones will be appropriate for everybody.

## Alzheimer's disease

Dementia is a devastating disorder that occurs in 5% of persons over the age of 65. By 85 years of age, almost half the population may have some degree of dementia. Dementia is estimated to cost the USA about 90 billion dollars per year. A small proportion of persons with dementia have reversible causes such as side-effects of drugs, depression, thyroid disorders, problems with vision and hearing leading to sensory isolation, normal pressure hydrocephalus (a condition of abnormal flow of cerebrospinal fluid that results in ataxia and incontinence as well as cognitive impairment), tumors, infection, or vitamin B12 deficiency. However, the majority of persons with dementia are diagnosed as having either Alzheimer's-type disease or dementia due to vascular disease (small strokes).

Alzheimer's disease is the commonest form of dementia. This disease was first described by Alois Alzheimer at the turn of the century and is characterized by pathological findings of plaques and tangles in brain cells. At the center of the plaques is an amyloid core. Experiments have shown that the amyloid beta-protein can produce amnesia in animals and can also be toxic to neuronal cells. The gene for amyloid beta-protein has been localized to chromosome 21. Persons with Down syndrome have an extra chromosome 21, overproduce amyloid beta-protein, and often develop Alzheimer's disease in their early forties.

Other studies have suggested a genetic association of Alzheimer's disease with chromosome 19 and apolipoprotein E. It is thought that one variant of this protein inhibits nutrient transport in neurons and results in the development of the neuro-fibrillary tangles.

A number of environmental factors have been associated with an increased propensity to develop Alzheimer's disease: head trauma, not smoking, low education level, depression, and possibly exposure to aluminum.

Numerous studies have demonstrated a deficit in the brain's cholinergic nervous system in persons with Alzheimer's disease. Other neurotransmitter systems such as the somatostatin and glutamate systems are also disrupted. Much recent attention in drug development has focused on the development of drugs that block the breakdown of acetylcholine in the synaptic space between nerve cells. The prototype of these drugs is tacrine (Cognex). This drug was studied in 653 patients with Alzheimer's-type disease. Over half the patients withdrew from the study because of side-effects, primarily liver abnormalities and gastrointestinal complaints. A small improvement in cognition was demonstrated in those who completed the study, but it is unlikely that this small effect is worth either the cost of, or potential danger associated with the drug.

Many of the advances in care for persons with Alzheimer's-type disease have focused on enhanced social care. Persons with Alzheimer's-type disease should wear a medic alert bracelet so that they can be identified if they wander away and get lost. Caregiver burnout is extremely common, and special support programs for caregivers are essential. Respite programs to allow caregivers time off are key. Depression is a common problem in caregivers and may need active management. Particular attention needs to be paid to driving by Alzheimer patients. In an attempt to improve care of institutionalized patients, special care units have been developed to provide increased care for the Alzheimer patient. There are few data to support the notion that such units are either cost-effective or actually improve quality of life for the patient.

Our understanding of Alzheimer's disease has improved impressively over the last decade, and we appear to be on the brink of being able to design rational drugs to treat the disease.

## Polypharmacy

Older persons are major consumers of drugs. When a person comes to a physician with a complaint, both physicians and public have been educated to expect that the solution should involve the writing of a prescription for a drug. Drugs clearly can produce side-effects and because of changes in the pharmokinetics and pharmacodynamics of drugs in aging patients, older persons are more vulnerable to these adverse effects of any drug. Older persons also are often on multiple drugs and drug–drug interactions are a common cause of problems in older persons.

In geriatrics, it is recognized that deterioration in patient functional status is often precipitated by a drug. In some cases the drug is not necessarily producing a bigger effect, but rather the same magnitude of effect may be far more devastating if the baseline has shifted. For example, a drug that produces a small detrimental effect on memory in a young person may have a devastating effect on functional status in an older person whose cognitive function has been lowered to a borderline level.

Many drugs are tested in middle-aged persons and relatively healthy older persons. Yet they are then marketed and used predominately in older persons with multiple problems. An example would be the increased renal failure seen in older persons using angiotensin-converting enzyme inhibitors to lower high blood pressure.

### Focus on function: The prevention of frailty

In geriatrics the major focus is on improving and maintaining function rather than on disease per se. The major factors causing deterioration of function are frailty, polypharmacy, mental changes (depression and dementia), and nutritional problems, all superimposed on a background of social problems. Older persons living below the poverty level have consistently been shown to have poorer function and greater morbidity and mortality. The poor outcomes of some of the minority groups in the USA appear to be associated more with poor resources rather than with genetics or other explanations.

Aging is associated with decreased strength, balance, mobility, flexibility and endurance that place the older person at risk for developing functional deterioration. While some of this deterioration of function is inevitable, some of it can be prevented by appropriate exercise regimens. A major danger to the older person is immobility. Despite this, many health care professionals have enforced immobility in older persons by placing them in physical restraints. Such restraints have been demonstrated to increase agitation, decrease fluid intake and produce dehydration, increase injuries and lead to bedsores.

The major causes of falls in older persons have been identified as decreased muscle strength, balance problems and polypharmacy. Also, drops in blood pressure either when standing or following a meal can result in a fall. All of these situations are treatable. It has been demonstrated that even 90 year olds living in a nursing home can enhance lower limb strength when they undertake a weight lifting program. There is increasing evidence that exercises aimed at enhancing balance will improve an older person's stability. The Chinese martial arts exercise, Tai Chi, appears to be particularly useful in improving balance. Exercises should also be undertaken to enhance endurance and flexibility and to maintain an upright posture.

Falls in blood pressure on standing (orthostasis) are not rare in older people. While there are many causes of orthostasis in the older person, it is often due to treatable causes such as overmedication, poor nutrition, low salt diet, anemia, dehydration or failure of the adrenal cortex to produce cortisol (Addison's disease). The fall in blood pressure following a meal (postprandial hypotension) is secondary to carbohydrate content in the meal and can often be corrected by increasing the fat content of the meal or eating multiple small meals. Persons who become dizzy when standing are at higher risk of subsequent disability than those who drop their blood pressure but do not become dizzy.

A major cause of frailty in older persons is depression. While depression is less common in older individuals than in younger individuals, the diagnosis is much more commonly missed by physicians. In a study, my colleagues and I found that internists failed to make the diagnosis of depression in 95% of persons over 70 years of age with this disorder. Depression can lead to malnutrition, isolation, muscle weakness, incontinence, a dementia-like syndrome, infections and death. Depression is easily screened by using the Yesavage Geriatric Depression Scale developed by Jerome Yesavage, MD, at Stanford University. Most older people with depression can be treated, usually with slightly lower doses of the classical antidepressants used in younger adults. Electroconvulsive therapy ("shock treatment") is highly effective in older persons, particularly those who are suicidal or who have severe malnutrition.

## Nutrition

In older persons weight loss leading to undernutrition is a very common problem. Being underweight in older persons is much more likely to lead to death and disability than being overweight. This is presumably because being malnourished interacts with aging to produce major deteriorations in immune function that can mimic those seen in patients with acquired immunodeficiency syndrome (AIDS). Depression is the commonest cause of malnutrition in older persons. Numerous other treatable causes of malnutrition exist in older persons, and these are easily remembered by the MEALS-ON-WHEELS mnemonic:

**M** edications
**E** motional (depression)
**A** lcoholism, anorexia tardive
**L** ate life paranoia
**S** wallowing disorders

**O** ral problems
**N** o money (poverty)

**W** andering and other dementia related behaviors
**H** yperthyroidism
**E** ntry problems (malabsorption)
**E** ating problems
**L** ate life paranoia
**S** hopping and meal preparation problems

Studies exist that have demonstrated that vigorous caloric supplementation can decrease mortality in all older persons and improve rehabilitation following hip fracture.

There is much current enthusiasm for the role of free-radical scavengers such as vitamin E in decreasing disease-allocated morbidity and mortality in older persons. Free-radicals have been particularly implicated in the pathogenesis of atherosclerosis, Parkinson's disease and some cancers. Available data suggest, however, that vitamin users do not live longer than nonusers of vitamins. Insufficient data exist to pinpoint which particular group of older persons may benefit from vitamin use, though evidence is accumulating that many nursing-home residents are vitamin deficient.

With aging, the optimal cholesterol level for survival appears to be higher than when there is younger. In women over 70, for example, ideal cholesterol levels may be as high as 240 to 300 milligrams per deciliter. Lower cholesterol levels have been associated with cognitive impairment and depression. Failure of adequately communicating this information to older persons has created an ethical conundrum, as older persons are more likely to be caught up in the cholesterol lowering mania that has gripped the American public.

## Other factors

At the turn of the century, 50% of the United States population were dead by 61 years of age. Today, the life expectancy for white women is 79.7 years and for African-American women is 76.2 years. White men can expect to live to 73.0 years and African-American men to 68.1 years. However, life expectancy is greater in 15 other countries of the world, with Japanese women living 2.4 years longer and Japanese men 3.1 years longer than their white American counterparts.

The reasons for the poor performance by the USA in the longevity stakes are multifactorial and include poor prenatal care and social violence, particularly among younger African-American men. Other factors identified for the poor life expectancy include a lifetime overconsumption of food (only Greeks eat more), excessive alcohol and tobacco use, an infatuation with automobiles that results in a decrease in physical activity, and an excessive use of high technology medicine, much of which has been poorly tested in older persons.

In addition, most countries that are in the top ten of life expectancies have higher intakes of fish than the USA. Japanese ingest 6.9% of their intake as fish, while American ingest only 0.8%. Fish contains eicosapentanoic acid, which decreases the stickiness of platelets in the blood and thus reduces the propensity to develop myocardial infarction. In the Netherlands, eating fish four times a week has been shown also to reduce the risk of myocardial infarction.

Another factor in poor outcomes for older persons in the USA may be the obsessive desire of administrators to reduce the length of stay during hospital admissions. Many western European countries with longer life expectancies than that in the USA tend to have lengths of hospitalization nearly twice the eight-day average of the USA. Japan has been reported to have much longer lengths of hospitalization. Despite the longer stays in other countries, the total cost per hospitalization in the USA is vastly greater.

Carefully controlled trials in the USA have demonstrated that a group of specially chosen frail older persons benefit from a six-week to three-month prolonged hospitalization for intensive rehabilitation after an acute illness. Benefits include increased life expectancy, decreased nursing home admissions, and improved quality of life. Over a year, figures show that, despite the increased period of hospitalization in what have been termed Geriatric Evaluation and Management Units, the total cost of care for individuals admitted to these units is less than that for those who do not receive this specialized "high touch" care. Despite this information, few of these units have opened outside of the Veterans Affairs hospitals, and a number of those initiated were closed because of the low reimbursement available for this effective method of care.

## Final thoughts

Presently, knowledge in geriatrics is increasing at an exponential rate. Unfortunately, most of this knowledge base has still to be adequately digested by primary care physicians. Studies have shown that the average primary care physician often fails to diagnose common treatable geriatric problems. There is a major need for increased education on geriatric problems at all levels. The lack of a significant number of geriatric subspecialists to undertake this training is a major roadblock to the future appropriate care of senior citizens.

Health care reform needs to recognize the scarcity and importance of geriatricians and provide protection for and nurture the growth of this subspecialty at the expense of the more procedure-orientated subspecialties such as cardiology and gastroenterology. There is also a major need for increased financing for the education of primary care physicians in geriatrics, both during medical school training and in postgraduate education, utilizing models such as the highly successful American Geriatric Education Centers that are sponsored by the Bureau of Health Professionals.

Available knowledge to help older persons achieve a good quality of life, to prevent diseases and overcome disabilities, is extremely large. The success of medicine during the next decade will depend as much on the ability of health professionals to apply these simple, "high touch" techniques as avoiding inappropriate high technology interventions in older people.

## Suggestion for further reading

Abrams, W. B., Beers, M. H. and Berkow, R. *The Merck Manual of Geriatrics*. Whitehouse Station, NJ: Merck Co. Inc., 1995.

Doress-Worters, P. B. and Siegal, D. L. *The New Ourselves, Growing Older*. New York: Simon & Schuster, 1994.

Hayflick, L. *How We Age*. New York: Ballantine Books, 1994.

Morley, J. E. and Solomon, D. H. Major issues in geriatrics over the last five years. *Journal of American Geriatrics Society* (1994) **42**: 218–25.

Perry, H. M. III, Morley, J. E. and Coe, R. C. *Aging and Musculoskeletal Disorders: Concepts, Diagnosis and Treatment*. New York: Springer-Verlag, 1993.

# Ethics and aging

GEORGE J. AGICH

EDITORS' SUMMARY

George J. Agich, PhD, groups his discussion of ethical issues in aging
under three categories. First, there is the struggle to develop a personal
meaning for growing old. Society focuses on being "actively retired," so
Agich explores the many meanings of autonomy and maturity, which may
not include such activity. Second, there is the problem of personal identity during loss of cognitive function, and the moral status of such persons.
Finally, there are the problems of intergenerational justice and the distribution of health care. Agich argues tellingly that the real problem of
justice occurs in a society that does not value aging, since the focus is on
distribution of care, rather than on the inherent personal and moral qualities of elderly persons.

There are many ethical issues associated with aging, but they can be conveniently
grouped around three themes. First, there are questions that relate directly to the
aging self and the personal meaning of growing old, such as how does one retain
personal autonomy and dignity in the face of the loss of capacities to act, to
remember, and even to think clearly? And, does loss of independence diminish
our fundamental worth as individuals? Second, there are questions that involve
our obligations to elders, especially frail elders. Are individuals ravaged by the
diseases and afflictions of old age, such as the loss of cognitive capacities associated
with Alzheimer's disease, less worthy of our attention and care? Or, do these
individuals retain moral worth that transcends their frailties and demands special
care and attention? Third, are demographic trends that show a marked increase

in the numbers of the very old relative to the rest of the population a cause for concern as this group consumes a disproportionate amount of resources compared to, for example, the very young?

## Personal meaning of aging

The first question concerns how an aging individual can retain a sense of self-worth as he or she inevitably experiences the anxieties associated with growing old and facing death. This question is especially pressing as the aging elder experiences the loss of vital capacities and becomes more dependent on others for help with various activities of daily living. This question is particularly poignant in Western society, which prizes individual autonomy and independence. This cultural ideal permeates popular images and expectations of old age. Promoted by groups such as the American Association for Retired Persons (AARP), the ideal of "active retirement" is seen more as an obligation and definition of the best kind of life during old age rather than the terminological contradiction that it is. Given the prevalence of these ideals, however, it is easy to see why the image of the frail elder who is not fully self-reliant is so abhorrent. Any sense of diminishment in the capacity to function independently is quickly viewed as a tantamount to subservience and inferiority, which, by extension, implies a loss of *personal* worth. Because we prize our individual autonomy so much, we are loathe to accept the dependencies that all too frequently accompany the aging process.

Philosophers, however, are quick to point out that autonomy has a rather wide range of meanings that include not only the common notions of self-rule, self-determination and freedom of choice and action, but also notions of individuality, independence, and absence of coercion and external influence. There are also important everyday senses of autonomy such that to be autonomous means to have a developed set of beliefs, preferences or values with which one identifies. Simply put, this means that an autonomous person is an individual who is at one with his or her own personal history and sense of self. There are important practical senses in which one can be decisionally autonomous, i.e. capable of rationally deciding for oneself, yet incapable of executing those choices without the help of others. Nonetheless, autonomy as a cultural ideal only reluctantly admits such distinctions.

The dominant view is that autonomy is most clearly expressed in the concept of independence, a concept that has been a perennial feature of American society. Even in the early days of our country, the French historian Alexis de Tocqueville noted the peculiar tendency of the Americans to draw apart and to keep to themselves. More recently, sociologists have characterized American society as constituting a "lonely crowd," as being engaged in the collective "pursuit of loneliness," and in manifesting an aversion to dependence of all kinds. This latter attitude towards dependence assumes that any form of dependence involves a degree of degrading submission and corresponding loss of self-worth. With increasing impact around the world, this cultural notion is spreading. The loss of self-reliance that necessarily occurs in the course of aging as one's powers diminish is thus a particularly pressing problem, since self-reliance and independence constitute one dominant cultural measure of personal worth.

Aging individuals naturally undergo periods of self-examination and self-questioning in the course of their aging. Important life events such as retirement, the death of one's friends and coworkers, or the marriage of children or the birth of grandchildren constitute occasions for asking "Who I am?" Similarly, the loss or impairment of capacity encourages reflective self-examination. The question, "Who am I?" is probably more a universal existential question that recurs in all stages of human development than it is a question unique to old age, but in our society it acquires a pressing poignancy because of the ideals of self-reliance and independence.

Aging has traditionally meant the attainment of practical wisdom and, by tradition, elders have been accorded a special place of honor as the living repository of the memory of the family or community. Even when being old does not bring with it exalted social standing, being old can bring a personal sense of worth as the elder reflects on the store of experiences and memories that provide a perspicacious view of the world and events. Such a sense of personal insight or wisdom, however, is frustrated by the lack of social acknowledgment. Rather than hold elders in esteem for what they are, namely people who are older and therefore more experienced than the rest of us, Western society seems to value elders precisely in terms of the ways that elders emulate the young and achieve the passions that drive the young. The ideal of active retirement sees the value of retirement as a time that affords the opportunity to pursue the activities of youth, e.g. sports, hobbies, and leisurely pursuits of various sorts. Paradoxically, to be old is valued only insofar as the old can enact our culture's idolization of youth. As a result, elders experience rather dissonant cultural messages about their status, messages that question the very meaningfulness of being old as such.

To properly understand the worth or value involved in being old requires that we first recognize that the mature self does not need to be a stalwart, robust individual. Rather, maturity and self-worth are consistent with various kinds of dependency, including loss or diminishment of capacity. While self-worth is consistent with dependence, the question of the personal meaning of one's life is a question that cannot be given a general answer. It is a question that is made more or less difficult by social attitudes, but inevitably each individual must forge or find personal meaning. This fact has led many gerontologists to look carefully at the role of narrative and story-telling in aging. In particular, it has been argued that our lives are themselves like narratives or stories that we tell as we enact them in the world. So, to understand the stories of elders, we can better understand the various ways in which elders successfully define their personal worth and what approaches create problems for the elder.

In acting in and experiencing the world, we can see ourselves as characters in the stories of others and in our own story. The sense in which there is but one "true" story is itself controversial, because some thinkers have argued that we enact a variety of roles in our everyday lives, each of which involves different ways of telling a story about ourselves. The problem of finding a personal meaning in old age or of defining a sense of self-worth might be less a task of finding one's own *true* story, but of maintaining the vital capacity to make a story of one's life. Viewed in these terms, the autonomy of even frail elders is significantly larger than the common culture would admit. The scope of autonomy is large because the possibilities inherent in the narrative unfolding of one's life are themselves quite extensive. Within this range of alternative stories of life, one can see the importance of others: others as intimates – as family members and friends and acquaintances – and others as strangers. In interacting with people in everyday life, the elder both reveals and discovers possibilities for action. "*Action*" in this context does not need be restricted to the image of a healthy person making his or her way against the resistance of the world, but rather includes the image of an individual who simply appreciates and understands the immediate circumstances. It is not in some romantic or epic drama that one expresses most fully one's autonomy, but in the everyday life in which each of us as human beings live. In daily life, then, the story of "who I am" is told by the elder no matter how frail or incapacitated. So long as the elder maintains cognitive and imaginative capacities, the process of making a world for himself or herself continues, a process in which one's personal story is revealed, if not discovered.

## The moral status of cognitively impaired elders

Generally speaking, as a proportion of the population, elders in America, have relatively high economic and political power when compared with other age groups. Nonetheless, growing old even in America is fraught with difficulties. Central among these difficulties is the fact that there are strong social and cultural messages that imply that personal dignity and worth are themselves diminished as one loses the capacity to act, to remember, or even to think clearly. This is spelled out well by Stephen Post (this volume), when he tackles our society's undue emphasis on rationality.

Thus, even healthy elders display prejudice toward nursing homes and other long-term care facilities as places for "the old," even where there is no disparity of chronological age between the perceiver and the perceived. The perception that aging inevitably means loss of personal dignity and worth is contradicted by a reality that elders are in the majority of cases cared for by family and friends in various ways. Of those individuals in nursing homes, the majority have medical problems that require specialized care or are individuals who have no family. Nevertheless, the question of meaning arises not only for the individual elder, but perhaps most distressingly for family and friends as they see their aging relative or friend changed significantly by the diseases and afflictions of growing old.

It is not surprising that the cultural ideal of independence has a contrary corollary, namely the icon of the nursing home, a place where frail individuals are seen as being utterly dependent on others. The image of the elder in such an institutional setting is frequently an image of a person without a sense of self or identity. Affected by Alzheimer's disease or other dementias, such patients appear confused, cannot remember from one moment to the next, and manifest sometimes striking changes in personality and behavior. The difficult question posed by demented elders for caregivers is whether such individuals retain any moral value or status such that the special effort and attention that their care demands is worthwhile?

This question is particularly paradoxical in our society given our fascination with independence. As independent individuals we determine for ourselves our way of life including myriad choices in daily activities of living. Our preferences for food and dress, for example, are intimate expressions of who we are. Yet, the capacity for such choices seem absent in elders who manifest cognitive disorders. They appear to be confused not only in their own experience of the world, but they exhibit a confusing image of themselves through their behavior and presentation. Add to this the loss of control over bodily processes and we have the image of an institutionalized elder who is disoriented, disheveled, dressed in mismatched clothes, drooling, and possibly soiled with urine or feces. As a result, it is hard to sustain the belief that such individuals are dignified or worthy of the care they so desperately need. As a result, some of our anger is rightly directed against the institutions and individuals providing care for these elders, but a good deal of anger is still left over for the frail elder herself.

Such a severely compromised elder is not recognizable as his or her former self. Even when family members and friends struggle to see in such severely compromised individuals vestiges of the former person, questions arise about the emotional and economic price of caring for such persons. Additionally, as members of society we are concerned about the expenditure of social resources, particularly for care at the end of life. Coupled with the growing movement supporting the refusal of death-delaying medical interventions and even unwanted life-saving medical interventions, it is natural to question whether patients so severely compromised "deserve" our care. This concern is two-fold, involving the question of whether such severely compromised elders are persons who deserve our attention and care and a question about what is the appropriate or just allocation of resources. I take up the first of these questions in the present section leaving the second question for discussion in the final section of this chapter.

Is an elder who no longer recognizes family or friends, who does not know where he or she is, and who recalls only isolated fragments from the past still a person who deserves our care? Questions such as the meaning of personhood seem remarkably philosophical, yet they are questions with which families struggle on an everyday basis as they watch a loved one undergo profound memory and cognitive change. The changes these individuals undergo are not isolated, but eventually affect their sense of continuity with the past, their recognition of loved ones, and behavioral and personality changes. Behavioral changes include bizarre and unusual behaviors that family members, usually spouses, struggle to hide and deny. In a real sense, such elders become different persons. We are no longer able to share a common past with them and their behaviors are sometimes completely out of character. In such circumstances, one can really ask whether such cognitively impaired elders still retain any sense of personal identity. Since our ethical obligations for care and concern arise from our concrete relations with persons, the elder's loss of a sense of personal identity raises emotional and psychological, philosophical and religious, and practical questions that seem to weaken our obligations to them.

Seeing a parent or spouse act in ways that are not only "out of character," but dangerous or embarrassing brings forth a range of psychological responses including defense mechanisms such as anger and denial. There is also frequently a sense of guilt: for not having noticed changes earlier, for having denied changes noticed, or for angrily reprimanding an elder for behaviors that are later attributed to an underlying pathology. Such reactions inevitably color and make the ethics of caring for cognitively impaired elders all the more difficult. A parent or spouse who, when in character, was not particularly kind, but was mean-spirited or even abusive is not likely to excite pity or genuine compassion from children or spouse. Yet, as we shall see, ethical problems are not confined to the fact that the elder is vulnerable and so presents occasions for past scores to be settled, but that the elder is importantly no longer herself such that settling of scores carries with it a profound injustice. At the other extreme, guilt over past failures to reciprocate love, for example, can sometimes motivate children toward unrealistic caregiving roles. Children and spouse of an elder might accept the burdens of care believing that their obligations are unqualified even in the face of other competing obligations. This is particularly true for families simultaneously facing the economic and emotional burdens of caring for children and frail parents.

The changes that cognitively impaired elders exhibit also involve philosophical and religious questions about the nature of the self, of personal identity and of the nature and worth of personhood. Noting that an individual who undergoes profound changes that alter his or her identity is, truly speaking, a different person clashes with the rather human observation that an elder may not be much of a person, but he or she is still "my parent." The philosophical question of personal identity is relevant for cognitively impaired elders who undergo profound changes of character or personality. Such changes thus prompt the question of whether the elder is the same person.

There is good reason to believe that the identity of a person is tied up with that individual's values and beliefs. Who that individual truly is involves the experiences, memories, habits, beliefs, and values with which that individual identifies. Should a drastic personality alteration occur such that an elder no longer identifies with his or her past values, but rather assumes a new set of beliefs and preferences, then we have grounds to suggest that a different person is present. Children may wish to maintain the prior preferences or values of the elder, values which he or she might not even be able to remember much less to identify with, whereas the direct caregivers might insist on honoring the elder's present wishes or values. The conflict that arises is less about the wishes or values themselves than about who the elder truly is. The question that needs to be asked is about who the *elder truly is*, not as a basis for shirking responsibilities to the elder, but to realize that our responsibilities are shaped precisely by who those others are. If my father is no longer the man I knew to be my father, my father nonetheless is *someone*, perhaps a stranger, yet my obligations to that someone are still the obligations of a son. These obligations are not just to a past father who no longer exists, but to an individual who is my father, though my father is now a different *person* from who he was before.

The deep problem that makes these cases so difficult is the age-old philosophical problem of personal identity. This problem has significant practical ethical implications that are only partly acknowledged in the bioethics literature. For example, bioethics has defended the use of advance directives as a way to ensure that patient autonomy is respected. The use of directives in America such as Living Wills or Powers of Attorney for Health Care allow individuals to express their wishes about life-sustaining treatment ahead of time. Numerous American states permit various kinds of surrogate decision-making, usually by family members who can make decisions based on their direct knowledge of the patient's own beliefs or wishes. The question that has not been adequately discussed is whether the past beliefs and values or expressed wishes of an elder carry more moral weight in the present than the present beliefs and values when they apply to a very different person. For example, an elder who indicates that she never wants to live in a nursing home, that she would rather die than leave her own home, might now not even recognize the home as her own, indeed might experience it as a place of fear and dread. Should that individual's previous wishes carry complete moral force such that caregivers struggle to maintain him or her in surroundings that are alien and foreboding to the elder's *present* self?

The standard answer is to say that a patient's expressed wishes must be honored. Patient autonomy, after all, is a foundational principle for most biomedical ethics. Yet, the very notion of autonomy is what raises the dilemma. If the elder is a significantly different person, one who autonomously expresses preferences, then why should a former self impose choices on a present self. While bioethicists are only now beginning to address these questions, families face them every day. One new way of looking at the question is to recognize that autonomy cannot just mean a capacity of choice or decision-making that involves cognitive expression of reasons for the choice. Few seriously compromised elders retain the capacity or inclination to communicate their preferences much less the rationale for their preferences. Nonetheless, even such elders manifest preferences in their everyday world: they prefer certain foods, certain clothes, certain persons. Good caregivers can readily identify such preferences. In fact, good caregivers look for them and intuitively adjust their care to the elders' own idiosyncrasies. That means that the elders' comfort and sense of security in their present world, in their present identity is highly relevant as a basis for care. Thus, the moral status of the elder is preserved even if the elder is a different person from the one we previously knew!

## Justice and aging

Even if it were clear that frail elders deserved special attention and care because of their significant moral worth, questions such as "Who is obligated to care for such old people?" would still arise. Is the expenditure of resources for the old ethically defensible when the use of scarce resources might limit the resources available for the care of children, for example? Such intergenerational issues of justice are not just academic, but are everyday practical dilemmas faced by the so-called "sandwich generation" families that find themselves facing the economic and emotional costs of simultaneously raising young children and caring for aged parents.

If we are lucky, we will all grow old before we die. To paraphrase Winston Churchill, being old is not so bad, given the alternative. Of course, such a view is not only ironic, it is downright pessimistic if aging is taken to inevitably involve a serious diminishment of one's capacities. To be sure, such a view of aging has been a dominant one, yet one that has been resisted as advances in living standards, medical care, and social provisions for retirement have meant that being old is no longer synonymous with dependence or destitution. In fact, intergenerational comparisons suggest that the elderly are relatively quite well off. Given that the need for medical services increases with age, other things being equal, inequality in the access and availability of health care services for the very young versus the very old provide a convenient focus for questions of intergenerational justice.

The fact is that most elders at some point before they die will spend some time in a nursing home or in a hospital. Costs associated with such health care are an understandable concern for older citizens, especially as health care costs continue to rise. Thus, aging has profound implications not only for one's personal state of health (and correlatively for one's need for health care services), but also for one's economic well-being or in countries with national health plans, social well-being. So it is no surprise that the question of the nature and extent of familial and societal obligations towards the frail old have become increasingly the focus of public discussion, a discussion that reveals the deep ethical ambivalence toward old age that characterizes our culture.

While it is understandable, even prudent, for individuals to set aside resources for their use in old age and to demand a share of societal resources, questions of justice exist about the distribution and use of resources across generations. For example, does justice require that expensive and exotic medical technologies be widely available for the old? Does it matter if increases in society's expenditure for care at the end of life diverts resources from education, medical care, or other social services for the young?

In an important sense, these questions are not simply questions about what is owed to one group versus what is owed to another, but involve a broader question about the nature of a just society. In a just society, a society that is well ordered, each individual and group of individuals would have a place that would allow those individuals to flourish. To have a place means that a fair share of resources will be available to each according to his or her needs. The problem, of course, is the universal problem of scarcity. Needs generally exceed available resources. Consumption of goods and resources by one group inevitably means that less is available for other groups. This fact is nowhere more evident than in terms of the employment of highly expensive and exotic medical technology at the end of life. The important ethical issue is whether life is so precious that each of us should cling to it irrespective of our experienced quality of life? For many Americans at least, the apparent answer is yes. Serious illness in the old, at least for most Americans, will involve costly medical care, all too frequently culminating in intensive services at the end of life. Such a practice results from many forces, but the net effect is not only to deny our finitude, but to make a special claim on resources for the end of life. Is such an extravagant claim on scarce resources justified in terms of social justice?

Clearly, we all age. From the point of justice this is a critically important point, because in contrast we do not change sex or race. Thus, to treat a member of a different race or sex differently creates an inequality that raises questions of justice. However, if the old and young are treated differently, an inequality may not necessarily be produced, since individuals will pass through each age group as they pass along the life span, enjoying each age group's specific burdens and benefits. If the old and young were treated occasionally and arbitrarily in an unequal fashion, then different persons would be treated unequally and a problem of justice or fairness would occur. But, if as a matter of social policy the old are treated one way and the young are treated another, and this treatment regularly occurs over the whole life of all members of a society, then all persons will be treated the same way and so no issue of unfairness need arise, because no inequality is produced. Thus, inequality by age is not itself a problem of justice.

The resources that the old (age greater than 65) consume are approximately 3.5 times the rate (in dollars) of consumption prior to age 65. However, prior to age 65, working people pay not only for their own actuarially fair costs of goods and services, but also for the costs of goods and services for the elderly and children. The problem of justice lies in the way that resources are spent on health care for the old – specifically, lavishing life-support for the dying, but withholding other kinds of services such as personal care and social support services that may be critical to well-being of elders when their lives are not immediately threatened. It is quite clear to almost all observers that our health care system is not effectively designed. It could do a much better job of matching services to needs at different ages of our lives and thus be far more effective and efficient than it is at present. Beyond efficiency, however, our health care system could be more responsive to the actual quality of the elder's life. Other nations provide far more generous social support and in-home nursing services than does the USA. As a result, the frailties associated with aging are not the cause of dramatic personal dislocation for elders or their families. There is no strong pressure to admit the elder to a hospital or to institutionalize the elder in a nursing home, because neither the elder nor the family are left to their own devices by society at large. Instead, social support exists for in-home care, transportation, and help with activities of daily living, which collectively contribute significantly to the elder's quality of life and sense of self-esteem.

Some thinkers (Daniel Callahan, for example) have argued that if one were to attend carefully to the actual experience of old age then one would see elders who are themselves engaged actively in communicating the sense of personal, familial, and social history and values to younger generations. This positive function that is the true and socially important activity of the old itself points to the obligation that the advanced old (perhaps beyond 85 years of age) be willing to accept death as "*appropriate.*" Because the old, in this view, have a purpose, namely to continue the tradition and to pass along community and family values and beliefs, their role gives them a special status and a special obligation. Elders are morally obligated to participate in the social role of making room for the next generation. This is actively undertaken by communicating important values and beliefs to the young, but also by knowing one's place as an elder who must surrender the stage to another. Thus, it would be unjust to support, much less demand, life-extending technologies and research for the very old. Such a demand would be quite selfish and contrary to the role and obligations of being old.

Such a recommendation, however, is greeted with horror in some quarters. A recommendation like this is seen as smacking not only of rationing, but as a denial of actual reality of aging in contemporary society. In America, the old are simply not honored for their age. As a culture, we increasingly value the young. We value the activities of the young. We value old age insofar as it provides opportunities to pursue those kinds of activities. But to be old and frail is not valued. Even if one is old, frail, *and* wise, it does not matter. Old age as such is not worth much in its own terms. Unfortunately, this response reveals our cultural prejudices about aging, but it is a reality with which any account of ethics and aging must come to terms. Many other cultures not yet influenced by Western technology value old age; indeed, they honor it. This does not mean that individual elders are honored or assumed to be honorable, but rather the very fact that they have lived into old age bestows an honor. The striking injustice of Western society and culture is its inability to find a place for old age that acknowledges it for what it is. As a result, treatment of the issue of justice is usually confined to the issue of the consumption of resources, rather than the personal and social qualities that are inherent in one's status as "old." The most prominent injustice in our society associated with aging, and the injustice that simple changes in social policy will not easily eradicate, is the injustice that fails to appreciate the old for what they are, old people. They are people who have survived and lived times past. Without appreciating this simple truth, our society will continue to struggle with ethical questions in the field of aging.

### Suggestions for further reading

Agich, G. J. *Autonomy and Long Term Care.* New York: Oxford University Press, 1993.

Callahan, D. *Setting Limits: Medical Goals in an Aging Society.* New York: Simon & Schuster, 1987.

Daniels, N. *Am I My Parents' Keeper? An Essay on Justice Between the Young and the Old*. New York: Oxford University Press, 1988.

Jecker, N. S. (ed.) *Aging and Ethics*. Clifton, NJ: Humana Press, 1991.

Moody, H. R. *Ethics in an Aging Society*. Baltimore, MD: Johns Hopkins University Press, 1992.

# People with dementia: A moral challenge

STEPHEN POST

EDITORS' SUMMARY

Stephen Post, PhD, discusses the difficult moral problems of dealing with persons with dementia. He presents some narratives from persons who had early dementia and struggled to find meaning. The experience is filled with frustration, fear, and loss of control. Caring for demented persons requires filling in for the missing structures of meaning, Post argues, and their dignity rests not so much on rationality, as on our commitment to promote whatever well-being might be possible. He concludes with practical guidelines for dealing with dementia as part of a philosophy of care.

In a culture that since the Enlightenment has so highly valued rationality and economic productivity, the life of the person with dementia, because it lacks such social-cultural worth, is easily characterized in the most negative terms, threatening to remove it from the protective moral canopy of "do no harm." Therefore, a fitting response to the epidemic of dementia, most often of the Alzheimer's type, requires us to enlarge our sense of human worth to counter an exclusionary cultural emphasis on rationality, efficient use of time and energy, ability to control distracting impulses, thrift, economical success, self-reliance, "language advantage," and alike.

My intent in this chapter is to present some experiences of dementia, the first of which was written by a person with early dementia. I present these experiences in order to underscore the importance of maintaining respect for the personhood and subjectivity of people with dementia. Through listening attentively to the experience of dementia, we outsiders can appreciate it as a search for meaning amidst a terrible anxiety about the future dismantling of abilities. This anxiety defines the most difficult existential phase of the illness that concludes when the person actually forgets he or she forgets.

## The struggle to adjust

Before radical self-forgetfulness is reached, there is a protracted period that may last for several years, during which the person with dementia depends on structures of meaning in order to make sense of his or her condition, sometimes against a religious background. People with dementia are meaning-seeking in the same way that we all are and their struggles to make sense of loss are akin to our own.

To present the picture of people with dementia as meaning seeking, I will rely on several autobiographical accounts. The first story – only lightly edited – was told by a woman named Jan who was diagnosed with probable Alzheimer's disease at age 40. In some pedigrees, early-onset dementia is common. Jan is now 43 and still highly conversant, although there are some days when she is too mentally confused to engage in sustained dialogue. Her meaning structure takes on a theological form.

### *Jan's Snow Flake*

It was just about this time three years ago I recall laughing with my sister while in dance class at my turning the big 40. "Don't worry Jan, life begins at forty," she exclaimed and then sweetly advised her younger sister of all the wonders in life still to be found. Little did either of us realize what a cruel twist life was proceeding to take. It was a fate neither she nor I ever imagined someone in our age group could encounter.

Things began to happen that I just couldn't understand. There were times I addressed friends by the wrong name. Comprehending conversations seemed almost impossible. My attention span became quite short. Notes were needed to remind me of things to be done and how to do them. I would slur my speech, use inappropriate words, or simply eliminate one from a sentence. This caused not only frustration for me but also a great deal of embarrassment. Then came the times I honestly could not remember how to plan a meal or shop for groceries.

One day, while out for a walk on my usual path in a city in which

I had resided for 11 years, nothing looked familiar. It was as if I was lost in a foreign land, yet I had the sense to ask for directions home.

There were more days than not when I was perfectly fine; but to me, they did not make up for the ones that weren't. I knew there was something terribly wrong and after 18 months of undergoing a tremendous amount of tests and countless visits to various doctors, I was proven right.

Dementia is the disease, they say, cause unknown. At this point it no longer mattered to me just what that cause was because the tests now eliminated the reversible ones, my hospital coverage was gone and my spirit was too worn to even care about the name of something irreversible. I was so confused and felt so alone and I didn't want to hear their advice that the support I so badly needed was available at the Alzheimer's Association.

I was angry. I was broken and this was something I could not fix, nor to date can anyone fix it for me. How was I to live without myself? I wanted Jan back!

She was a strong and independent woman. She always tried so hard to be a loving wife, a good mother, a caring friend and a dedicated employee. She had self-confidence and enjoyed life. She never imagined that by the age of 41 she would be forced into retirement. She had not yet observed even one of her sons graduate from college, nor known the pleasures of a daughter-in-law, or held a grandchild in her arms.

Needless to say, the future did not look bright. The leader must now learn to follow. Adversities in life were once looked upon as a challenge, now they are just confusing situations that someone else must handle. Control of *my life* will slowly be relinquished to others. I must learn to trust – completely.

An intense fear enveloped my entire being as I mourned the loss of what was and the hopes and dreams that might never be. How could this be happening to me? What exactly will become of me? These questions occupied much of my time for far too many days.

Then one day as I fumbled around the kitchen to prepare a pot of coffee, something caught my eye through the window. It had snowed and I had *truly* forgotten what a beautiful sight a soft, gentle snowfall could be. I eagerly but so slowly dressed and went outside to join my son who was shoveling our driveway. As I bent down to gather a mass of those radiantly white flakes on my shovel, it seemed as though I could do nothing but marvel at their beauty. Needless to say he did not share in my enthusiasm, to him it was a job; but to me it was an experience.

Later I realized that for a short period of time, God granted me the ability to see a snowfall through the same innocent eyes of the child I once was, so many years ago. Jan is still here, I thought, and there will be wonders to be held in each new day, they are just different now.

Jan experiences frustration, fear, loss of control, and anger, but she is able to adjust to her circumstances with some success. In my conversation with Jan, she emphasized two key factors that make adjustment possible. First, her husband Ken and her sons discuss with her any limitations on her privileges, and she is able to negotiate mutually satisfactory resolutions with them. For example, Jan no longer drives, but on the condition that family members provide her with transportation. She no longer walks across the street alone because she is confused about the meaning of the red, yellow and green lights, but this is on the condition that others escort her routinely. Second, Jan refers to the love she feels from Ken and her sons as essential to her quality of life. She once said that it is better to speak of quality of lives than of quality of life.

### Robert's faith

In another autobiographical account from a person with Alzheimer's disease, Robert Davis found peace in his perception that his condition was a test of faith like Job's. But later, in more confused moments, even this spiritual framework was lost and he experienced profound fear and forsakenness, a "dark night of the soul." His hope was for an imminent death to take him home quickly rather than experience further self loss.

In this account by a man trying to find meaning in subtle and then more dramatic changes occurring within his mind, the story of the disease is conveyed from the inside. The existential story must be told powerfully for the undemented to properly appreciate the struggle for meaning. In *My Journey into Alzheimer's Disease*, he says "My brain may be dying, but in my spirit Christ has healed me, and I can say with the songwriter, 'It is well with my soul'," and later, "The private emotional relationship with the Lord that I enjoyed is distorted and does not comfort me now. When I pray, I often pray in silent blackness of spirit." He became periodically forgetful even of a cherished system of meaning to which his life was devoted. But in the last analysis, it was his faith, coupled with the support of his wife Betty, that allowed him to gain enough perspective on his condition to write a meaningful book with Betty's help.

These accounts of dementia highlight individuals' endeavors to retain hope and meaning. Both remain persons, with their gratifications and frustrations, their own unique background, and their own unique destiny. There is a benevolent point in the progression of dementia when all anxiety and embarrassment are forfeited in favor of grateful amnesia. Family and friends become as strangers while the familiar and the foreign lose the elasticity of their boundaries and become as one.

When the capacity to seek meaning in the midst of decline itself gives way to dementia, as it will in the more advanced stages of illness, then the experience of the person must be understood in relational and affective terms. Here the challenge to set aside the values of hypercognitive culture is upon us.

## To forget that one forgets: Quality of life

At a meeting of the local Alzheimer's Association, a support group for individuals affected by Alzheimer's-type disease, and their families, I conducted a discussion of the moral meaning of quality of life in the context of dementia. There was much concern expressed by the group that only people with Alzheimer's and their family caregivers are in a position to speak thoughtfully about quality.

Sharen, for example, told the story of her father. Dad held onto his identity until the very end, she said. He wore his favorite cowboy hat in the shower, slept with it on his head, and never let it out of sight. Somehow he knew that there was something special about this hat, that it was somehow connected with who he was. While dad could no longer talk much, and never coherently, he could still play a pretty good game of cards long after he forgot Sharen's name. Dad has never read books much anyway, remarked Sharen. Her conclusion was strong: people of intellectual capabilities would not appreciate dad's many moments of joy. Of course, there were down moments for dad, like there are for everyone. But his quality of life cannot be adequately evaluated by intellectuals, who are not a jury of his peers. The lesson is that we outsiders must be careful about judging quality of life for those with dementia.

### *Mrs G*

Take the case of Mrs G She came from an old family of distinction. I visited her in the Alzheimer wing of a quality nursing home.

> She carries an old book under one arm as she walks slowly down the corridor. I greet Mrs G but she says nothing. However, she shows a picture to me and seems to smile but I am not quite certain of this. It is a James Audubon print book. They tell me she always has it open to the same page and points to the same picture, a bluebird. (I think a bit sardonically for a moment that at least with dementia novelty requires only one book and only one picture.) I guide Mrs G to a table and we sit. I ask her how her children are. She does not respond, although she again appears to smile. She seems to say "sky," but who knows?

She has a certain graceful charm and a slight smile. They say that habitual mannerisms and demeanor are so ingrained that they are the last things to go. Are these graces just the simulacrum of the self, a kind of deception that suggests more of Mrs G is there than meets the eye? Slowly Mrs G rises and walks away, a little tear in her eye. She seems to have emotions left, and emotions are a part of well-being. The ability to experience emotions has not been lost, and in this sense Mrs G is as fully human as anyone else, or maybe even more so.

Because of the human propensity to treat with indignity those who supposedly cannot experience indignity, it is vital to assume a continuing self-consciousness and a correlative sensitivity to the manner of treatment afforded by others.

## Ethics of loyalty

In cases of profound and terminal dementia, it may be difficult to convince the observer that epistemological agnosticism (doubt) is all that empiricism (experience) will permit. Solicitude is then best understood in terms of loyalty. Throughout the experience of dementia from the struggle for meaning to profound loss of capacities and abilities, solicitude is tied to filial or conjugal loyalty. It is loyalty that precludes moving people with severe dementia "into the house of the dead." Here, a solicitude grounded in bestowal rather than appraisal comes to the fore.

In an article on "Seeing and knowing dementia," David H. Smith describes the loyal care provided for his mother-in-law, Martha. A woman from an apartment down the hall from Martha called to say that his mother-in-law was getting lost and behaving strangely. Yet "Social graces remained: smiling at a visitor, laughing with the crowd, responding briefly and politely in conversation." In looking back on several years of caring for Martha prior to her death in a deep coma following a seizure, Smith raises the question of "identity, status, or ontology. How much do demented persons matter, and why do they matter?"

Smith rejects the notion that personhood, the basis of moral considerability, is measured by continuing moral agency. He cites biomedical ethicist H. Tristram Engelhardt Jr, who argues that people "are persons . . . when they are self-conscious, rational, and in possession of a minimal moral sense." By this narrow definition of personhood, Martha had ceased to be a person. Smith asks, "But what follows? That she need no longer be respected? That she was no longer part of the family? That it was incoherent to continue to love her? That she could make no reasonable claim on the resources of forbearances of the larger society"? Smith's conclusion is that the narrow personhood theory of moral considerability is an "engine of exclusion" that can lead to "insensitivity" if not "wickedness."

## Toward a moral philosophy of dementia care

Dementia strikes at the idolatry of rationality. It is incumbent on us to see the glass of that humanness as half-full rather than as half-empty, and certainly rather than as empty. Caregivers can interact with the person caught in the anxious struggle for meaning, they can respect subjectivity and intentionality even when the means of conveying thoughts is eroded and when some days are more lucid than others. Caring need not be characterized as an ongoing funeral.

Enhancing well-being and making the most of what strengths are still present is key. Understanding communication methods is important to enhance rapport and diminish anxiety. Above all, quality of life for the person with dementia is always partly subjective and is largely a matter of emotional adjustment facilitated by interactions and environment. Quality of life is always a self-fulfilling prophesy. If we think that there can be no quality of life because of cognitive deficits, then we will probably not do the things that can enhance quality. In the process, we consign those with dementia, retardation, and a host of other brain-related conditions to neglect.

Now obviously many people with Alzheimer's disease dementia do not adjust emotionally to their plight, nor do all family caregivers find caring sustainable. People with the disease can manifest severe behavioral problems and even violence, at which point family caregivers can be overwhelmed and a nursing home placement may in some cases be called for. Yet I have seen and heard about many people with dementia who seem to have a profound appreciation for the little things in life, from gazing in delight at a colorful flower to whistling with the birds while walking on the path around the nursing home. They seem to transcend their cognitive impairment.

## Practical conclusions

Thus far, I have attempted to enter into the experience of dementia to underscore the continuing respect due those with Alzheimer's disease. The subjectivity of the affected individual must be taken with utmost moral seriousness. In conclusion, I will quickly highlight some maxims that are consistent with respect for the personhood of those with dementia.

### *Diagnosis*

The diagnosis of probable Alzheimer's disease should be sensitively communicated by the physician to the affected individual and his or her family. Disclosure of the diagnosis can be beneficial when support groups (such as the Alzheimer's

Association) are available to affected people as well as to family caregivers, and when counseling and other services can be provided to facilitate emotional adjustment to the diagnosis. With diagnostic disclosure comes the responsibility to direct the affected individual and family to available resources.

### Driving

Diagnosis of Alzheimer's disease is never itself sufficient reason for loss of driving privileges. The person whose driving privileges are to be limited because of dementia should be an active participant in the decision. Appropriate limits to driving and other activities of daily living can often be delineated and mutually agreed upon through open communication between the affected person, family and health care professionals.

### Exercise competency

People with dementia should be allowed to exercise whatever competency for specific tasks and choices they retain. Competency should be specifically assessed for particular tasks. A person may lack competencies to drive, handle financial affairs or live independently in the community, but have the abilities to make authentic decisions about place of residence and medical care. In almost all cases, judgments of competency in health care settings for medical decision-making can be made without the need for legal proceedings.

### Social modifications

The best approach to problem behaviors relies on social and environmental modifications, and on creative activities, thereby preserving independence and self-esteem. Physical and chemical restraints should not be substituted for social, environmental, and activity modifications. Behavior control drugs should be used cautiously for specified purposes. An individual profile of the person with dementia should be available to institutional caregivers, highlighting an interactive and activity-based care plan known to be most effective for the individual.

### Planning ahead

Family members, Alzheimer's disease-affected people, and health care professionals should sensitively discuss and plan for a good death, supported by appropriate documentation. The disease will eventually result in the affected person's death and is therefore a terminal chronic disease. Many people with families want to entrust treatment decisions to loved ones who will act in their best interests, and this should be supported. Alzheimer's disease should be legally acknowledged as a terminal illness, thereby removing any doubt about the right of affected people to refuse any treatment by advance directive should they become incompetent to make medical decisions.

### Suggestions for further reading

Brody, E. M. *Women in the Middle: Their Parent-Care Years.* New York: Springer Publishing, 1990.

Davis, R. *My Journey into Alzheimer's Disease.* Wheaton, IL: Tyndale House, 1989.

Foley, J. M. and Post, S. G. Ethical issues in dementia. In *Handbook of Dementing Illnesses,* ed. J. C. Morris, pp. 3–22. New York: Marcel Dekker, Inc, 1994.

Mace, N. L. and Rabbins, P. V. *The 36-Hour Day.* Baltimore, MD: Johns Hopkins University Press, 1991.

Melnick, V. L. and Dubler, N. D. (eds.) *Alzheimer's Dementia: Dilemmas in Clinical Research.* Clifton, NJ: Humana Press, 1985.

Office of Technology Assessment. *Losing a Million Minds: Confronting the Tragedy of Alzheimer's Disease.* Washington, DC: United States Government Printing Office, 1987.

Smith, D. H. Seeing and knowing dementia. In *Dementia and Aging: Ethics, Values, and Policy Choices,* ed. R. H. Binstock, S. G. Post and P. J. Whitehouse, pp. 44–54. Baltimore, MD: Johns Hopkins University Press, 1992.

Thomasma, D. C. From ageism toward autonomy. In *Too Old for Health Care? Controversies in Medicine, Law, Economics, and Ethics,* ed. R. H. Binstock and S. G. Post, pp. 138–63. Baltimore, MD: Johns Hopkins University Press, 1991.

Wickler, D. (ed.) *Medical Decision Making for the Demented and Dying.* The Milbank Quarterly (1986) **64**, Supplement 2.

Zgola, M. *Doing Things: A Guide to Programming Activities for Persons with Alzheimer's Disease and Related Disorders.* Baltimore, MD: Johns Hopkins University Press, 1987.

# Personal dying and medical death

STEVEN MILES

EDITORS' SUMMARY

Steven Miles, MD, proposes that the most important aspect of dying is a moral one, the confrontation with one's own mortality. For the family this is also the most important value, placing the dying person in a continuum of family and personal values. He contrasts this with the impersonality of medical death that all too often occurs to people today, dying in institutions under the care of health care providers who are also strangers. When this happens, the priorities of medicine and medical care seem to trump the personal and family values, creating needless crises at the bedside of the dying person. There is an enormous gulf between the two value systems, one to a large extent created by technology's need to be delivered in institutions. Miles challenges us to think of ways to overcome such a crisis.

How will I die? In this society, at this time, the answer to this ancient question has acquired layers of institutional complexity. Simultaneously basic comfort for the personal crisis posed by death and dying has become, for many, less accessible. There may be a link between the complexity of medical death and the seemingly inadequate resources for confronting death at a personal level. Perhaps the new technical cloak for death is a tattered attempt to shelter people and a society that lacks a more fundamentally consoling shelter. If so, medical death is confusing and officious because it is simply another face of a confused society. It is also possible that the complex legal and technical enterprise of modern medical death is the problem and not a symptom. If so, the legal and technical complexities of medical death are avoidable barriers to a meaningful and personal dying.

These diverging possibilities raise difficult social and personal questions. How should highly technological, regimented, and impersonal health care institutions support persons and families during medical care before death? How can people draw strength from personal sources of meaning as they or their loved ones die in such institutions? This chapter can hardly begin to resolve such profound issues; it can only orient a reader to places where personal quests for solace can be found in medical care and explore overcoming barriers to that necessary task. I contrast medical and personal death in the case below, and in my discussion.

## A case of a grandfather

My grandfather was debilitated, bedbound and skinny when he died. He had widespread prostate cancer which, depending on how one looks at it, may have killed him. Or maybe he died of cancer-related anemia, or of his decision to refuse blood transfusions or of malnutrition, or of his refusal to eat or drink after cancer took his hunger away, or of our decision to not put a feeding tube down his nose and into his stomach. The cause and course of his official death is named in his medical record and a death certificate which list the cause and effects of each of the diseases, conditions, and treatment decisions preceding his death. At 87, my grandfather died of "natural causes" but not of "old age."

My grandfather would have filled out his death certificate quite differently. He was not a student of the biology of aging but he had raised his family and sent them into the world, built and retired from his law firm, married and buried his wife, put aside his beloved fishing rods, and long retreated to memories and set opinions. He closed his story in the home he would not leave on the first anniversary of his life-long wife's death. As a grandson-physician, I attended his medical and his personal death. I reluctantly became his physician when his doctor refused to leave his kingdom to come to my weakened grandfather's castle. I was obliged by blood and empowered by my status as a first-year doctor to order the morphine for his pain.

## Medical death

Western society is making over death in a new "managed" image: "medical death."

- Medical death is the outcome of disease, rather than the end of a natural life. It occurs in, and (judged by the degree of control) "belongs" to, health care institutions, hospitals, intensive care units, and nursing homes rather than to personal communities.

- The course and conduct of medical dying is largely governed by laws and expert policies rather than by the neighborly moral conversation over the back fence, around the kitchen table, or in places of worship. The roles of priests and physicians as bystander and chaperone have been reversed.

- Medical caregivers will use the patient's values to guide treatment decisions but are reluctant to use them for the larger solace of a dying person or family (a family task which is deferred to the funeral eulogy). Until recently, during dying the life-long values were integrated into the care of the dying person and the family; after death the person was buried with the stark final simplicity of "dust to dust."

- The role of the family has changed. Once, they were intimate attendants who provided the physical care and spiritual escort of a dying loved one. Today, they are often replaced in the former role by professional strangers. In the latter role they have become supplicants who bear the dying person's values to be blessed by a doctor's orders to permit life to close by withholding an otherwise nearly automatically provided life-prolonging treatment.

For families, medical dying means to patients an event to be managed by others and attended by intimates, rather than one to be lived through or learned from.

Dying does not rest easily in its new, managed construction. The older traditional version, "personal dying," surfaces, sometimes volcanically, like a submerged continent rising on ancient geologic forces. Hospice, a family-centered and less intrusively technologic caregiving, is entering the health care system. A priest may help a dying patient assert the plea to "die in peace" and reject radiographic scans and chemotherapies. Euthanasia, often an attempted escape from a medical death that a patient has lost confidence in or derives no comfort from, is now a pressing social issue. Even so, on the dominant cultivated landscape of medically managed death, these events are too often misunderstood as quaint or medically unreasonable, or even neglectful.

"Medical" and "personal" dying profoundly differ in terms of the experience and expectations of dying persons, their communities, and health care institutions:

- The fact that life is cut short by disease is an intrusive and unacceptable accident that can obscure the fact that we all must come to terms with closing of lives.

- The rules for managing medical death rest on the impersonal foundations of medicine and law that only imperfectly and crudely correspond to the personal founts of meaning that patients and families draw on.

- The replacement of families with strangers in the care of the dying person's body is a form of familial estrangement that is so profoundly alienating, that we can barely talk about it.

The power of the hospital to cast personal dying as quaint or medically ignorant suppresses personal values, emotional truths, and family ways and thus forever shapes the story and meaning of death, which is handed down from generation to generation within families. To recover the possibility of a personal death in our health care system, we must remove the protective cover of habit and ordinariness that suggests medical dying is the only way that things can be.

## Death from disease

The idea that life is ended by a specific disease, rather than by its natural mortality, which can take many forms, is powerful. It spurred research leading to the identification and treatment of literally thousands of acute and chronic diseases. It discredited neglect of avoidable suffering and premature death and disability. The life-prolonging and pain-relieving value of this idea has endowed healers with respect, resources, and power. It also obscured the fact that the end of life is not centrally about the physiology of dying but about the problem of mortality.

Hippocrates once admonished doctors to avoid persons who were overmastered by disease. Only a decade ago, doctors shrugged their too broad shoulders and said, "There is nothing more that can be done." Today, medical writers can be criticized for saying that someone is hopelessly ill or that care is futile. Today, the last stages of life – no matter how far advanced – can be medically treated, however briefly and uselessly, with life-prolonging technology. So, medical death does not arrive as the manifestation of mortality but as the last event in the course of modifiable biological processes.

Ironically, even as medical science has banished mortality, it has also made the approach of death telescopically visible from a distance. Few persons die young. Far fewer die suddenly. The vast majority of people bear medically recognized and treated signs of impending death in the form of the gradual onset of a chronic disease such as cancer, acquired immunodeficiency syndrome (AIDS), dementia, or meteor showers of bits of cholesterol to the blood vessels of their heart. To the greatest degree in history, we peer forward to the scientifically recognized sequence of chronic disease even as the natural and personal mortality, which lies at the end, and beginning, of that process, is obscured.

## Dying in health care institutions

We die under medical care. In the USA, 85% of deaths take place in hospitals and nursing homes. Hospital death is far more common. This a big change; fewer than 50 years ago, most deaths were at home. Though death often occurs in health care environments, it is surprising how often health care providers try to avoid caring for a dying person. Nursing homes often transfer patients to hospitals as the end draws near. Surgeons maintain some people on life support until the inevitable but delayed death can no longer be attributed to surgery. Oncologists or cardiologists will often return dying patients they have treated over a prolonged struggle with an illness to family doctors for "continuity" of care. Patients and families are often surprised by, and can fight, such events. But life's last hours can be frightening even when the end is welcome. Sometimes the overwhelmed family of a patient in a home hospice summon aid for an emergency hospitalization in the face of a sudden crescendo of pain or apparently difficult breathing. A nurse I know once wisely called this "dying at home but taking the last breath in a hospital."

Thus, it is medical institutions, not just a mortality-denying science, that sets the stage for medical death. Institutional habits and history drive events and largely determine what choices will be available and, more importantly, how those choices are understood. For example, visiting rules can change the comfort and shape of how the dying person interacts with friends and family. Most of these rules are not supported by research. The steady pressure of institutional settings can be resisted if families and patients are not too tired or overwhelmed.

The fact that dying takes place under medical care can result in the morality of medical decisions about the end of life becoming hopelessly confused with the pressing need for a personal solution to address, and accept, mortality itself. Each of dozens of decisions to provide, change, or stop any medical machine, pill, or feeding tube can be named as a decision to postpone death, hasten its approach, or allow it to come in its own time. Three-fourths of hospital deaths follow a decision to withhold or withdraw treatment. Half of intensive care deaths occur in such a manner. It is even higher in nursing homes. End-of-life medical decisions occur on the contested terrain of medical and personal death where precious life meets mortal life. It is rarely a banal fight between respecting life or casually accepting death.

The medical paradigm has the upper hand in defining and directing end-of-life health care. Decisions about medical treatments are bewildering when considered as technical minutiae and mind-boggling in the full combinations and sequences of a complete treatment plan. Family or patients who try to think like doctors, especially when the doctor unfairly presents only one option, are quickly disempowered. Furthermore, treatment proposals arise from a scientific paradigm that respects life, biologically explains, and therapeutically acts and that lacks an account of, or solace for, the necessity or inevitability or meaning of death. The decisions about my grandfather's transfusions, hydration, bedsores, nutritional intake, chemotherapy, and home-based nursing care were a vast muddle to his family. He saw the task more simply: to keep his life in its own context as it came to its natural, and inevitable, close. His was a personal triumph over medicalized death.

## Struggling for personal death under end-of-life medical care

Personal and medical death must take place together. The techniques for easing pain, explaining and relieving the hard breathing without shortness of breath of life's last hours, or for postponing death that is far before its necessary time are too valuable and too life-affirming to forego. However, medical machines are aimless if not constantly held accountable to the context of a personal story about how life was valued, and how a person bowed, with retained dignity, in the face of mortality.

The fact that we search for a solace to the problem of personal dying on a medical stage has lead to two kinds of difficult struggle in hospitals and nursing homes. Some successful personal deaths occur when a person is assertive enough and strong enough to plow and plant their own values in the wintery, but nurturing, fields of medical dying. Others, compliant to the rituals of a medical death, somehow find comfort in the struggle or the quest that is the premise for the use of aggressive medical interventions up to the moment of death. Too often, however, it is too arduous for the dying to assert or find personal solace in medical end-of-life care. It need not be this way.

### Suggestions for further reading

Aries, P. *Western Attitudes toward Death: From the Middle Ages to the Present*, translated by P. M. Ranum. Baltimore, MD: Johns Hopkins University Press, 1974.

Becker, E. *The Denial of Death*. New York: The Free Press, 1973.

Broyard, A. *Intoxicated by My Illness*. New York: Fawcett Columbine, 1992.

Friedan, F. *The Fountain of Age*. New York: Simon & Schuster, 1993.

Nuland, S. B. *How We Die*. New York: Alfred A. Knopf, 1994.

Tolstoy, L. *The Death of Ivan Ilyitch. The Cossacks: Happy Ever After*. London: Penguin Books; New York; Viking Press, 1960.

# Stopping futile medical treatment: Ethical issues[1]

NANCY S. JECKER and LAWRENCE J. SCHNEIDERMAN

EDITORS' SUMMARY

Nancy S. Jecker, PhD, and co-author, Lawrence Schneiderman, MD, have distinguished themselves in a series of articles regarding medical futility. Here they summarize their view by arguing that there are two kinds of medical futility, quantitative and qualitative. In the first, the likelihood of benefiting the patient is vanishingly small, perhaps one chance in a hundred. In the second, the quality of benefit is so small that the healing aims of medicine would not be achieved. They point out that the goals of medicine are subverted in futility cases, and that health professionals, too, have values that should not be destroyed by compelling them to "do everything possible."

Physicians and other health professionals sometimes find themselves continuing aggressive medical procedures well beyond the point at which such measures would be useful. The impetus for this may come from a variety of sources. Perhaps the patient, fearful of death, desperately seeks every conceivable way to avoid it. Alternatively, a family member or loved one may implore the health care team to "do everything possible" and "spare no expense" when a patient is unconscious or delirious and unable to speak for himself or herself.

Sometimes it is physicians or other members of the health care team who insist on nonbeneficial medical treatments because they cannot let go. The reasons for this may lie in the emotional bonding that takes place between a desperately ill patient and dedicated providers, or may lie in misguided notions of ethical or legal obligation. Perhaps a physician has already "saved" the patient from the brink of death on several occasions, and feels that giving up now would mean that these previous efforts were of no avail. Or perhaps members of the health care team regard death as a personal failure, to be avoided at all costs. Perhaps, too, education and training has inculcated in health providers a bias toward action; therefore, standing by and "doing nothing" in the face of death seems inconceivable. Finally, in this day when the practice of medicine has been opened up to public scrutiny and ethical dilemmas sometimes become media spectacles, pressure to continue extreme life-prolonging treatments may come from outsiders with political, religious, or other motives. These and other attitudes give rise to situations in which medical interventions are used despite overwhelming evidence that they will not benefit the patient.

One example of nonbeneficial medical treatment is mechanical ventilation (the respirator) employed on a patient in a permanent vegetative state, a condition characterized by permanent unconsciousness. Because the patient is utterly incapable of appreciating, and will never appreciate, ventilator assistance as a benefit, the treatment strikes us as unambiguously futile. Almost as clearly futile, in our opinion, is attempting cardiopulmonary resuscitation on a terminally ill cancer patient whose death is imminent and who is experiencing unremitting pain and discomfort or is under heavy sedation. In this case, the harms of the procedure (more misery from invasive intubation, broken ribs, organ damage including possible brain damage) must be weighed against the potential benefits (a few more hours or days in the intensive care unit). Another example of futile treatment is tube-feeding in a patient with multiple organ failure whose condition is so irreversible that the patient will never leave the intensive care unit of a hospital.

In all of these instances, although the intervention produces an *effect* on the patient's body, it fails to provide a *benefit* to the patient as a whole. If the goal of health care is to heal (from Old English "to make whole") the patient (from the Latin for "suffer"), then such interventions are inappropriate – or futile. We argue that not only are health professionals not obligated to provide nonbeneficial treatments, they are obligated *not* to provide such treatments, because they violate ethical standards of the health care professions. The purpose of this chapter is: (a) to clarify the meaning of medical futility; (b) to illustrate, through actual case examples, the application of medical futility; and (c) to urge increased efforts to provide more appropriate, care-oriented treatments when life-sustaining measures are futile.

## When is an intervention futile?

We define a medical treatment as "futile" in situations where either the *likelihood* of benefiting the patient is so vanishingly small as to be unrealistic or the *quality* of benefit to be gained is so minimal that the healing goals of medicine are not being achieved. We refer to these two situations as quantitative and qualitative futility, respectively.

Informed and well-intentioned people may disagree about the specific levels at which quantitative and qualitative definitions of medical futility can be drawn. For example, people may set different cut-off points regarding how low the odds of success must be for a treatment to be futile. This is not surprising because where one sets the cut-off point for quantitative futility involves a value judgment. What counts as an acceptable chance? Likewise there may not be unanimous agreement about what qualities of outcomes are poor enough to qualify as futile. Again, the judgment of futility requires a value decision.

Some may prefer to wait for "absolute certainty" before making judgments about futility. As pointed out by philosophers as venerable as John Stuart Mill and as contemporary as Karl Popper, however, absolute certainty is an absolute impossibility. We can never be certain about the presence of causal connections in life, much less in medicine. We can draw conclusions only in a common-sense way after making empirical observations. In formulating a specific approach to medical futility, therefore, one should look for a common-sense account. We propose calling an intervention quantitatively futile if experience has shown that it has less than 1 chance in 100 of benefiting the patient. With respect to qualitative futility, we identify several exemplar cases. The most obvious we have already cited: the patient in permanent vegetative state – of whom it is estimated that perhaps as many as 35 000 in the USA alone are being kept alive by artificial nutrition and hydration. There can be no treatment of any benefit to such a patient. Another situation is the patient whose illness confines him or her permanently to the intensive care unit, where the magnitude of benefit is dubious because treatment requires the patient's entire preoccupation to the exclusion of other life goals.

One problem with applying this definition in practice is that health providers and family members sometimes interpret the patient's plea to "Do everything!" to mean doing everything possible to prolong life – even at great cost in suffering and dignity. An interpretation more consonant with the ethical traditions of health care however, is that it is incumbent on physicians and other health professionals to do everything possible to benefit patients, including maximizing comfort and dignity when life prolongation is neither reasonable nor desired. The idea that health providers are obligated to prolong biological life for its own sake is a modern idea, an idea without precedent in health care's ethical traditions. Hippocratic physicians, for example, shunned the use of futile treatments in order to distinguish themselves from charlatans. Hippocratic physicians were expected to recognize when a disease "overmastered" known medical treatments and to refrain from using such treatments. The emphasis of Hippocratic medicine was to restore the patient's health where possible, and to alleviate the patient's suffering. Sadly, in modern times, health care teams sometimes become preoccupied with life-prolongation efforts and overlook the duty to care for patients. It is important to emphasize that although *treatment* may be futile, *caring* for patients and patient's loved ones never is.

## Are health care providers obligated to use futile interventions?

When a patient or family member insists on continuing futile treatments, the health care team has no obligation to provide such treatments. After all, as we have said, the goal of health care is to benefit the patient. When treatment ceases to provide a benefit, there is no clear reason to continue with it. Furthermore, providing futile treatments often violates the health care provider's duty not to harm patients because such treatments are often invasive and burdensome, only adding pointlessly to the patient's pain and discomfort. Finally, the use of futile treatments should be avoided because it wastes society's resources. Typically, interventions that are futile are also costly, requiring the use of expensive medical technologies, the attention of multiple health care providers, and the use of hospital beds that other patients may need.

Some have argued that patients have a right to be autonomous or self-determining agents and that denying futile treatments to patients deprives them of this right. Yet, in response, it can be said that although autonomy guarantees a competent adult patient the right to refuse any medical treatment, it does not guarantee the patient a right to obtain any medical treatment. Whereas the former, negative right, simply asks others not to interfere, the latter, positive right, requires actions on the part of others who also have a right to be autonomous and self-determining. Patients cannot ethically compel physicians, nurses, and other health care workers to provide treatment that violates the health care provider's own conscientiously held personal beliefs, or the standards of the health profession to which the individual belongs. No other profession labors under an unlimited mandate of the people it serves. A lawyer, for example, is obligated to provide a vigorous and competent defense for a client accused of murder. But a lawyer is not obligated to do *anything* the client wants – phone the governor, phone the President – since the lawyer is responsible also to the reasonable standards of the legal profession.

## Examples of medical futility

To illustrate medical futility's meaning and ethical implications, we now describe two real cases involving the use of futile interventions. The first was reported in the popular press and concerns an intervention, cardiopulmonary resuscitation, that in this case was qualitatively futile because the patient will never be conscious or appreciate any benefit from treatment. Thus, treatment serves merely to prolong biological life, without benefiting the patient as a person.

### Case 1: Prolonging Life[2]

Baby K was born with a condition known as anencephaly, in which most of the brain – except for rudimentary portions of the lower brain – are missing. The doctors caring for Baby K know that Baby K will never think, hear, see or indeed perform any of the activities that we associate with a living person, except for so-called "vegetative functions," such as breathing (with the help of a machine), digestion, and urine formation. Now 16 months old, Baby K is placed on a mechanical ventilator whenever Baby K has trouble breathing, following the patient's mother's wishes. Although the patient lives at a nursing home, the patient has been taken to a hospital in Fairfax, Virginia, at least three times with severe respiratory problems.

The child's mother has insisted that doctors "do everything" for Baby K, and has stated that she wants God, not the health care team, to decide when her child will die. Yet Baby K's doctors have argued that mechanical ventilation will do nothing to replace absent brain tissue and therefore ameliorate the child's underlying condition. The hospital where Baby K is being treated requested court permission to deny emergency treatment.

The child has spent about 120 days at the Fairfax Hospital, at a minimum cost of $174 000, paid for by State Medicaid and private insurance.

In response to this case, some may argue that biological life itself is sacred or has intrinsic value, and therefore Baby K's life should be continued. Yet even if this position is correct, again we emphasize the ends of medicine (and other health professionals) are not to keep bodies alive but to benefit patients. Doctors' and nurses' commitment is to persons, not to biological life as such. Nor do medicine and nursing function simply as instruments to carry out others' (family's or patient's) wishes. Rather, as professions, medicine and nursing uphold certain standards and values as their own. Continuing to provide futile treatment to Baby K undermines the integrity and goals of the health care team. Furthermore, futile treatment for Baby K expends society's finite resources, and may violate principles of justice in the allocation of society's resources.

In contrast to case 1, consider the following case involving a young man in the terminal stages of acquired immunodeficiency syndrome (AIDS). One of us (L.J.S.) was involved in an ethics consultation involving this patient. Members of the health care team were attempting to decide whether or not to recommend terminating life support measures for the patient, and shifting the emphasis of treatment to palliative and comfort measures.

## *Case 2*

A 36-year old male bartender is admitted for the third time to the intensive care unit with complications from AIDS, including Kaposi's sarcoma (a form of cancer), pneumocystic pneumonia, idiopathic thrombocytopenia purpura (a severe blood clotting disorder) and disseminated mycobacterium intracellulare avium (a tuberculosis-like infection). Over the several days in the intensive care unit he developed failure in several organ systems, including the lungs, kidneys, liver, heart, and became progressively obtunded (out-of-it). Although the man's partner at first insisted that "everything be done" to keep him alive, it soon became apparent to the physicians involved in the case that, from previous experience with patients in this condition, the patient had no realistic chance of survival.

Recognizing the futility of life support efforts, the physicians requested an ethics consultation. The outcome of discussion among members of the health care team was a decision that treatment objectives should be changed from "full code (do resuscitate), full care" to "no code (do not resuscitate), comfort care."

Members of the health care team did their best to persuade the patient's partner to accept the patient's impending death and assist with efforts to contact family members and friends. After extended discussions, the patient's partner agreed. The health care team used a morphine drip to keep the patient comfortable, and pillows to prop up the patient to make his breathing easier. The patient's partner, family members, and friends, recognizing that death was imminent, visited the patient and said their farewells. The patient died a few days later.

In sharp contrast to case 1, case 2 reached a point of moral and emotional closure for the health care team and for those who were close to the patient. Rather than using futile treatments as perhaps a symbolic show of love and concern, more fitting measures were used to convey caring. Although death is always an occasion for grief and sadness, in this instance medicine did not fail to offer solace in the face of it.

## What should health care providers do when life-saving measures are futile?

The foregoing discussion suggests what doctors and nurses should be doing when they can no longer cure or ameliorate disease because death is imminent: they should offer comfort to patients and loved ones. Rather than abandoning the patient or those who are close to the patient, health care providers should instead assure their continued support, and should use low technology, care-oriented measures that offer relief and dignity to the patient. Simple acts, such as holding a patient's hand, moistening a patient's lips, talking and listening to a patient and loved ones, are very much a part of health care as are the latest high technology machines and gadgets. Sadly, modern health care has sometimes lost sight of this, and providers have sometimes neglected their obligation to help patients and families in these ways.

Rather than viewing the end of life as a medical failure, we submit that health professionals should instead regard death as a planned-for and natural event. Rather than understanding the impending loss of a loved one as an occasion to defy death, we have argued that family and friends should instead help patients and each other to face death with courage and dignity.

### NOTES

1. This paper is adapted from proposals and arguments we have developed in greater detail elsewhere, most notably in the following previously published manuscripts: (with Albert R. Jonsen) Medical futility: Its meaning and ethical implications, *Annals of Internal Medicine* (1990) **112**, 949–54; Futility and rationing, *American Journal of Medicine* (1992) **92**, 189–96; Ceasing futile resuscitation in the field, *Archives of Internal Medicine* (1992) **152**, 2392–7; Medical futility: The duty not to treat, *Cambridge Quarterly of Healthcare Ethics* (1993) **2**, 149–57; Futility in practice, *Archives of Internal Medicine* **153**, 437–41; An ethical analysis of the use of "Futility" in the 1992 American Heart Association Guidelines for Cardiopulmonary Resuscitation and Emergency Cardiac Care, *Archives of Internal Medicine* (1993) **153**, 2195–8. The ideas contained in this chapter are also developed in more detail in Jecker, N. S. and Schneiderman, L. J. *Wrong Medicine: Doctors, Patients and Futile Medicine*. Baltimore, MD: Johns Hopkins University Press, 1995.
2. Adapted from the *Seattle Times*, 12 February 1994.

### Suggestions for further reading

Beauchamp, T. and Childress, J. *Principles of Biomedical Ethics*, 3rd edn. New York: Oxford University Press, 1989.

Brody, H. *The Healer's Power*. New Haven, CT: Yale University Press, 1993.

Callahan, D. *Setting Limits*. New York: Simon & Schuster, 1987.

Cassell, E. *The Healer's Art*. Cambridge, MA: MIT Press, 1989.

Churchill, L. R. *Rationing Health Care in America*. Notre Dame, IN: University of Notre Dame Press, 1987.

Macklin, R. *Enemies of Patients*. New York: Oxford University Press, 1993.

President's Commission for the Study of Ethical Problems in Medicine and Biomedical and Behavioural Research. *Deciding to Forego Life Support and Sustaining Treatment: Ethical, Medical and Legal Issues in Treatment Decisions*. Washington, DC: Government Printing Office, 1983.

Weir, R. F. *Abating Treatment with Critically Ill Patients*. New York: Oxford University Press, 1989.

# The Sorcerer's Broom: Medicine's rampant technology†

ERIC J. CASSELL

EDITORS' SUMMARY

Eric J. Cassell, MD, suggests that medical technology has become like the Sorcerer's Broom, taking on a life of its own almost beyond the control of practitioners. His argument is built on observations about the lure of technology: it causes wonder and fascination, it gives immediate results, it is unambiguous and clear, it is certain, it is self-perpetuating, and it confers a false sense of power. All of these qualities disturb the essential features of caring for persons rather than diseases. Cassell closes with an appeal to educate physicians to be more comfortable with ambiguity, since the problem of technology is not with the techniques, but with the persons designing and using them. The mistake lies in believing that technology reduces the patient's distress to a controllable "problem" without entering the personal world of the patient.

If one element can be singled out as the engine of the medical economic inflation now occurring everywhere in the world, that element is the seemingly irresistible spread of technology into every level of medicine – irresistible to doctors, patients, and nations alike. Yet, evidence that technology is a problem is everywhere in medicine. In intensive care units the world over, the technologies of monitoring, organ support, and resuscitation are used where

† An amended version of a paper originally published in *Hastings Center Report* (1993) 23(6):32–6. Reproduced by permission. © Hastings Center.

appropriate, i.e. related to the aims and purposes of the sick person, but also where inappropriate, as defined by the capabilities of the technology and the consequent expertise of physicians rather than – or even contrary to – the best interests of the sick person.

The thesis of this chapter is that, like the broom in Disney's *The Sorcerer's Apprentice*, technologies come to have a life of their own, not only because of their properties but also because of certain universal human traits. Technologies come into being to serve the purposes of their users, but ultimately those users redefine their own goals in terms of the technology. Like the twentieth century, during which they have come to dominance, technologies as a class are reductive, over-simplifying, intolerant of ambiguity, and democratic. Purposes and goals are ideas that are often slow to disseminate. Democracy has only gradually spread over the world. The transistor radio spread in a decade. It would be simple to say that technologies are just things, neutral in and of themselves, employed solely at the will of their users, but that statement is not true. Technologies are developed, manufactured, and marketed for purposes that are meant to match the intentions of their potential users. Thus, their deployment will not change until we understand the singular category they occupy in medicine and specifically train physicians to control technologies as well as to employ them.

To revisit the long history of the debate about the wonders or dangers of technology is neither necessary nor useful except to acknowledge its existence and the literature it has engendered from Goethe's *Faust* to Aldous Huxley's *Brave New World*. Nevertheless, systematic concern about technology is largely a child of the second half of the twentieth century. The famous 1911 edition of the *Encyclopedia Britannica* has no entry on technology; the 1991 edition devotes 32 pages to the subject. The issue no longer is "pro" or "anti." There is no going back to nontechnological medicine – who would want to? The issue today is how to solve the difficulty epitomized by poet and essayist Ralph Waldo Emerson's observation, "Things are in the saddle and ride mankind." Technology is not the problem. It is the relationship of technology and those who want to use it that is problematic. If this is not solvable, our entire project is a waste of time.

What is technology? The definition of technology presents problems for which dictionaries are no help, because the term "technology" can be used in a manner so broad as to defeat understanding. Thus, any tool employed in a craft could be said to be that craft's technology. In this case the word technology would be virtually the equivalent of the word "means" in the phrase "means and ends." Thus, in medicine any means a physician employs to meet diagnostic or therapeutic goals could be labeled technology. Taking a case history or doing a physical examination would, by that definition, be technologies (be extension from technique), and so, too, would stethoscopes, ophthalmoscopes and scalpels.

However, in this discussion, by technology I mean modalities and instrumentalities that greatly extend the power of human action, sensation or thought independent of the particular user. In addition to the instruments and devices usually considered as technology, we should include, for the sake of understanding, high power medications – whether cardiac, anti-microbial, psychotropic or whatever. These medications greatly extend our therapeutic power, and it is our power that technology expands.

Technology is not science. Technology and science are frequently lumped together – as in "sci-tech" – but they are distinct. Science is not my topic. The topic is to see what there is about positron emisson tomography (PET) scanners, magnetic resonance imaging (MRI), angioplasty, endoscopy, automated chemistry machines and so on – the whole wondrous parade – that poses problems for medicine.

Technologies are reductive and oversimplifying. Much of their hold on medicine, however, is a result of two historical reductive steps. The first step was reducing the problem of human illness – with all its intricate physical, social, emotional and cultural aspects – to the biological problem of disease. For example, tuberculosis with its complex determinants and consequences was reduced to the problem of the anatomical and microscopic manifestations of the disease and their relationship to the patient's symptoms and physical manifestations. This step was a natural evolution from all the attempts to understand illness that preceded the discovery of diseases in the early nineteenth century in France.

The history of the medical profession and its dominance is tied to knowledge of the body. Diseases were initially defined as physical entities with unique anatomical (later biochemical) characteristics and unique causes. These two characteristics permitted precise definitions of diseases. Precise definitions in addition to anatomically (or biochemically) discernible characteristics finally permitted the productive entrance of science into medicine.

The second reductive step follows from the scientific investigation of diseases. Here the findings of science become the accepted picture of the disease, further oversimplifying the problem. Tuberculosis again provides an example. The scientific discovery of the disease agent, the tubercle bacillus, completes the simplification, as though the presence of the organism were virtually the equivalent of the disease. These definitions, identifiable characteristics, scientific investigations, and consequent technologies perpetuated the oversimplification of human illness, some of whose consequences bring us together. However, the presence of a certain circularity should be acknowledged. Disease definitions permit the entrance of science. Science increases knowledge of the disease by employing technologies and promoting the development of further technology. These technologies come about because of the scientific understanding of the disease and reinforce the original picture of disease that started the cycle. This circle also contains the values that direct the technologies toward the facts which support the values. Breaking out of such a circle is one of our tasks – but not an easy one.

What I have just noted may be the way technology entered medicine, but this knowledge will not end technology's almost autonomous growth. We will not solve the problem of technology without providing other solutions or defenses against the human characteristics that lead to our difficulty. I discuss here six such characteristics, wonder and wonderment, the lure of both the immediate and the unambiguous values, the avoidance of uncertainty, and the human desire for power.

## Wonder and wonderment

The first hold that technology has on us I call wonder and wonderment. When I lecture at other institutions, I am frequently taken on a tour of the place. Once, in Pittsburgh, I was shown their new cardiac catheterization laboratories – four of them! Why? Did they think I'd never seen a cath unit? Why didn't they take me by somebody's office (whispering, so as not to disturb) and say, "There's one of our smartest doctors?" Because everybody loves the new and the shiny, especially when they do fantastic or seemingly inexplicable things that enthrall us. Wonderment must be reduced to bring the world back into order. So people have to figure out what this new wondrous thing is, how it works, and, of course, how to control it. I could demonstrate this process in a minute in any audience of physicians. If I were to put on a table a device that looked different and had a screen and strange keyboard, people would soon start poking at it, manipulating the keys to bring up the control system, find out what it is, and how it works. Wonder is not easily put aside and is quickly reawakened – one taste leads to a desire for more.

Wonder and wonderment cause physicians to use and overuse technology. They like to see it in action – and they want a new model as soon as possible. Wonder may seem a childish motivation in a very serious pursuit. It *is* childish; that is one of its attractions. Wonder helps to solve the problems of boredom, absence of meaning, and loss of motivation. The human body is wondrous and so is its psyche. In defined circumstances, some doctors love to take the body apart and others love to pry into the psyche. Yet, surgeons are socialized as are psychiatrists never to cut into the body or mess in the mind unless in the patient's good. We know that curiosity – an aspect of wonder – is not easily held in check, but much time in medical education is spent successfully socializing doctors to hold their curiosity in check. This says it can be done.

## The lure of the immediate

The second reason for technology's hold on physicians is that technology roots us in the immediate. The numbers on the readout, images on film, dexterity required for deployment, technical complexities, tubes, wires, plugs, valves, needles, gauges, mirrors, focusing devices, and on and on exist in the here and now – the immediate moment. And these things are immediate in another related, but perhaps more important sense; they are unmediated by our own reasoning – the thing in and of itself. Computer jargon even has a name for it, WYSIWYG, "What You See Is What You Get." The user doesn't have to reason from one output to another; each is distinct.

How different this is from the patient. Look at a patient, see only the here and now, and you have missed the truth of a sick person. Any one moment of life – in an intensive care unit or a nursing home – contains only one bit of the importance of something much larger. A human life is a trajectory through time, the historic route of a society of complex parts, as the philosopher Alfred North Whitehead explained. Sick persons, all persons, are difficult to understand. The doctor in attendance is also part of a society of complex parts pursuing a historic route that interacts with the patient.

God bless the immediate – no need to get caught up in all that complex sick person stuff! That is why, given a complicated human question in the care of the sick, we doctors love to start talking about physiologic parameters, calling up diseases or planning tests. For example, the attending physician and medical students stood outside the room of a dying patient whose suffering could not be controlled. Did they speak about her suffering or what to tell her or do for her? No, they read her test results and X-ray films – irrelevant to her present problem but much simpler and more immediate. Medically they have been trained to do this. Why could they not be trained equally to care for her suffering?

Why is not the examining hand on the abdomen just as immediate as looking at a readout or computer-generated image? Because the hand is not just a hand or sensations in the fingers; it is a doctor feeling responsible for the approximation of what fingers disclose of an unseen reality – and what it means. Why is not the same true of the image on the film? It can be true, it *should* be true. The physician viewing the image *should* be reasoning from the information contained in the image about what must have come before and what will follow. And, then, how that information fits with what he or she knows of the patient and the patient's interests, desires, purposes, fears, and concerns. But as technology gets more advanced it becomes more autonomous, telling you directly what it means in immediate terms – like the computer-generated electroencephalograph interpretations. Or, a specialist, whose sole job is to interpret the image, tells you what it means in unmediated terms. As we all know, less often than they used to physicians read their X-rays (even with the interpretation in hand) or go to the pathology department when the biopsy is being read or question the precision, accuracy or validity of the automated chemistry report. The reason is not that doctors are lazy: doctors have come to accept these technologies and their output as the equivalent of what is being tested.

Some specialties in medicine have always lived more in the immediate than others – surgeons are the best example. The open wound, flowing blood and exposed viscera are more immediate than the evolution of a drawn-out illness. The special attraction of the immediate is one of the reasons that surgeons are different from internists (or the other way around).

The system of answers that we teach physicians about diseases are ill-suited to frame the longer-term, larger questions raised by the sickness of the person to whom the monitors are connected. Science has ruled out of court the information from values and aesthetics by which we live our lives, allowing only brute facts. One of the advantages of the immediate is that it provides answers – information – while more relevant understanding requires deeper reasoning and greater involvement from doctors as persons. Understanding cannot operate separately from the reasoner as can the computer. Immediacy, and its lesser requirement for reason, facilitates a detachment from a patient's suffering. Thus, we can be in thrall by technology as much by the seeming advantage of the immediate as we are by its wonder.

## The lure of the unambiguous

The third aspect of technology, unambiguous values, keeps it employed sometimes even when inappropriate. Watch the movie of a coronary arteriogram. It is like a Western where you can quickly tell the good guys from the bad guys. The values are clear and unambiguous. A good coronary arteriogram, with adequate dye in the vessels, is anatomically clear. You can compare it to those taken previously and subsequently. A good coronary artery is open and a bad one is obstructed, although there are criteria for degrees of good and bad. A good obstruction is short with adequate run-off and not so tight that an angioplasty balloon will not pass through it or that it cannot be bypassed. Cardiologists, although they may disagree with one another about details, are absolutely clear and straightforward about such things. When they are not, they make new criteria to remove ambiguity.

Virtually all technology is marked by similarly unambiguous values. In fact, lack of ambiguity is essential to good medical science. If cardiologists at Cornell cannot speak the same language and mean the same thing by technical terms as cardiologists at Stanford, Oxford, and the Hôtel de Dieu, international research is impossible and progress in medical science will be impeded. So, on the face of it, the unambiguous seems reasonable – except that we physicians do not generally know how or when to abandon it. Many, in fact, most of life's simple pleasures are also unambiguous. We generally know what is good and bad in behavior, food, wine, and sex. However, the development of sophistication in nontechnological pursuits involves appreciation of complexity and ambiguity. Sophistication in technology, I believe, goes in the other direction. More sophisticated means less ambiguous; the better the piece of equipment, the clearer the values.

But good or bad as measured by technology are not necessarily the same as good or bad for patients. Coronary artery disease and its technologies are a case in point. Imagine an instance, common enough, where a middle-aged man without symptoms wants to join an exercise program. He is required to have a treadmill exercise test. In this example, the test (following the usual Bruce protocol, which approximates to no exercise you have *ever* done) is positive by published criteria. He is advised that these unambiguous criteria often indicate coronary artery disease and should be followed by a thallium stress test. The test (in this case) is also positive, and he is advised that he should have a coronary arteriogram. (In many instances the thallium scan is considered redundant – the patient goes directly to the arteriogram.) The arteriogram shows significant obstruction of a coronary artery. Subsequently, a coronary artery angioplasty is done to reduce the obstruction. This scenario is extremely common in the USA – and increasingly so elsewhere, although no quality evidence indicates that the outcome of this chain of events makes a positive difference in the life of such a patient. The relationship between what is considered good and bad in test results and what is best for the patient is at the very least obscure and at the worst, just plain wrong.

The scenario demonstrates that because available technology permits visualizing the major coronary arteries, atherosclerosis of these vessels, which can be demonstrated unambiguously, has come to be taken definitionally as the equivalent of coronary heart disease. Coronary heart disease is a more complex entity than mere atherosclerosis of the major coronaries, although the two are often associated. For example, an autopsy will commonly show old people with coronary arteries so filled with the calcium deposits characteristic of advanced atherosclerosis that one wonders how blood ever gets through them. Why did these people have coronary artery disease but show no evidence during life of loss of everyday function due to heart disease? Because they did not have heart disease. Unfortunately, physicians infrequently attend autopsies nowadays and thus are not exposed to this common phenomenon. Conversely, sometimes one sees a patient with clear-cut signs of coronary heart disease, but little evidence of obstructed arteries.

As I suggested previously, human sophistication is marked by tolerance for ambiguity. Sophisticated technology removes ambiguities by narrowing the field of difference between the good and the bad, so that ultimately one test result is taken to be good and another result bad. Thus, the state of the coronary arteries became accepted as the equivalent of a disease of the heart itself in the circumstances I have described. Whitehead's fallacy of misplaced concreteness writ large.

We must not forget that technological measures of value, even as they achieve a life of their own, are derived from human values. When medicine's priorities (another word for values) are too simplistic, they will be represented by a technology that also exemplifies simple values, as when a part is given priority over the whole – sustaining an organ but losing sight of what is best for the whole person. We value the preservation of structure over function, and value the body over the person, survival over maximum function, and length of life over quality of life.

The development of technology is not an event, but a process. Technology is invented to solve problems arising out of the pursuit of medical values. Technological values, however, foster and reduce the ambiguity in medical values, and that process leads to a new stage of technology. The result stifles the sophistication necessary for physicians to tolerate the ambiguity that would inevitably follow attempts to break out of the circle. (Were one to fault modern physicians for lack of sophistication they would most likely dismiss the criticism by pointing to the sophisticated equipment they use.)

So, to wonder and the lure of the immediate, we add unambiguous values as a reason why technology runs doctors rather than vice versa.

## The pursuit of certainty

The next reason for the dominance of technology is medicine's pursuit of certainty. Uncertainty is the central problem that physicians confront. It is doubt that grays hair. In an essay written many years ago, "Training for uncertainty," Renee Fox identified two reasons for uncertainty: first, defects in the knowledge of the individual physician and, second, the inadequacies of the profession's knowledge. Even if I, impossibility granted, knew everything medicine knew, I would not know everything. Uncertainties would remain. But in an ideal world of complete knowledge, in this view, we could be certain. Unfortunately, as Drs Sam Gorovitz and Alisdair MacIntyre pointed out 20 years ago at a Hastings Center meeting, two other roots of uncertainty can never be removed. First, every decision, small or large, is made about the future. All medical decisions are about the future, a future that starts an instant from the present. Second, uncertainty can never go away because all of science, medical and any other, is about generalities, and every patient is a particular individual and necessarily different in some respect from the general. It is in the nature of the uncertainties of clinical medicine that the more important the knowledge required by the decision, the less tolerable is the uncertainty. Since physicians commonly must make decisions that have profound implications for the lives of others, uncertainty is a constantly disturbing factor in medical practice.

## Technology and uncertainty

Technology would not produce problems in relation to uncertainty if it did not, in fact, frequently reduce uncertainty, sometimes dramatically. Probably because of the change produced by effective technologies, I believe that doctors are no longer trained in the management of uncertainty in the fashion first described by Renee Fox. As a consequence, they tend to utilize any diagnostic or therapeutic technique that promises to reduce uncertainty, leading to a sort of Gresham's law of technology: whatever technique promises greatest certainty, even if inappropriate, will diminish the use of techniques associated with greater uncertainty. Hard facts drive soft facts into hiding, which in turn drives softer facts into oblivion. Technologies produce hard facts.

All the wonder, the dislike of ambiguity and the fear of uncertainty that afflict doctors are present among patients. And the stakes are highest for the latter. In the current medical world of advanced countries, patients have a significant voice in the choice of diagnostic strategies and treatment. They are now generally knowledgeable to an unprecedented degree. Not surprisingly, their knowledge is greatest about new technologies and treatments, details of which fill the pages of newspapers, magazines, and health promotion newsletters. It is fair to say that many patients believe that the test rather than the physician makes the diagnosis, and the drug rather than the physician effects the cure. Consequently, patients have been an active force in the increasing deployment and dominance of technology.

## Technology is self-perpetuating

Employing one technology frequently leads to the use of another, as demonstrated by the function of computers in neonatal intensive care units. Commonly, each "bed" in a neonatal unit has its own computer to analyze and display the physiological state of the infant. The requirement of computers for digital information encourages the proliferation of instrumentation that produces such data. Another example is the effect of automated blood chemistry machines in making redundant the manual skills of technicians. Other automated laboratory examinations become necessary because technicians no longer do the tests by hand. The results consequent on the use of one technology frequently raise questions that can apparently be answered only by other technologies. Computerized tomographic images of the central nervous system may introduce doubt that only magnetic resonance imaging can resolve.

Doctors who have mastered a technology tend to use it as often as possible, not necessarily for reasons of profit, but because they love their skills and technologies. As noted earlier, problems tend to be redefined so that a technology becomes appropriate when it might otherwise not be. A saying that makes the point has become popular among physicians, "To the person with a hammer, everything is a nail."

## Power

The final reason for the inappropriate use of technology is the power it confers on physicians and their institutions. Although the meaning of power seems self-evident, some further explication is required. The power to act is basic to human existence and is employed to control or influence events. We exist for ourselves and for others in our actions; i.e. when we act, we simultaneously create ourselves and our world. The scope and effectiveness of our actions in both self-creation and influence in the world are determined by the degree of our power. Since we are social beings, virtually all of our actions take place in a world of others, and our power is relative to the power of others. Thus, my ability to act among others is partly dependent upon permission to exercise my power by those more powerful or my desire to exercise my power in relation to those less powerful. Frequently the word hierarchy is used to refer to social ranking according to power. Power does not reside in us only as individuals but also by virtue of our acknowledged place in society – in our social status. Thus, hierarchy may be status-dependent rather than the result of self-generated power. Power relationships, which also exist among and between animal groups, are dynamic. It is difficult to exaggerate the importance of the exercise and experience of power.

Even in sophisticated societies, the ability to do things better than others confers power. Possessions confer power because they bestow status – material wealth is the most obvious example – but so does access to objects of superior efficacy. In fact, changes in one's access to things that contain superior efficacy in themselves may alter one's status.

There is little doubt that one of the attractions of technology is its ability to confer status and rank on individuals. Medical power is demonstrated when infection is treated, blood pressure is lowered or pain relieved. Every therapeutic and diagnostic act is a demonstration of efficacy and thus of power. The therapeutic effectiveness of the relationship between patient and doctor is dependent, in part, on a belief in the physician's individual and institutional power over the forces of nature. In previous epochs, the physician's power came not only from his or her shared knowledge of the body and disease but also from the personal development of knowledge about the sick and sickness and demonstrated effectiveness in the diagnosis and treatment of patients. Personal power of this sort takes many years to develop and is inevitably a result of the ripening of the medical self. Technology confers power on individual doctors with much less personal involvement. The modern tendency toward specialization encourages this more easily gotten power because it narrows the amount of knowledge necessary to exercise it.

Technology also gathers to itself personnel and space that exhibit power and tend to be self-perpetuating. Intensive care units and transplant units are the perfect examples. In medicine, as elsewhere, technology engenders special training, which furthers the world view of technology, which further increases political power. Employing or having access to technology also garners social power or status from lay people, the press, the university, or the hospital trustees. In like manner, technology confers status on hospitals and other medical institutions. The example I gave earlier about being shown around shiny cardiac catheterization laboratories when I visited a Pittsburgh hospital can also be used to exemplify a hospital possibly showing off its power as demonstrated by the fact that it had more such laboratories than its neighbors.

Technology would not confer power on doctors and the profession of medicine if it were not seen by the larger society as having power in itself. It erroneously appears to free the patient from the necessity of depending on the individuality and individual skills of the physician. (This point is reflected in uniform fee schedules that pay for a particular physician's act – office visit, surgery and etc. – as though doctors dispense a uniform technology rather than a personal individual service.)

## Knowledge at a distance

In our daily lives, we are accustomed to confronting much of our world in its representation, rather than in itself – in photographs, recordings, radio, movies, and television, resulting in a widened perspective and scope of knowledge about things distant and close to us with which we are all familiar but, especially for doctors, problems. Technology represents a kind of knowledge. In fact, it epitomizes the twentieth-century ideal of knowledge – scientific, objective, and existing seemingly separately from humankind. In medicine, the scientific knowledge and subsequent technology developed in response to the challenge posed by sickness and suffering has assumed an actuality more convincing than the reality of sick persons themselves.

Consider this common situation. A patient has severe pain in the hip and the doctors can find no evidence of disease. With each negative test, increasing doubt is raised about whether the patient is truly in pain. Then a radionuclide bone scan is done showing cancer in the hip bone. The patient will now be believed. Why is the celluloid rectangle with fuzzy black dots more believable than the patient's pain? The usual answer, that the pain is subjective, will not hold water. The pain may be subjective, but the report of pain can be evaluated. Further, we are of a piece; we cannot have severe pain without it being reflected in other aspects of our physical, social, and psychological selves. A person with severe pain moves, acts, thinks, feels, displays emotion, and relates to others differently than the same person pain-free. All of these features – apparent to others or that can be evoked – are objective.

Objectivity alone is not the issue. The way we would know that the person really has pain does not meet the ideal of medical scientific knowledge developed over the last 150 years. Scientific knowledge, surely not the only way to know things, has come to be accepted as more actual than patients or their pain or suffering. Medicine's technology also produces representations of patients' original reality that are another reality in themselves. For example, electroencephalograms, X-ray machines, monitors, computer-assisted tomography (CAT) scanners, MRI machines, and PET scanners are all imaging devices that distance physicians from the sick person. Their focus of interest is inevitably drawn away from the patient and onto the part or the disease – out of the context of the whole patient and the patient's lived world. To return medicine's focus to the sick person, physicians and commentators and critics of medicine have largely depended on moral injunctions. It is an uphill struggle because the problem is based, in part, on the nature of medical knowledge itself and is shared by twentieth-century humankind.

## Conclusion

Technology holds sway over medicine and its public because of its capacity to induce wonder, root us in the immediate, remove ambiguity, increase certainty as well as its self-perpetuating character and its enhancement of power. Since this is not well known, it is hardly surprising that technology, by itself inert and useless (although beckoning for attention through its inherent purposes), should be blamed for the troubles it brings rather than blame being placed on the doctors who use it, the public who love it or the narrow knowledge on which it is based. Medical technology's form and character arise from medicine's focus on disease and pathophysiology as the arena in which the origins and solutions to human sickness are to be found. The values on which it is based come primarily from the spectrum of pathophysiological and anatomical criteria for disease and normalcy, now largely defined and perpetuated by the technology. Our task, it seems to me, is to stop blaming, regulating and complaining about technology – without which modern medicine is unthinkable – and start working towards a solution based on understanding, as we have done with so many other problems.

The search for new goals of medicine can be one step in such a task. I do not believe we are seeking goals new under the sun, but rather should be trying to return medicine and doctors towards a focus on persons sick and well and on their suffering. Conversely, no change in the ends and purposes of medicine is possible without bringing technology under control. Toward this end we must learn how to teach doctors, who are in themselves the primary instruments of diagnosis and treatment, to tolerate uncertainty, accept ambiguity, deal with the complex, and turn away from mere wonder. Accepting these assignments and redirected goals and following them as far as they lead will be a sufficient task for the next decades.

### Suggestions for further reading

Fox, R. C. Training for uncertainty. In *The Student Physician: Introductory Studies in the Sociology of Medical Education*, ed. R. K. Merton, G. G. Reader and P. L. Kendall, pp. 207–41. Cambridge, MA: Harvard University Press, 1957.

Gorovitz, S. and MacIntyre, A. Toward a theory of medical fallibility, *Hastings Center Report* (1975), **5**, 13–23.

Toombs, K. Illness. In *The Meaning of Illness*, pp. 31–50. Boston, MA: Kluwer Academic Publishers, 1992.

# Modern technology and the care of the dying

RONALD E. CRANFORD

EDITORS' SUMMARY

Ronald E. Cranford, MD, explores the changes in health care delivery that have occurred since the Second World War. He contrasts how people died before the war with how they might die today. The contrast is powerful. Before, people died at home from "old age," surrounded by family, friends, and a physician who knew them well. Today, they are usually found in an institution, surrounded by high technology and the specialists familiar with that technology, but strangers to the person and his or her values. The constant growth of medical technology creates new choices and, as Cranford clearly establishes, with new choices come new ethical dilemmas. The future will bring even more of these challenges than the past.

Life and death seemed so simple in the early twentieth century. Prior to antibiotics, respirators, resuscitation, dialysis, intensive care units, and paramedics there was little control over the dying process. Of course, many people died premature deaths. Diseases now treatable were invariably fatal: pneumonia and other infectious diseases, kidney failure, respiratory insufficiency, heart attacks. At the turn of the century, life expectancy in the USA was 47 years, in contrast to 76 years at the present time, even longer in other countries. Much of this increase in life expectancy is due to improved public health measures, e.g. improved sanitation, better-quality drinking water, and vaccinations, but dramatic advances in medical technology and specific therapies have contributed significantly to this increased life expectancy and improved quality of life.

In this chapter, I examine some of these technological advances, show how they have impacted on life expectancy and quality of life, and illustrate some of the ethical dilemmas resulting from these recent advances in medicine.

## The old and the new

Let us contrast a typical death bed scene of the 1940s with one of the 1990s.

### Dying in the 1940s

An elderly female, in her seventies, is dying at home. She is dying in the home she has lived in for the past 30 years. Friends and family gather at her bedside – sad at her passing away, but remembering the good old days when she was lively, healthy, and full of life. Several generations of family are present: children, grandchildren, perhaps even great grandchildren.

Over the last few years, the patient's health has dwindled. She is afflicted with arthritis and a heart problem. Her mind is not quite as sharp as it used to be. But no one really questions what she is dying of, including her physician; she is dying of old age, of natural causes.

The family physician, who has known her, her husband, and many of the other family members for several decades, comes to the patient's home each day: seeing the patient, examining her heart and lungs with his stethoscope, sitting at the bedside while concerned family members look on. The doctor tells the family the obvious, "There is really nothing we can do at this stage. She is just dying of old age." The family is sad, but understands. During the patient's last few days, she is surrounded at the bedside in her home by her close family, friends, and a family practitioner who knows her and her family's values well.

### Dying in the 1990s

An elderly female, in her late eighties, has been readmitted to an acute care hospital for the sixth time in the last 15 years. The patient has resided in a long-term care facility for the last seven years after the family was no longer able to care for her at home. She has had multiple small strokes, at least one heart attack, and two episodes of pneumonia – all of which were vigorously treated at the time by specialists in neurology, cardiology, and infectious disease.

Prior to this most recent illness, the patient's memory had seriously deteriorated, and she no longer recognizes her children. Because of her rapidly deteriorating mental and physical condition, she has been admitted to a medical intensive care unit attended by specialists in the fields of cardiology and neurology. This is the first time these specific specialists have ever seen this patient. They know nothing about the patient's previous personal history and life style. The specialists talk to the family whom they have never met before.

In the medical record are listed numerous specific medical diagnoses: arteriosclerotic, cerebrovascular and coronary artery heart disease, emphysema, mild peripheral vascular disease, and probable Alzheimer's disease or multi-infarct dementia. No mention of old age or natural causes.

Finally, after several days in the intensive care unit, the patient has a cardiac arrest, but the patient is not resuscitated because the family and doctors have agreed to a DNR (Do Not Resuscitate) order.

These two scenarios may seem rather dramatic and disturbing, but they do reflect the reality of dying in the modern day with our emphasis on technologies and specialists. What were some of the changes, in medicine and society, that caused such a sharp contrast in these death-bed scenes?

## Medical advances in the latter half of the twentieth century

To fully understand the current scene, it is important to briefly review some of the major advances in modern medicine and to appreciate how relatively recent these advances are. The 1940s and the Second World War, mark an important turning point in modern medicine. Antibiotics were a rare commodity prior to the war. Sulfa drugs had been used for quite a few years, but the use of penicillin was only in its infancy. The dramatic response of, for example, the infectious disease pneumonia to drugs such as penicillin were considered nothing short of a miracle by physicians in those days. Penicillin was in such short supply that it was given in extremely small doses compared to the modern-day amounts.

The widespread use of respirators during the polio epidemics of the 1940s and 1950s was directly linked to the knowledge and experience gained during the Second World War in the use of respiratory support systems for pilots at high altitude. A knowledge of respiratory physiology was extremely important for the Air Force as the pilots flew at higher and higher altitudes for longer periods of time during the war. The iron lungs used in the immediate post-war period for polio patients were a rather crude form of pulmonary support system. During the last 30–40 years, however, there have been dramatic improvements in the use of ventilators and intubation. Many of the polio patients who died then, would have been saved today with our more current pulmonary support systems.

Other important advances have been the resuscitation techniques developed in the 1960s and 1970s. The technique of basic life support (closed cardiac massage and mouth-to-mouth ventilation) can sustain patients for a brief period of time until advanced cardiac life-support systems (defibrillators and numerous cardiac drugs) are used to reverse abnormal heart rhythms. It is hard to imagine a more dramatic life-saving technique than cardiopulmonary resuscitation (CPR).

With the advent of these advanced life-support systems for primarily cardiac and respiratory abnormalities came the development of intensive care units of all varieties, beginning with medical intensive care units, then cardiological, pulmonary, neurological, nephrological, pediatric, and neonatal units. In the last 20 years, the rise in the use of intensive care units has been nothing short of phenomenal.

The early 1960s also witnessed the start of developments in renal dialysis and transplantation, both major advances in treating disease. But modern technology is costly. When the United States Congress was asked to pay for all renal dialysis in 1973, the cost was estimated to be approximately $135 million a year. Just ten years later, the actual cost for this program was $2.3 billion on an annual basis. Another major advance has been the development of different ways of providing nutrition and hydration into the body of patients no longer able to eat or drink in the normal manner.

## The trade-off: The good and the bad

How have these advances improved our health, and how have they also produced ethical and social dilemmas never dreamed of before?

Modern medicine has given us control over the death and dying process. With this control have come choices of when it is appropriate to use or not use these new therapies, and these choices result in ethical dilemmas. So, the more control we have, the more choices, and the more choices the more dilemmas.

Consider the following two scenarios, which illustrate the good and bad of modern medicine.

### Case 1

A middle aged male with a history of mild coronary artery disease suddenly develops severe chest pain and falls to the ground unconscious. A bystander trained in basic life support applies closed cardiac massage and mouth-to-mouth resuscitation, while a companion calls 911 (a United States emergency number). Paramedics soon arrive;

using techniques of advanced life support, the paramedics immediately apply ventricular defibrillation, and the heart rate is restored to a normal rhythm. Direct communication is maintained with the nearest hospital emergency room. The patient is maintained on life-support systems while being transported to the hospital and then again in the emergency room. Because resuscitation was applied promptly, the patient leaves the hospital several days later with no physical or neurologic after effects. Even though the patient does have coronary artery disease, he lives another 15 years in good health.

## Case 2

All the facts of case 2 are identical to those in case 1 except that the people witnessing this arrest are not trained in basic life support. Therefore, no resuscitative measures are begun until the paramedics arrive. By this time the patient has had a full cardiac arrest and lack of blood to the brain for perhaps as long as 15 minutes. The paramedics are successful in restoring the basic heart rhythm, but the patient is in a coma at the time of admission to the hospital. During the hospitalization, the patient evolves into a vegetative state from which he never recovers. The patient lives another 15 years in a long-term care facility, fed through a feeding tube in his stomach, when he finally succumbs to pneumonia.

In the USA every year 25 000–50 000 patients are successfully resuscitated from cardiac arrest. These patients go on to lead normal healthy lives for many years. In these cases, cardiopulmonary resuscitation is not only life-saving, but also helps to preserve a good quality of life for patients for many years thereafter. In the days prior to the widespread use of cardiopulmonary resuscitation, patients with a sudden cardiac or respiratory arrest simply died.

With our modern medical treatment modalities, life no longer seems to have a natural end; nobody dies of old age or natural causes any more. Even when patients in their eighties die, family members feel like they have been cheated of a few extra years of life. No wonder many doctors still view death as a defeat!

## Creatures of modern technology

Today we have various clinical syndromes resulting from medical treatment that hardly ever existed before ("creatures" of modern technology): brain death, where the entire brain is destroyed, but the patient can be kept "alive" on artificial life support systems; locked-in syndrome, where the patient is severely paralyzed, but has normal awareness; and advanced dementia, usually of the Alzheimer's type, where the last five to ten years of the patient's life may be spent in a mindless existence.

Perhaps the best paradigm for creatures of modern medicine is the persistent vegetative state – as in the cases of Karen Ann Quinlan and Nancy Cruzan discussed below. Patients in this condition experience a loss of the functions of the higher brain while the lower brain (brainstem) is intact. So, even though these patients have no awareness of their condition and no ability to think, feel, or speak, their eyes are open and may move in a random, aimless fashion, and they have sleep–wake cycles. These patients may live for years: Karen Ann Quinlan lived for ten years like this, and Nancy Cruzan for eight years. This syndrome is the best example of the disparity between the quality and the quantity of life produced by modern medical technology. The patient has no mental functioning or conscious awareness at all by current criteria, yet can live for prolonged periods.

In the USA it is estimated that there may be as many as 15 000–35 000 patients in the vegetative state – adults and children. The annual costs for maintaining patients in this condition are estimated to be between $1 billion and $7 billion. Usually, the only medical treatment sustaining these patients is artificial nutrition and hydration (the feeding tube).

With increased life expectancy, Alzheimer's disease has become a much more common syndrome in the elderly. It is estimated that there are in the USA three to five million patients with Alzheimer's-type disease; many of these in the later stages of the disease. It now seems evident that as many as 25% to 50% of all patients living into their eighties will develop Alzheimer's disease. How aggressively should these patients be treated when we know they will invariably develop a progressive deterioration of the mind to the point where they may even become totally amented – that is without any consciousness at all, in other words, in the permanent vegetative state? Can Alzheimer's, then, be regarded as a slow-motion terminal disease?

## The future

Given these trends in modern medicine, what can we expect in the future? Is it ever possible in modern times for patients to experience a "good death"? I believe that our ability to prolong life will continue to outstrip our ability to meaningfully cope with these modern-day dilemmas. The costs will continue to be astronomical. People are often looking for wonder drugs or miracle cures that, somehow, can achieve our immortality.

The one area where we have not made a great deal of improvement is in the treatment of neurologic diseases. Will there be effective medications for diseases such as Alzheimer's or severe strokes in the near future? I seriously doubt it. Will brain transplants be the answer? Clinicians have been somewhat successful in treating Parkinson's disease, a severe neurologic disease causing progressive impairment of movement, by transplantation of fetal brain cells into a specific area of the brain, the substantia nigra. But there is an enormous difference between transplanting brain cells into a specific area of the brain and transplanting cells of the cerebral cortex from one patient to another. The personality of the individual lies in the cerebral cortex. Therefore, to transfer this vital area from one patient to another would be to change aspects of the personality of the individual receiving the "transplant" to that of the person "donating" the cerebral cortex. Thus, even if brain transplants were more effective in the future than they are today, something I greatly doubt, it would still present ethical dilemmas beyond our wildest dreams.

Lastly, physician-assisted suicide and euthanasia will probably become more common. Much of the recent impetus in the USA for these practices arises directly from the conviction that modern medicine keeps desperately suffering and hopelessly ill patients alive far beyond any reasonable limits. One state, Oregon, has officially legalized physician-assisted suicide by prescription. The requirements are modeled on Dutch practices where euthanasia is still illegal but tolerated under controlled conditions. As medicine continues to advance, so too will our ethical dilemmas concerning the end of life.

### Suggestions for further reading

Breo, D. L. *Extraordinary Care*. Chicago: Chicago Review Press, 1986.

Colen, B. D. *Hard Choices: Mixed Blessings of Modern Medical Technology*. New York: G. P. Putnam's Sons, 1986.

Hoefler, M. *Deathright: Culture, Medicine, Politics, and the Right to Die*. Boulder, CO: Westview Press, 1994.

Humphry, D. *Final Exit: The Practicalities of Self-Deliverance and Assisted Suicide for the Dying*. Eugene, OR: The Hemlock Society, 1991.

Malcolm, A. H. *This Far and No More*. New York: Times Books, 1987.

Nuland, S. B. *How We Die*. New York: Alfred A. Knopf, 1994.

Quill, T. *Death and Dignity: Making Choices and Taking Charge*. New York: W. W. Norton and Co. 1993.

Starr, P. *The Social Transformation of American Medicine*. New York: Basic Books, Inc., 1982.

# Care of the dying: From an ethics perspective

T. PATRICK HILL

EDITORS' SUMMARY

T. Patrick Hill, MA, argues that the focus of the doctor–patient relationship moves from cure of disease to the broader commitment to care for the whole person when that person is dying. What formerly might have been considered optional qualities in that relationship now become essential. The relationship itself moves from a more medical to a more ethical focus. This is because the dying person suffers a loss of connections to everyone around him or her, to loved ones and to society itself. There is a loss of dignity within a loss of self-esteem. The act of caring therefore becomes a communion of persons testifying to the inherent dignity of the dying person's life, and for that very reason, of the lives of the caregivers as well. It is especially important to maintain this focus on personal dignity, and the requirements of informed consent that emerge from such dignity, rather than on technological strategies that might prolong life in an unacceptable condition for the patient.

In a society lacking recognizable social rituals that embrace the reality of death, dying becomes an embarrassment to all involved – to the person dying, to family members, and to health care professionals. Institutions, whether medicine, law, or religion, similarly find themselves at a loss for language and corresponding behavior that would enable them to accompany the dying in the way they are expected to accompany the living.

The source of this embarrassment lies in the intellectual and emotional difficulty we experience today with what was called, in the Karen Ann Quinlan case (see Cranford, this volume), the "right to a natural death." In theory, we can see that medicine, law, and religion might offer mutual support for such a right. Today, medicine, with systematic application of advanced technology, can and does prolong biological life. The law secures each person's right to live until the life span ends naturally. Religion, while reminding us of the limitations of biological life as an end in itself, also insists on the sacredness of human life and its inviolability against any direct assault.

## Interdependence

Taken together, these three positions suggest an interdependence. Although the law would not assert a belief in eternal life it could not overlook the right of individuals to hold such beliefs and, accordingly, control, where possible the circumstances of their deaths. For its part, medicine would not be expected to cause natural death directly but neither should medicine prevent death from occurring when nothing can be done to restore some minimum level of human activity. The role of religion is not to determine biological death, but religion has every expectation of forming a correct conscience to accept death when death is inevitable. Religion always underlines human mortality not as a biological necessity but as a moral challenge to each person to make the best of his or her life.

In practice, however, interdependence has given way to independence, with the result that medicine, law, and religion have realigned themselves oddly as mutually countervailing forces. Medicine has succumbed to the technological imperative to treat and wherever possible to sustain biological life, even without expectation of cognitive or affective life; law has been used to assert a patient's right to die. Religion has given voice to fear that the so-called "right to die" is a depreciation of human life in general and, in particular, a major concession in the direction of euthanasia.

The resulting tensions from these trends make more difficult reconciliation of legitimate individual, professional, and social interests so as to provide an ethical construct for the care of the dying. In principle, there is no denying the obligation of the physician to treat patients, using whatever clinical interventions are indicated and reasonably available. Therein lies the perennial temptation of medical paternalism. Nevertheless, United States common law tradition affirms the right of the individual to refuse medical treatment, including life-prolonging treatment. If the emergence of patient autonomy has added to the complications of the clinical setting, it has also become problematic for the balance between an individual's stewardship of life and religion's belief in a divine dominion over life. Complicating matters further, is that individual stewardship can also come into conflict with the legitimate interest of the state in the protection of life.

That these tensions should be brought into such sharp relief now is not surprising. After all, we live in circumstances where the timing of death, once beyond our control, has increasingly come to the point where we can exercise clinically a considerable degree of influence. As we know, only when we *can* act do we question, "Should we?"

This unprecedented capacity for action brings with it a much more complicated challenge to some of our central moral assumptions regarding life and death. The experience has not been without disturbing contradictions. With our affinity for science and technology, we have welcomed the power both place at our disposal. At the same time, we become troubled with the dawning realization that the use of technology invariably intensifies the moral consequences of actions today. In that sense, we are beginning to appreciate that medicine is now different from what has gone before. For example, artificially delivered food and water now can mimic the natural digestive system and keep a patient in a persistently vegetative state alive. So we have been persuaded to reformulate the definition of death itself. As technology increasingly informs medical practice, medicine itself becomes profoundly problematic in ethical terms.

## Care of dying: An ethical task

Perhaps nowhere is this uncertainty more apparent than in the medical care of the dying. The interdependent model outlined above might work as long as anyone involved – the patient, the family, the health care professional, and society – could remain essentially passive in the face of the onset of death. With no ownership of the process available, death "simply happened." All could remain morally neutral onlookers. To cooperate was natural. That stance changed once the process of dying became increasingly manageable clinically, and claims to ownership, based on personal, professional or social interests, became reasonable at some point along its newly emerging continuum.

As a result, any realistic assessment of the ethical issues involved in the care of the dying must be made within the framework of this new set of circumstances and in direct reference to the resulting tensions. Now we enjoy greater clinical discretion over the circumstances of dying and the very timing of death itself. The first ethically indispensable questions must be: "Whose claims on that discretion come first?" and "Whose death is it?" We have now come to understand very clearly that when someone is dying, others have central interests that will be affected by the eventual outcome of the process and the manner in which it has been controlled. Consequently, now, with some measure of control over the process of dying, comes presumably a responsibility to exercise that control in a manner that recognizes and protects as reasonably as possible the interests of all those involved. Unlike the recent past, when the circumstances and timing of death were effectively beyond clinical capabilities, that responsibility is now positively exercised to enhance the ethical quality of the use of such capabilities and to further their moral results.

The question, "Whose death is it?" really asks "Who has final responsibility for these clinical capabilities, their ethical use, and the ethical quality of the results of their use?" The question is significant because it presumes a condition that hitherto has not been presumable, i.e. that the process of dying and death itself, even though inevitable, can be a moral enterprise conditioned by informed and free consent. In other words, the process of dying can be ethical because available clinical control over circumstances is possible. When a person is dying, a will to die can be exercised. The basis of this possibility is precisely that we are now able to invoke, in legal terms, a so-called "right to die."

Another consideration is significant at this point: if a patient may will to die, then presumably the patient may consider the qualities required in the process of dying so that death can be willed as something good. In other words, death, like life, has finally become a real ethical issue, and thus merits serious discussion within ethics. As a result, a good death has become an end to which we ought to aspire. In this sense, the wish for a good death or the wish to die with dignity, far from being an oxymoron, can be a moral matter, at the center of which is the person actively and responsibly engaged in dying in the way that he or she has been accustomed to be involved in living. However inchoate, the phrases "a good death" and "dying with dignity" are actually eloquent expressions of newly realized and heightened moral possibilities in our experience of life all the way up to and including the process of dying. In earlier times, life was too easily hostage to whatever fate the form of death might take. Today, at least within the clinical circumstances of acute and chronic illness, we can relinquish life in much more measured steps because in so many ways life has become ours to relinquish. Accordingly, nothing ought to befit our lives more than the manner in which we die.

If that premise is accepted, then the central ethical goal in the care of the dying is, paradoxically, to ensure that they live as well as possible until they die. The daunting question before us now is how to ensure this state and, at the same time, respect the interests of all involved. Precisely because distinct, on occasion conflicting, interests are involved, some have called for an environment "hospitable" to dying as an integral part of the natural continuum of human life. Distinctive to this environment, which would embrace families, the health care establishment, and society at large, is the working principle that the values of living should not be superseded by fears of dying. Consequently, within this environment would be the overriding expectations, shared by all, that the dying person, now a patient, is to be treated as a full human being until death occurs. This commitment includes universal recognition that no one has a greater stake in this expectation, for purposes of achieving a good death, than the person who is dying.

## Moral integrity of the dying person

Treating the dying patient as a human being rather than as an object is possible only when those providing treatment do so in a way that recognizes the physical and moral integrity of the person. In the treating relationship, for example, the physician may not provide treatment independently of the patient's consent, whether explicit or implicit, direct or indirect. The moral integrity of the patient, however, requires that consent come from a clear understanding of the medical condition, as well as the nature and anticipated benefits and risks of the proposed treatment. In addition, informed consent cannot be given without an understanding of any other treatments, their benefits and risks, that might reasonably be proposed as alternatives.

Failure on the part of the physician to take whatever steps are required for a patient's informed consent is a failure to acknowledge the inherent freedom of the patient, a necessary condition for behaving ethically. By definition then, the fundamental terms under which medical treatment is given and received are moral, so that without informed consent for competent patients, no matter how technically successful the results of the treatment may be, the patient–physician relationship is ethically unacceptable.

With this appreciation of the ethical implications of informed consent for the patient–physician relationship, another premise of this relationship – benevolence – takes on even greater ethical significance. Benevolence is a psychological attitude that wishes the good of everyone. In the patient–physician relationship, benevolence should inform all of the physician's clinical decisions so that, when a treatment is provided, intentions as well as actions produce benefit and do no harm. For the patient, benevolence provides justification for the trust placed in the physician, with the expectation that good, not harm, will result.

## Patient–physician relationship

Ideally, two results will emerge from a patient–physician relationship that has been characterized by informed consent and benevolence: truthful communication and personal communion. Desirable in all patient–physician relationships, both are indispensable when a physician treats a dying patient. That fact alters substantially the balance in the partnership, increasing the patient's dependence on the physician and, thereby, proportionately increasing the physician's responsibility to the patient. Perhaps nowhere does the ethical nature of the relationship between patient and physician appear so clearly as here.

Truthful communication at this point entails sharing clinical information that will enable the patient, the family members, and all the health care professionals involved to acknowledge intellectually and emotionally that the patient is dying. For the patient in particular, but also for the family, this sharing requires unusual sensitivity. As some point out, truthful communication occurs on a continuum, bordered at one extreme by benevolent deception that makes liars of everyone and on the other by insensitive bluntness that is destructive. Truthful communication constructs an environment hospitable to dying and gives all parties involved permission to function within that spirit of hospitality. As a result, rather than asking, "Should dying be acknowledged?" everyone, including the dying patient, is encouraged to consider *how* to acknowledge dying in a manner that leads eventually to a state of "personal communion."

Personal communion is difficult in a setting that has become, almost by definition, impersonal and anonymous. For many patients and health care professionals receiving and providing medical care have become impersonal experiences as the practice of medicine has come to be defined in terms of particular human organs and their diseases, and as technological capability has displaced direct human agency. Within the paradigm of curing we have been prepared to pay this price of successful medicine. But in the case of dying, where there is no cure, we must recognize the need for a different paradigm. The new paradigm, commonly called the paradigm of care, is built on the premise that proper treatment of the dying requires functioning within a community of interests discernible through an act of communion on the part of all involved. At the center of this personal communion is the paradoxical recognition that with the prospect of death all persons involved must act to confirm the life of the dying person in particular, as well as to confirm what all of them share in that life as they care for it in its waning moments. To care for the dying is to care for one's own life too.

## The approach of death and personal communion

Disease, so long as it is not terminal, primarily threatens our activities. Death, as the negation of life, threatens our very being. As a result, when death approaches, undermining the patient's sense of being and the meaning that lies at its center, personal communion expressed through caring medicine can protect that meaning. Since the dying patient is beyond the services of curing medicine, it is not uncommon, for example, to think that the professional relationship between patient and physician is over. In fact, when we put to one side the curing dynamic in the relationship, we can see that a physician is never more a physician than when caring for a dying patient. Caring, not curing, is the ultimate objective of medicine. Practiced to its fullest capacity, caring medicine will always include the possibility of curing. The reverse cannot be said of curing medicine, i.e. that it will always include the possibility of providing care, particularly when cure is achieved by practice of technologically driven medicine.

To manage symptoms unavoidably encountered in the process of dying, physicians and other health care professionals practice palliative medicine. No less aggressive on its own terms than curative medicine, palliative medicine provides active care when cure or prolongation of life is no longer possible. No less skillfully than when treating a sick but curable patient, the physician must diagnose a range of physical symptoms as well as psychological symptoms. Proper care of the dying patient, accordingly, means managing the physical symptoms and supporting the patient and family psychologically through the terminal phase of illness. It is a very active process.

The symptoms of dying demonstrate the need for personal communion as a central feature of proper treatment, whether physical or psychological – the latter a term that, for the purposes of this chapter, includes both emotional and spiritual symptoms. These symptoms have a more radical significance for the dying patient than they would for an acutely ill but curable patient because they are a final, negative testament to a life, threatening to erode its inherent dignity, even its meaning.

That the physician chooses to treat these symptoms, without hope of cure, is nothing less than a declaration of the abiding dignity of the patient, symptoms notwithstanding. Health care professionals have pointed out that common to dying patients is fear of pain and other physical symptoms. But so too is fear of being cut off from the support of family, friends and caregivers. "Why me?" is a persistent question, as the patient searches for meaning with diminishing hope in a diminishing self-image. The need for meaning in the presence of impending death is just as clear in the fear and distress family members experience as they accompany the decline and eventual death of a loved one. That need is no less urgent for the health care professionals treating the dying patient. When the diagnosis is one of terminal disease, physicians and nurses, realizing that their clinical skills are no longer likely to cure the patient, may incorrectly conclude that there is "nothing further they can do" for the patient.

Quite the contrary. By means of personal communion, all persons involved in care of the dying recognize that impending death creates a community of interests and a community of needs, both of which can best be accommodated within an environment hospitable to dying. Here, as Elizabeth Latimer observes, the presence of patients who are well informed indicates truthful communication between them and caregivers; that of patients who are comfortable and reasonably alert suggests effective and balanced management of pain and other symptoms; that of patients at peace within themselves suggests that they and their families are receiving appropriate psychological support. Being at peace suggests that health care professionals have recognized the differences between curative and palliative medicine. They know how inappropriate curative medicine is for dying patients because it may prevent them from living well until they die. Finally, being at peace in the presence of impending death, suggests that patient, family, and health care professionals have had the moral awareness to reconfirm the life of the dying person by reconfirming the personal and professional meaning of what has been shared in clinical care. Since that meaning can no longer be realized in curing, where will it find realization?

The physician's positive obligation of beneficence and negative obligation of nonmaleficence (doing no harm), still prevail, if anything with greater force, in the case of the dying patient. What was a matter largely of clinical judgment, in the case of a curable patient, becomes, in the case of the dying patient, fundamentally a matter of ethical judgment. Now that recovery is no longer possible the focus of clinical attention must move from treating the disease itself to caring for the patient as a person. Nothing else will do because this commitment depends on the community of interests at the center of which is the meaning of a dying person's life.

## Suggestions for further reading

Cassell, E. J. The relief of suffering. *Archives of Internal Medicine* (1983), **143**, 522–3.

Culver, C. and Hanover, C. M. *Ethics at the Bedside*. New Hampshire: University Press of New England, 1990.

Dubose, E. R., Hamel, R. P. and O'Connell, L. J. *A Matter of Principles?: Ferment in U.S. Bioethics*. Valley Forge, PA: Trinity Press International, 1994.

Hill, T. P. Pain management: Theological and ethical principles governing the use of pain relief for dying patients. *Health Progress* (1993), Jan.–Feb., 30–9.

Hill, T. P. Freedom from pain: A matter of rights? *Cancer Investigation* (1994), **12**(4), 438–43.

Hill, T. P. and Shirley, D. *A Good Death: Taking More Control at the End of Your Life*. Reading, MA: Addison-Wesley, 1992.

Ladd, J. *Ethical Issues Relating to Life and Death.* New York: Oxford University Press, 1979.

Latimer, E. What is in the hospital care of dying? *General Palliative Care* (1991), **7**, 12–17.

May, W. F. *The Patient's Ordeal.* Bloomington, IN: University Press, 1991.

Moller, D. W. *On Death Without Dignity: The Human Impact of Technological Dying.* Amityville, NY: Baywood Publishing Co., Inc., 1990.

Ramsey, P. *The Patient as Person.* New Haven, CN: Yale University Press, 1970.

Wanzer, S. H., Federman, D. D., Adelstern, S. J., Cassell, C. K., Cassem, E. H. and Cranford, R. E. *et al.* The physician's responsibility toward hopelessly ill patients. *New England Journal of Medicine* (1989), **320**, 844–9.

Young, E. W. D. *Alpha & Omega.* Reading, MA: Addison-Wesley, 1989.

# Euthanasia and assisted suicide

PIETER ADMIRAAL

### EDITORS' SUMMARY

Pieter Admiraal, MD, PhD, an anesthesiologist and specialist in palliative care for cancer patients, presents a review of the social and cultural attitudes about euthanasia and assisted suicide. Throughout history it was sometimes approved and sometimes forbidden, approved because it was a way out for suffering, terminally ill persons, or for reasons of dignity, forbidden because it violated the rule against killing and letting God instead of the self be master of one's life. In the last and this century many physicians supported it and legislation was proposed but never approved. The Nazi brutalities were not euthanasia, but they tended to put on hold any legislative initiatives. These are only now starting again, beginning in the Netherlands and now also in the USA.

## A short history of euthanasia

The word euthanasia originated in Greece. It comes from two Greek words, *eu* (good), and *thanatos* (death). In its most neutral form, then, "euthanasia" means a good death. Since most of us would want a good death for ourselves and for others, this cannot be the aspect of euthanasia that engenders so much dispute today. Instead, that dispute comes from social, cultural, and religious values that come into conflict about duties we have to die well and duties others have to assist us.

Conflicts of this sort have always surrounded the idea. Plato and Socrates regarded suffering as a result of painful disease to be a sufficient reason for stopping life through suicide. We all know that Socrates died by taking hemlock, not because of a painful illness, but for a noble reason, to uphold the very rule of law that had condemned him to death. Plato's and Socrates' views on the matter diverged from those of Aristotle, who argued that suicide was not courageous and was an offense against the state. Pythagoras and Epicures also condemned suicide. Yet in some city-states of Ancient Greece, suicide was approved. Magistrates kept a supply of poison for anyone who wished to die. Perhaps it was against this widespread acceptance that the Hippocratic physicians took an oath to "give no deadly drug," as they were part of a reform movement of physicians influenced by Pythagorean ethics.

The Stoics, another later branch of Ancient Greek and then Roman philosophy, accepted suicide as an option when life was no longer acceptable for any serious reason. To the Romans, suicide for halting life during painful terminal illness was acceptable. Due to the Stoic influence, the idea of dying well was a *summum bonum*, the highest good, and part of a noble life: "A good death gives honor to a whole life," said Epictetus, a Roman Stoic thinker. In Rome, people were permitted, sometimes expected, to commit suicide to escape from disgrace at the hands of an enemy, or scandal (as it is in Japan even today), or as an alternative to public execution (as Field Marshall Rommel, a German hero, was given the option by the Nazis when it was learned he had plotted to kill Hitler).

With the advent of Christianity in the Roman Empire, this viewpoint waned. Under the growing influence of this religion, and its acceptance as the official religion of Rome at the time of Constantine, suicide was no longer acceptable. The rule against killing had its origin in the Christian view of the Commandment, "Thou shalt not kill," and the pacifism of Christ and the early Church. Life was seen as a gift from God over which persons had to take ordinary care. St Augustine, for example, argued that suicide was against the Sixth Commandment, against killing, and that life and suffering were divinely ordained for the individual. The moment of death was in God's hands, and to usurp it was a sinful act of pride, a denial of God's power over human life. Persons who committed suicide were usually buried outside the city walls, at a crossroads, where that cross might ward off the devil seeking the troubled soul of the individual.

Growing intolerance of suicide continued. In AD 553 the Council of Orléans officially denied funeral rites to anyone who had killed themselves, and in 693 the Council of Toledo announced excommunication for attempted suicide. From early times, then, persons who were successful at suicide could not be buried in the churchyard. This intolerance culminated in the thirteenth century with St Thomas Aquinas. Aquinas, the great theologian, argued that suicide was the most dangerous of sins and against not only Divine Law but the law of nature (self-preservation). His views certainly reflected an unbroken tradition of Christianity about suicide and any assisting thereof.

Yet perfect unanimity is hard to maintain, and probably never existed. Some softening of views occurred during the Renaissance, after the grisly fourteenth century when the plagued wiped out a third of the population in Europe. Death was everywhere. It touched everyone. Intensive study of Ancient Greek and Roman sources and their cultures rehabilitated earlier "pagan" values. During the Reformation, for example, Luther favored euthanasia and seemed to cause his own death when he went for an extended walk knowing he had heart trouble. Sir Thomas More, the famous saint and martyr for the Catholic Church and adversary of the English King Henry VIII, published in 1516 in his *Utopia* the need for voluntary euthanasia for terminal patients. This was an acceptable measure for control of suffering in the ideal society he envisioned.

The idea continued to take hold among serious thinkers and physicians. In 1742 in Scotland, philosopher David Hume's essay "Of Suicide" was published in *Essays, Moral, Political and Literary*. In 1794 Paradys, a physician, recommended in his *Oratio de Euthanasia* (an "Oration concerning Euthanasia") an easy death for an incurable and suffering patient. Interest continued high in euthanasia later in the nineteenth century, both in Europe and in the USA. Books on euthanasia and euthanasia societies existed in both places over a hundred years ago, with draft legislation proposed as well. Karl Marx, in his "Medical euthanasia," criticized physicians who treated diseases instead of patients, and pledged a better alleviation of suffering. Schopenhauer said, "as soon as the terrors of life ... outweigh the terrors of death, a man will put an end to his life." Thus, right to the end of the nineteenth century and into the early twentieth, physicians spoke openly about the possibility of euthanasia. Because of a Prussian equivalent of a national health plan, there was even open discussion a hundred years ago of the need for the state to provide euthanasia for persons who became incompetent to request it (Adolf Jost, *Das Recht auf dem Tod* [The Right to Die], 1895).

## Action in this century

In 1931, a health officer in Leicester, England, named Dr Millard, published his Voluntary Euthanasia Legalization Bill. He made the following specific proposals:

1. An application for a euthanasia permit may be filed by a dying person, stating that he has been informed by two medical practitioners that he is suffering from a fatal and incurable disease, and that the process of death is likely to be protracted and painful.
2. The application must be attested to by a Magistrate, and accompanied by two medical certificates.

3. The application and certificates must be reviewed by the patient and relatives as interviewed by a "euthanasia referee."
4. A court will then review the application, certificates, the testimony of the referee and any other representatives of the patient. It will then issue a permit to receive euthanasia to the applicant and a permit to administer euthanasia to the medical practitioner.
5. The permit would be valid for a specific period, within which the patient would determine if and when he wished to use it.

In its requirements for consultation and deliberation, the resemblance to the rules the Dutch use to become immune from prosecution during administration of euthanasia is remarkable. This bill was discussed for five years, but was defeated by the House of Lords in 1936.

A year prior to that, in 1935, the British Voluntary Euthanasia Society was founded, the first in the world. Shortly thereafter, a similar one was started in the USA, the Euthanasia Society of America (1938). In 1939 this society proposed a bill to legalize euthanasia in New York State, but it was never introduced into the Legislature. During the ten years preceding the Second World War, several court cases dealing with assisted suicide and mercy killing occurred in the USA and Great Britain. Although all suspects in these cases were found guilty, the sentences in most cases were very mild.

What happened in Germany prior to and during the Second World War clearly had an impact on the euthanasia movement. Euthanasia, as we have been discussing it, was legalized by Hitler. A physician could help to put a patient "out of his or her misery" if they were suffering during a terminal illness. But this comforting measure was mixed up with others that were, in any event, terrible. Earlier, by 1920, two physicians, K. Binding and A. Hoche, published a book contributing to the concept of eliminating "valueless life" (*Die Vernichtung des lebensunwertens Lebens*). Arguments about the best way to kill "life not worthy of life," "wasted," or "worthless" lives were discussed and published. These lives were those of the physically or mentally handicapped, the senile, the retarded, and psychiatric patients not able to work after five years in an institution. Nazi party physicians spearheaded efforts toward "social hygiene," one step of which was to sterilize those with hereditary illnesses, something done in America already, and which China contemplates doing today. This was stopped in 1941 officially by Hitler after protests from the Churches. By then perhaps 90 000 patients had been sterilized.

During the war itself, due to Nazi nationalist and racial ideology, the most evil crime of the century took place, the Holocaust, in which 6 million Jews and 3 million gypsies, socialists, and protesters were exterminated. This became possible through total control by the National Socialist Party of the engines of the State, social and political control exercised by the SS and the Gestapo, and the German propensity to educate for total obedience to law and order.

Was this euthanasia? Of course not. It was murder, purely and simply. Thus the Nazi experience actually has nothing to do with the contemporary debate. And yet in another way it does. Everyone should be wary of the power of the state and of involuntary killing.

After the war the debate continued, but initiatives to include the right to euthanasia in the United Nations charter, for example, initially supported by Eleanor Roosevelt, Chairman of the United Nations Commission on Human Rights, were withdrawn as the revelation of Nazi atrocities increased. Parties on either side of the debate became more and more alienated from one another. New proponents emerged, and mainly religious opponents became more explicit in their condemnations. As we have seen, the Roman Catholic Church considered euthanasia, like abortion, to be against "the law of God," i.e. to be a deliberate and direct attack on human life. In 1956 the Pope reiterated this view to an International Congress of Doctors. But in 1957, addressing another international group of physicians, he accepted the possibility of eventual life-shortening use of drugs such as morphine to relieve unbearable pain, because there is no causal link between that effect (death) and the intent (to relieve suffering). The idea of "passive euthanasia" was therefore introduced into the mainstream of the debate, along with the distinction between using ordinary rather than extraordinary means to preserve life.

Meanwhile, in 1954, Joseph Fletcher, an Episcopal priest-theologian, published *Morals and Medicine*, one of the first books in "secular" bioethics. In the chapter, "Euthanasia: Our right to die" he contradicted the Catholic view. He was a lifelong proponent of euthanasia. Glandville Williams, professor of law at Jesus College, Cambridge, in 1957 published *The Sanctity of Life and the Criminal Law*, in which he stated: "the greatest of all Commandments is to love, and this surely means that euthanasia is permissible if performed truly and honestly to spare the patient and not merely for the convenience of the living." These two arguments in favor of euthanasia, that it is a right of an autonomous person and that it is a kindness or benevolence of caregivers, are the mainstays of the movement. They are addressed, head-on, by Leon Kass in this volume.

From the 1960s onward, two additional developments had remarkable influence on the debate. It would be hard to overestimate them.

First, there was an unprecedented revolution, an explosion of progress, in medical science, with major discoveries and inventions that had the cumulative effect of protecting life against almost all disease and to prolong life for a much longer time than possible before. Heart transplantation is only one technique among many, a brilliant inauguration of an age when, perhaps, only the doctor might decide not only when life begins (see p. 23) but also when it ends (see p. 163). Death and dying were increasingly denied.

The second influence was the attitude of patients themselves to this technology. A great disappointment occurred. Prolonging life could also mean prolonging suffering. On top of this, death, naturally, was not something one could ultimately escape. Witness that realization among younger persons with the advent of the epidemic of acquired immunodeficiency syndrome. It proves once again the vulnerability of human life.

One could almost predict the response. Patients wanted to decide for themselves about their own lives. The right of self-determination became a world-wide movement. Ethicists and religious leaders tempered their condemnation of euthanasia with concerns about unduly prolonging life and protracted dying. These concerns stemmed not only from compassion, but also from justice, the balance of marshalling our resources properly. At the same time, increased interest arose in providing care for the dying patient. The psychiatrist Dr Elizabeth Kübler-Ross published her book *On Death and Dying* (1969), describing the stages of a dying person, and suddenly the most popular course on college campuses and in medical schools became the one on death and dying. The subject of dying was no longer taboo. Most importantly, however, the attitude of doctors to their new-found power also changed. Different American and English polls in the 1960s, asking whether euthanasia were acceptable for doctors, proved that up to 60–70% of the respondents answered "Yes, if it were legal," and many admitted already performing euthanasia in extreme circumstances.

Opposition, however continued. Many doctors opposed all forms of euthanasia, even passive ones. The World Medical Association adopted 1968 resolutions opposed to euthanasia. Beginning in 1975 with the case of Karen Ann Quinlan, the court system became more lenient toward forms of euthanasia. Recall that Quinlan lapsed into a coma after taking a combination of drugs and alcohol, and was in a persistent vegetative state. She was placed on a ventilator. Counseled and supported by a parish priest and others, after three months the father signed a release, on the basis of her previously expressed wishes, to permit her physicians to turn off the respirator. The physicians and hospital refused. Her father went to court to be named her guardian to authorize discontinuance of all extraordinary means of sustaining vital processes. In appeal, the New Jersey Supreme Court agreed, and Karen was removed from the ventilator. The physicians involved planned, actually, to defy the order of the court, but were successful in weaning her off the respirator. She died about ten years later in a nursing home, never regaining consciousness.

Other supreme courts of states, and in one case, the US Supreme Court, have decided always in favor of individual directives against prolonging life, and in favor of passive euthanasia. The Quinlan case profoundly influenced the Living Will statutes enacted in 36 states, as well as Durable Power of Attorney laws in most states (where one designates a surrogate to speak for one's values about medical care during any time of incompetency). These laws have the effect of legally recognizing one's right to die with dignity, i.e. to die with one's values intact. In some states, the Living Will law was resisted most prominently by the Right to Life movement, itself originating among Catholics and fundamentalists (like Baptists) opposed to abortion. This group opposed Living Will legislation because it believed that persons were thereby authorized to commit suicide (by not accepting medical technologies that could prolong their lives). Such resistance was not a majority opinion in any case. The most extreme and fundamentalist wings of this world-wide movement today are the Human Life Alliance and the Club of Life.

In 1980 Derek Humphrey and Ann Wickett formed the Hemlock Society in America to help people to learn how to commit suicide painlessly in the case of terminal illness. That same year the World Federation of Right to Die Societies was formed from 27 groups in 18 countries. Thus, the debate moved to the present day from more academic circles to much more public and organized world-wide movements.

## The Dutch experience

To a large extent on the European continent, nothing changed after the Second World War. Almost all countries had been occupied by the Germans, most of their cities and infrastructure having been destroyed, and their Jewish inhabitants exported and killed. Everyone had had some terrible experience with the Nazis. In 1941, for example, Seyss-Inquart, the German Commander in the Netherlands, tried to coerce Dutch physicians to participate in the Nazi sterilization of the mentally handicapped and the extermination of the Jews. Without hesitation these physicians refused, despite the fact that over 100 were promptly shipped off to concentration camps.

The Dutch view of euthanasia as voluntary and as a good death was solidified by this resistance to its abuse. The starting point of more public acceptance of medically induced voluntary euthanasia occurred with the case of Dr Postma. In 1973 she was found guilty of mercy-killing her mother, but was only given a one-week suspended sentence and a year's probation for this act. Also in 1973 the Dutch Society for Voluntary Euthanasia was founded. It established a "member's aid service," giving members proper information about euthanasia, and it acts as a mediator between patient and doctor, although it would never distribute lethal drugs or offer physical help in dying. In 1980 this organization published

*Justifiable Euthanasia* (written by me), advising about the most suitable drugs and their proper administration for euthanasia. This publication was sent to 19 000 doctors and 2100 pharmacists in the country.

The Royal Dutch Medical Association (RDMA) was a major influence in the Dutch debate, and still is. In 1973 it had already issued a provisional statement on euthanasia: "Legally euthanasia should remain a crime, but that if a physician after having considered all the aspects of the case, shortens the life of a patient who is incurably ill and in the process of dying, the court will have to judge whether there was a conflict of duties which could justify the act of the physician." Updating this view in 1984, the same association published its requirements for doctors who assist in dying, to prevent prosecution. These were mostly taken from the Rotterdam Court standards of 1981 for non-criminal aid-in-dying. These rules have subsequently been confirmed in several court decisions. They are as follows:

1. The request to die must be the voluntary decision of an informed patient.
2. The request must be well considered by a person having a clear and correct understanding of his or her condition and of other possibilities. The person must be capable of weighing these options, and must have done so.
3. The desire to die must be of some duration.
4. There must be physical or mental suffering which is unacceptable and unbearable.
5. Consultation with a colleague is obligatory.

In 1990, the RDMA and the Ministry of Justice agreed upon a notification procedure about enthanasia with the following elements:

1. The physician performing euthanasia or assisted suicide does not issue a declaration of a natural death. He or she informs the local medical examiner by means of an extensive questionnaire.
2. The medical examiner reports the death to the district attorney.
3. The district attorney decides whether a prosecution of the physician involved should be inaugurated. If the doctor has complied with the five requirements listed above, the attorney will not prosecute.

In effect this constitutes immunity from prosecution, although the act is still against the law. The notification procedure acquired the force of law by being included under the Burial Act of 1993. To this point, then, the Netherlands is the only country in the world where euthanasia is accepted.

Following these guidelines, in 55% of cases the physician who performs euthanasia will offer the patient a drink with a lethal dose of barbiturates. The patient will die in a deep coma as a result of respiratory depression. About 70% of them will die within 3 hours. In most cases the doctor will shorten this period using curare, a drug that paralyzes all muscles. In 45% of cases, the doctor will give an injection containing both barbiturates and curare. The patient will die this way within a few minutes. These drugs are only available to doctors with a prescription. The physicians in question know their patient and the patient's value system, and are acquainted with the family. Most of these deaths occur in the family's home. These are important points, since no euthanasia should be legalized if it is not part of a comprehensive program of caring for the dying, such as that offered by hospice, and should be used as a treatment of last resort only.

## Future challenges

Recent opinion polls in almost all countries of western Europe and in the USA have proved that a majority of the populace, sometimes over 80%, answers "yes" to the question: "Is voluntary euthanasia in case of unbearable suffering acceptable for you, and should this be legalized?" But why has it not yet been legalized?" Only in the Netherlands has it been regulated, although it is still illegal. There are an estimated 3000 cases a year in the Netherlands, and it has been practiced there as part of a total commitment to caring for the dying for over 20 years.

There have been four separate attempts, two in California, one in Washington, and one in Oregon, to legalize euthanasia in the USA. Those efforts continue. Despite public support for the concept, each time these efforts lost to a narrow majority of voters. This demonstrates that, although the concept is fertile, making it legal is "iffy" for a lot of people. Strong campaigns against legalization have been mounted by the Roman Catholic Church, other religious bodies, state medical societies, and politicians and legislators.

There is no indication that the views of the Roman Catholic Church, Jewish bodies, Islamic groups, and fundamentalists, will change. They rest on the principle of "the sanctity of life." Some Protestant reformed churches have accepted euthanasia under the rubric of shortening unbearable suffering. The religious and medical groups in general find that there is no need to legislate for euthanasia, since adequate pain control and passive euthanasia can be used instead, and are already morally accepted. Meanwhile the growing secularization of all elements of social life in the world contributes to independent thought. While religious leaders do have an influence, it is waning. Surely legalization will occur in one or two places first, be tried "experimentally," and then will grow more popular when it is shown that abuses of this power are kept in check.

Further, throughout the world there are different systems of justice. Sometimes efforts will continue to legalize through public vote or through legislative action. At other times the court system and the decisions of judges and/or juries will assist in the growth of acceptance of euthanasia. Already in the state of Washington, a federal judge argued that a law restricting assisted suicide there was unconstitutional, since it infringed the rights of individuals to die as they wished. This decision will most certainly be taken on appeal to a higher court. In Michigan, a state that formerly had no law against assisted suicide, one was created explicitly to ban the actions of Dr Jack Kevorkian. He has already been tried under the law for assisting one person to die and was acquitted by a jury. However, he may be tried for assisting two others (he helped a total of 20 persons) before the law was enacted. This is a typically complex juridical point, since an appeals court in Michigan has declared the law passed to outlaw assisted suicide unconstitutional because it contains more than one provision (not because it outlaws assisted suicide itself). In England, a physician was tried and convicted of deliberately ending the life of a patient.

Generally speaking, courts and juries are open to euthanasia if physicians explain carefully enough how they might follow rules that could eliminate abuse. The official stance of most of the world's medical associations, like the churches mentioned above, still maintains that a doctor's duty is to preserve life such that he or she cannot end it even upon request of the patient. I see this official stance changing in the near future.

Perhaps the greatest fear of abuse is presented as the "slippery slope" argument: gradually voluntary euthanasia will slip over into involuntary killing of the demented and the mentally handicapped. This fear is based on the abuse of power that virtually defined the Nazi experience. But is that fear rational? Are our countries like Germany at that time? After 20 years of experience in the Netherlands there is no indication of a slippery slope occurring, although one case did happen wherein a depressed patient petitioned her psychiatrist for euthanasia, and, after consulting others, he finally complied.

Without question, the biggest challenge in the immediate future for the legalization of euthanasia will be to spell out barricades against potential abuse. Once these are written into legislative proposals, they have a good chance of being passed. Meanwhile, efforts to support the right of a patient to request assistance will increasingly be successful in the courts.

### Suggestions for further reading

Admiraal, P. Justifiable active euthanasia in the Netherlands. In
    *Euthanasia: The Moral Issues*, ed. R. M. Baird and S. E.
    Rosenbaum, pp. 125–8. Buffalo, NY: Prometheus Books, 1989.
Battin, M. P. *Ethical Issues in Suicide*. Englewood Cliffs, NJ:
    Prentice-Hall, 1982.
Brahams, D. Euthanasia: Doctor convicted of attempted murder.
    *Lancet* (1992), **340**, 782–3.

Brody, H. Assisted death – a compassionate response to medical failure. *New England Journal of Medicine* (1992), **327**, 1384–8.

Callahan, D. *The Troubled Dream of Life: Living With Mortality*. New York: Simon & Schuster, 1993.

Campbell, C. S. Religious ethics and active euthanasia in a pluralistic society. *Kennedy Institute of Ethics Journal* (1992), **2**(3), 253–77.

Campbell, C. S. Aid-in-dying and the taking of human life. *Journal of Medical Ethics* (1992), **18**, 128–34.

Cundiff, D. *Euthanasia Is Not the Answer: A Hospice Physician's View*. Totowa, NJ: Humana Press, 1992.

De Wachter, M. A. M. Euthanasia in the Netherlands. *Hastings Center Report* (1992), **22**(2), 23–30.

Gomez, C. *Regulating Death: Euthanasia and the Case of the Netherlands*. New York: The Free Press, 1991.

Graber, G. C. and Thomasma, D. C. *Euthanasia: Toward an Ethical Social Policy*. New York: Crossroads/Continuum, 1989.

Humphry, D. and Wickett, A. *The Right to Die: Understanding Euthanasia*. New York: Harper & Row, 1986.

Humphry, D. *Final Exit*. Secaucus, NJ: Carol Publishing, 1991.

Khuse, H. *The Sanctity-of-Life Doctrine in Medicine: A Critique*. New York: Oxford Press, 1987.

Kohl, M. Altruistic humanism and voluntary beneficent euthanasia. *Issues of Law and Medicine* (1992), **8**, 331–42.

Kushner, T. CQ Interview: Derek Humphry on death with dignity. *Cambridge Quarterly of Healthcare Ethics* (1993), **2**(1), 57–62.

Quill, T. E., Cassel, C. K. and Meier, D. E. Care of the hopelessly ill: Proposed clinical criteria for physician-assisted suicide. *New England Journal of Medicine* (1992), **327**: 1380–4.

Rachels, J. Active and passive euthanasia. *New England Journal of Medicine* (1982), **306**, 639–45.

*State Commission on Euthanasia: Report on The Cases of Euthanasia*. Den Haag, The Netherlands: Staatsdrukkerj en Uitgeverj, 1985.

Van der Maas, P. J., Van Delden, J. J. M., Pijnenborg, L. and Looman, C. W. N. Euthanasia and other medical decisions concerning the end of life. *Lancet* (1991), **338**, 669–74.

Wanzer, S. H., Federman, D. D., Adelstein, S. J., Cassel, C. K., Cassem, E. H., Cranford, R. E. *et al.* The physician's responsibility toward hopelessly ill patients: A second look. *New England Journal of Medicine* (1989), **320**, 844–9.

Welie, J. V. M. Euthanasia: Normal medical practice? *Hastings Center Report* (1992), **22**(2), 34–8.

Welie, J. V. M. The medical exception: Physicians, euthanasia and the Dutch criminal law. *Journal of Medicine and Philosophy* Aug. (1992), **17**, 419–37.

# Physician-assisted suicide: Progress or peril?

CHRISTINE K. CASSEL

EDITORS' SUMMARY

Christine K. Cassel, MD, a general internal medicine geriatrician, argues
that physician aid in hastening death does not necessarily violate professional standards, and sometimes enhances patient care during the dying
process. The centerpiece of her argument is that the procedures for helping dying patients during pain and suffering are sometimes indistinguishable from assisted suicide and euthanasia. Arguments against these actions
look duplicitous and self-serving when faced with clinical activities such
as giving high doses of morphine to make the dying unconscious or to
relieve their suffering while hastening their death. Cassel suggests that
some moral reasoning about caring for the dying actually disguises a
reluctance to face death, and helps to hide from public accountability
some of the actions physicians take at the bedside.

The ethical and legal issues of physician-assisted dying are much in need of discussion among the professions and the public. The debate about physician-assisted
suicide (PAS) is a perfect example of the tension between what one might call
the rule-based or principle-based approach to medical ethics and newer ways of
thinking about medical ethics or questions of values in health care. These ways
are more experimental, grounded in narrative and in case-based analysis, more
inductive than deductive, and, to use an older term, "situational." A lively discussion has already occurred about the conflict between these two approaches to

solving problems in ethics and doing research in ethics. The problem of the physician's role at the end of life and in particular the question of the active role of the physician in assisting with dying is one that really forces us to examine these theoretical questions. There is some clarity to be gained by understanding the differences between these approaches and the contribution of both to resolution of a complex and contentious question.

### Some critical questions

To begin, I pose some questions. What does it mean when ethicists claim that there is a strict proscription against physicians helping people to die; that is to say, assisting in suicide or active euthanasia? I know many excellent clinicians – eminent, highly respected leaders of medicine – who have assisted in dying and who will say, usually quietly, when you talk with them, "Well, of course there was that one patient or those two patients, and that was a special circumstance." We have a firm rule that says we do not do this, and yet almost everyone can think of exceptions that they might find morally acceptable on one term or another. So these are the main questions: "Are these doctors murderers?", or "Is the rule wrong?", or "Are we somehow wrong in thinking about the way we use or apply rules in medical ethics?" The arguments, pro and con, have been fairly well aired and codified by now.

The arguments in favor of allowing physician assistance in dying in some circumstances appeal to the right of the patient to decide about care at the end of life, claims to dying with dignity, and arguments from ethicists that *allowing* to die is morally the same as *causing* to die if one's intention is benevolent. The arguments against allowing physicians ever to hasten death intentionally are either Hippocratic and fundamental ("We don't do this; doctors don't kill") or pragmatic (worries about the slippery slope and what will happen to vulnerable patients where these practices might be abused). Interestingly, both of these arguments – pro and con – are concerned about better care of dying patients. This commonality needs greater emphasis because it affects the number of patients who are not receiving adequate comfort care in their last days, potentially far greater than those asking for euthanasia. The widespread lack of competent palliative care is related to a problem that rule-based ethics presents: what happens to decisions and clinical care when you come close to the line, close to the bright line of actively assisting in dying?

The doctrine of "double effect" emphasizes the difference between what we intend or expect (to relieve pain) and what happens (death is hastened) when we give morphine to a patient in great pain. But what is the result of having a clear and precise bright line that says "We give morphine to treat pain but not to hasten death?" Does the bright line guard against abuses, or on the contrary, does it create a barrier to the effective care of patients who are dying? I believe that the bright line is less effective as a barrier against abuse of the vulnerable than it is as a barrier to physicians' intense caring for their dying patients.

## Arguments against euthanasia

In the USA a great deal of progress has occurred in understanding the ethics of the relationship with a patient who has a terminal illness or a life-threatening illness. We have made great progress in recognizing that competent patients have the right to refuse life-sustaining treatment and to exert some control over their future, through the use of advance directives, even if they become incompetent. Although there are many cultural and important attitudinal barriers toward the use of advance directives, they certainly have increased our ability to talk about these decisions.

To some degree, standards of comfort care have also improved. The hospice movement has made enormous progress since the 1970s, when it began, in promoting the idea that it is not always the prolongation of life that is the most important goal of treatment and that in certain situations comfort care is a much more important goal. Although we need more progress in this area, we have less worry about the addictive characteristics of narcotic analgesics and other pain-killing medications. We have a little less worry about the double effect; that is to say, "If I give this patient adequate pain control might it depress respiration and hasten death?" In the case of a patient with a terminal illness, we have been taught that it is acceptable not to worry about hastening death if comfort is the primary goal. However, the recent debate about physician-assisted suicide has made people more, not less, nervous about that issue.

In spite of all this progress, hospice still remains a sort of "ghetto" in medicine. We transfer our patients to hospice programs and often never see them again. We turn to specialists who care for dying patients rather than accept that this ought to be part of what every physician does when his or her patient becomes terminally ill. In most academic centers, we make palliative care rather invisible. Community hospitals have actually done a much better job than most of our teaching hospitals at establishing hospice programs within the hospital setting. But even there, often house staff are transferred off the case when the patient has gone into hospice, as if there is nothing left to learn from or give to that patient. We thus make it invisible to the people who are the most important for the future – the trainees who are learning to become physicians.

We have learned this from hospice experts about most patients: thoughtful and competent comfort care *can* make the dying process gentle and free from suffering – certainly free from pain, but nevertheless perhaps not always free from suffering. In fact, there is no evidence that even all physical suffering associated with the terminal illness can be relieved. All too often we focus only on pain as if it were the only symptom that accompanies death, when in fact there are many other symptoms afflicting dying patients that are much more difficult to treat than pain. These sometimes are protean, and can be difficult to diagnose and to understand, but often are unremitting and terrible for the patient experiencing them: symptoms such as nausea, vomiting, shortness of breath, inability to handle secretions, and nightmares and episodic, terrifying delirium, sometimes caused by the very medications doctors give to treat the pain. Also, patients have incontinence, immobility, weeping sores that will not heal – all these require specific forms of treatment and surely contribute to the suffering of the person who is dying. They certainly contribute to the fears people have about the end of life. These symptoms as a source of suffering cannot always be abolished by medical treatments, unless we include medically induced coma as a therapy. Some people argue "Well, if we just treated pain well enough, then there would be no demand for assisted suicide or active euthanasia." I believe that is not true; there are things besides pain that are much more difficult to treat adequately.

Even if we could treat these symptoms adequately, and there are ways of doing so, there are some experts in palliative care who oppose the idea of active assistance in dying. They argue a principle of "Let them sleep, before they die," – a heavy sedation so that patients become unconscious and die of dehydration in that state. This is not considered active euthanasia. By some people it is considered, for physicians, to be more humane and more ethical than euthanasia.

## Death with dignity

But even if we accept that idea, from the patient's perspective there is still one very important consideration that is not addressed by any of these medical approaches. That is the problem of dignity. This is what the public talks about: death with dignity. It is an old idea. Nonetheless it is something that we all need to understand about the public debate. If we understand it, physicians will become better and get greater rewards in the practice of medicine. But dignity is much harder to achieve than relief of pain, especially in our modern hospitals, where in the USA 80% of people still go to die.

People are afraid of the symptoms that threaten personhood, not so much the pain or even physical suffering, but the loss of dignity and selfhood. The classic example here is a patient with acquired immunodeficiency syndrome (AIDS) anticipating AIDS dementia, having seen friends who have gone through this losing their very sense of self and dignity, a descent into a kind of absurdity or degradation, not being able to say goodbye on your own terms but totally dependent on others, without awareness or control. Whether one sees this in a religious or a secular context, it is important somehow to derive some kind of meaning from this most profound moment in life – death itself. We often erase that meaning with modern medical technology. Unfortunately, all too effectively we obliterate it. Death ought to be one of the most profound and meaningful experiences in human life. To be with someone who is dying and to be witness and companion I consider to be one of the most privileged moments of being a physician, comparable only to being present at a birth. And yet, how few times in modern medicine do physicians allow themselves to take part in the personal significance of death.

What the patients are asking is that physicians give this meaning back to them. In fact, modern ways of dying have made it harder to recognize and capture that meaning. This is described brilliantly by historian Philip Aries in his book *Western Attitudes to Death*. He describes how difficult it is in modern medicine to pinpoint the moment at which a patient dies. There are many "little" deaths along the way. So even if you wanted to have a ceremony or say something meaningful about a person's death it is hard to know at what point to do it. Indeed, Dr Lewis Thomas, who had a way of always capturing in a very clear and profound way what people are thinking, has said this too: "And so we have come to view death as a failure, as opposed to an inevitable and meaningful part of life, and because of this we have lost the old feeling, the feeling of respect for dying and all of the awe." He uses the term "awe," a word rarely heard in modern medicine. I wonder how often physicians feel that sense of awe in the presence of the end of the life of a human being.

Dignity is an important need and not something we can diminish by categorizing it as just a popular political slogan. People are asking for something concrete when they are asking for death with dignity. They are *not* necessarily asking for suicide or euthanasia, but in some way see these as the only dignified alternatives, since they fear medicine will take over their death. Addressing this need for dignity might be more profoundly significantly human than all the other caring that physicians do. Many patients are more willing to put up with pain if they can just have dignity. Thus, I think that we need to include that in our goals in the treatment of patients at the end of life.

## Arguments for physician assistance

It is not surprising, then, that public opinion surveys demonstrate that there is enormous support for the active involvement of physicians in either euthanasia or assisted suicide. The National Opinion Research Center at the University of Chicago has done a general social survey every year since just after the Second

World War. They have asked, in one way or another, "Do you feel that in the face of terminal illness, a physician should be permitted to actively help a patient hasten their death, if the patient requests it?" A majority of people have agreed with that statement. Since 1977 the number has gone from 62% to 75% as most recently reported (1991). This general trend of more and more people wanting to have this option persists, despite the actions of Dr Jack Kevorkian in Michigan. Even then, there is enormous public approval for what he has done. Even the courts have been unable to convict him. People would like to have this option, even if they are unlikely to use it. The polls demonstrate that people do not like their options in our modern medical world when they face death. What they see as their options are the intensive care unit or abandonment. They do not see death on their own terms, a death that respects their own meaning at the end of their life, nor do they possess confidence that their suffering, whatever form it takes, will be competently relieved.

It was Kevorkian and his initial activities in 1990 that prompted many scholars to begin thinking seriously about this issue. The first patient he assisted in suicide, Janet Adkins, was a 53-year-old woman with a rather mild case of early Alzheimer's disease. When she died, most experts in medical ethics immediately stood up and said "We don't do this; this is not acceptable physician behavior." I could not help but think, "It's more complicated than that." To believe it is more complicated is not to approve of Kevorkian or what he is doing. Some ethicists, including Diane Meier, agreed that there needed to be some medical voice saying, "This issue deserves deep consideration and discussion within the profession." She and I wrote an article published as a "Sounding board" in the *New England Journal of Medicine*. We called for a more thoroughgoing discussion among physicians and ethicists, going beyond reliance on a simple rule.

In response to that publication we received dozens of letters from physicians – not sent to the *Journal* – but to us personally. These letters thanked us for raising the issue, and many of them related personal experiences. One after another these stories began to unfold, either about physicians who had helped a patient to die and wanted to talk about it, or about a physician who had refused and now regretted it. It was a fascinating tapestry of medical stories. One of the stories that came out of this interchange was the eloquent account of "Diane" by Timothy Quill that was published the following year in the *New England Journal of Medicine* (1991) and that led to his summons to appear before a Grand Jury in Rochester, New York. He had to face possible indictment for actively assisting in the suicide of his patient, Diane, suffering from leukemia, who took an overdose of barbiturates that he had prescribed for her.

These events suggested something about the power of the individual case, the "story" or the narrative analysis of the situation in exploring the boundaries of right and wrong. If you listen to informal conversations, even in the elevators at major university medical centers, the power of the story emerges. When Quill's article came out many people commented that even though they disagreed with his action, they would like to have that man for their own doctor! So even if one is opposed in principle to physician assistance in dying, there is a quality about the sympathy, compassion, and the relationship that makes people think "Well, if it were me, I would want to have that option for myself." This observation could be seen as a kind of hypocrisy, if we say that we do not condone the practice, that it is not consistent with medical ethics. Yet many physicians feel that when the time comes they will have a little stash of morphine or barbiturate and they will be able to take control of the end of their own lives. We are not willing to offer to our patients an option that we feel that we might want to make use of ourselves.

More candour and more acknowledgment of uncertainty in this area is important. Many ethicists point to the Hippocratic tradition, arguing an ancient proscription against a physician's actions causing death. It is true that the Hippocratic oath says this, "I will neither give a deadly drug to anyone if asked for it nor will I make a suggestion to that effect." Very clear and unambiguous. This statement was especially important in Hippocrates' time as he distinguished his "scientific" creed of medicine from other more magical systems prevailing then. The historical context is very important. Physicians now often turn to Hippocrates to support something that they believe. Yet there are also many things in the Hippocratic oath that physicians reject these days. For example, there is the proscription against doing surgery, a proscription against abortion, even a proscription against taking "fees" for teaching medicine! We have chosen to leave these Hippocratic rules behind in the name of medical progress and social change. It is important not to undermine the fundamental importance of this life-and-death issue, but one should be clear that, just because a statement is in the Hippocratic oath, it is not automatically a rule that we always have to follow. We should look instead to the basis of this rule.

If one looks to the next line, in fact, one learns more about the underlying reasons. It says "In purity and holiness, I will guard my life and my art." This section does not say anything about patients and what patients need. It emphasizes guarding of professional integrity, as important to the public respect of the guild of practitioners, then as now. The ethical analysis now about why doctors should or should not assist in dying has to do with the same concern that, if doctors begin to allow themselves to actively assist patients to die, this will erode the integrity of the profession. We have to "guard our life and our art." I certainly have enormous respect and love for the profession and would not want its integrity to be eroded, but we should at least realize that this claim is not based on patient care values. It is a claim based on a very strict and ancient notion about the profession and about what its first principles ought to be. In some ways it is more concerned with the purity of the profession than it is with the needs of the patient. This need for purity and unwillingness to engage in complex or ambiguous situations could be seen as "excessive scrupulosity." Hippocrates' writing, especially the rule-based approach, is deeply rooted of course. But it is not always reflective of the real and lived values of our contemporary society, our individual relationships, or even our commitments to healing and solace.

As we look at what actually happens in the care of dying patients, we should consider the possibility that a fundamental construct of medical ethics actually obscures an important reality about the choices that face us. The construct I refer to is the notion of double effect – and the related importance of intention or expectation – that has been very useful in medical ethics. If one cares for a patient who has cancer that has spread to lung and bone and is suffering from terrible pain, morphine is the drug of choice. The only way to treat that patient's pain adequately is perhaps to hasten his or her death. Morphine may reduce respiratory drive especially if the lung function is already impaired, but, if this is the patient's preference, then it is morally accepted. The physician does not intend for the patient to die, but does intend to treat the pain. Hastening death is an unintended side-effect. In fact the double effect doctrine has been very useful in persuading physicians to treat pain adequately in the dying patient.

Yet this doctrine might have outlived its usefulness, or at least has had some unintended side-effects itself. It has created an illusion that we do not intend or expect death in such cases. Because of that, it has allowed us not to confront death, not to take part in it, not to understand the issue of dignity, and finally, not to obtain rewards that we might get from actually participating in helping a patient to die in a meaningful and dignified way. In the double effect doctrine, we must believe that we are not intending death, which can begin to seem like denial at best, or self-delusion at worst. We fool ourselves into thinking that death is an accident, or as Lewis Thomas would have said, "a failure." We were really just trying to treat the pain and it was a failure of medicine that the patient died in the meantime. The following clinical examples, which are not bizarre or unusual ones, demonstrate this problem.

### *Case 1*

Discontinuing ventilation in a patient who has end-stage lung disease is a decision that physicians have to make all the time. In a patient who is conscious, the primary symptom is not pain. Probably the most frightening symptom of terminal illness is suffocation – dyspnea, air hunger – and the best treatment for it is morphine. One of the reasons for morphine's effectiveness is that it suppresses the drive for air, so in fact the patient lives a shorter period of time because he or she is not struggling to stay alive. In that patient to whom we give morphine once we discontinue the ventilator, how is it possible to say we are not intending that person to die? What kind of an illusion is this? Do we really believe that the patient's demise is just an unfortunate side-effect of the treatment of the dyspnea? Using this construct reinforces our own denial and makes it possible for us to continue to practice medicine thinking that we are not dealing directly with death, and confronting this most challenging and sometimes personally threatening fact of human mortality.

## Case 2

Another example is the discontinuation of hydration and nutrition in a person in a persistent vegetative state. Consider the case of Nancy Cruzan and the thousands of people like her who did not want to be kept alive in a persistent vegetative state. The Supreme Court has supported the concept that nutrition and hydration are medical treatments that the patient, or family by proxy, can refuse. But there is no double effect here. When you discontinue nutrition and hydration in a patient in a continuous vegitative state, how is it possible that you are not intending that person to die? We usually rationalize it by saying we are respecting our patient's right to refuse treatment, rather than embrace a right to die.

## Case 3

Consider the next case from an ethics committee. A competent patient is dying of widespread swelling ovarian cancer. She had draining, foul smelling wounds; she could not move because of massive ascites (swelling), and became delirious and hallucinated with the morphine they were giving her. She hated it. She requested an overdose saying she was ready to die. The committee decided they would recommend sedation so that she would become unconscious, then allow her to die of dehydration. The question here too, is whether death was intended, and whether it matters morally. Whose needs are being met by this decision? Is it the patient's need or is it some need of the committee members to think that they somehow did not actively create that death. There are numerous instances of this sort when we make rather bizzare decisions and we need to ask ourselves are we doing it for our good or the patient's good.

## Helping patients to die

As a parable, perhaps a little extreme but illustrative nonetheless, I relate a story from anthropology. A pre-literate nomadic people called the Tiwi revered aged people until they physically could no longer keep up with the tribe, in which case it became impossible to keep them around. When the old person became frail or sick, the family performed a ritual called "covering her up." Her family in loving ceremony – while the rest of the tribe went on ahead – stayed behind with the elderly person so they could bury her in the ground up to her neck. Then they would leave. They did not kill her – nobody killed her. She was still alive the last time they saw her. Of course she was too weak to get out by herself and she would inevitably die. I was struck by that story and I wondered how many of end-of-life decisions are rather like that. They usually are not as cruel because the person is not conscious, but in the same way we reassure ourselves that we are not actively causing the patient's death. Yet our actions are just as directly related to that outcome as were the families of the Tiwi elders.

The strict ethical boundaries we have lived with for the last 20 years, which have helped us to deal with ethical problems at the end of life, may now be more of a barrier to effective care than they are moral guidelines that really help us along the way. Especially now physicians are so worried about the legal ramifications of active euthanasia and assisted suicide, they may back away from the dying patient, fearful that aggressive comfort care may be construed as hastening death. The more and more public this debate becomes, in fact, the less good we are at taking care of the symptoms of dying patients because of the worry about legal ramifications.

The "slippery slope" argument is a legitimate concern. Ethicists point out the social risks if we were to legalize physician-assisted suicide. What might happen, then, in a country such as the USA where not everyone has health insurance, many people have inadequate insurance, and there is a lot of discrimination against patients with certain kind of disease (for example, AIDS), and patients of certain socioeconomic and other groups. How would we prevent vulnerable people from being victimized by euthanasia? Another social risk that is raised is a concern that people might stop trusting their doctors, if they were worried that "You never know when they might kill you!" On a more fundamental basis, another concern is the possible loss of the reverence for life, which would be a tremendous loss.

There are serious concerns, but they are not inevitable consequences of a more open policy on physician-assisted dying. It is at least possible that, if we were openly to acknowledge that there are some circumstances when it is morally acceptable to assist a person to die, the social forces might work in the other direction. The abuse of this practice with "disvalued" people could be prevented by limiting it to competent patients. In fact, one wonders now that there is a much wider acceptance of foregoing life-sustaining treatment in the first place, whether such practice is in fact being abused already. Especially since we do not talk openly about it, we do not have any sort of accountability in this arena. We often make these decisions informally, and it may be that we are already discriminating against vulnerable populations. If we were to make the practice more open, more accountable, and have to defend it to one another, we might do a better job of preventing abuses. We cannot know which would occur without doing the experiment.

Ethicist Dan Brock argues that the motives are much more important in ethical analysis than the act itself, not whether or not we expect or intend the patient to die, but, much more significantly, whether the motive is benevolent and compassionate or not. If we could really stipulate this factor, it would be the best safeguard against abuses. What about trust of physicians? Perhaps it is true, as Leon Kass and other physicians argue, that we might erode the trust in the profession if we were actively to help patients to die. Nevertheless, it may be that physicians are already losing it, because patients think doctors do not understand the issue of personal dignity. They do not think they can get what they want from the physician at this most important time – at the end of life – when they are most needy. Janet Adkins might not have chosen to kill herself so early if she had trusted her doctors to do the right thing when her Alzheimer's disease became so advanced that she would not have wanted to live anymore. The reason she felt she needed to kill herself was that she did not trust her doctors to do what she wanted on her terms.

Finally, let me turn to reverence for life. It is not at all clear to me that by hiding the reality of death from ourselves we are in fact generating greater reverence for life. It may be that, if we allowed ourselves to participate in this most profound moment and to deal with our own mortality in a way that doctors do not do now, it might give a renewed appreciation for the meaning of life as well as for death. I do not believe that continuing to draw a bright line against physician-assisted suicide ensures or even encourages improvement in end-of-life care. I have not seen this improvement yet.

Alongside all the debates, the reality is that hastening of death is already happening. Some physicians are occasionally finding themselves in a situation where assistance in dying seems the right thing to do. Earlier I mentioned my experience in getting all those "confessional" letters from physicians. Quill's article led to many more letters sent to him that he turned into a book! He relates fascinating accounts of physicians struggling with this issue, many of whom have assisted patients in dying and would do so again.

We have to look at what happens in the real world. Examining the law in this area we find the same double standard. Modern courts seem to enact the same ambivalence that one sees within the medical profession. In every case in the USA when a physician has been indicted or has had to stand trial for causing the death of a patient, the physician has been acquitted. This is true even in cases that involved direct injection. An obvious interpretation of this phenomenon is that the jury understands the human dimensions of this issue. In every single case when the doctor knew the patient, when the patient was terminally ill, and when this was clearly being done in a merciful way, with what Brock would call the right motive or the right intentions, the people who make up the jury seem to say that the law is wrong. That what *this* doctor did in *this* case is not culpable. So we have these laws on the statute books that we do not really respect, but that make the physician nervous about even discussing this issue with a patient. Then we have to ask if respect for the law, too, is at stake here. Are we better off with a covert process, in which we rely on the courage of the physician who might be willing to commit an illegal act for a patient for whom they have a long-standing relationship and continue to maintain the social fiction that "We don't do this" because it is somehow better for the purity and holiness of the physicians' art? Would it not be better to make it a more open, strictly limited but more account-able practice – risking losing the intimacy of it perhaps, but gaining in disclosure and scrutiny of this most important subject?

## Suggestions for further reading

Aries, P. *Western Attitudes Toward Death: From the Middle Ages to the Present*. Translated by P. M. Ranum. Baltimore, MD: Johns Hopkins University Press, 1974.

Brock, D. W. Voluntary active euthanasia. *Hastings Center Report* (1992), **22**, 10–22.

Brody, H. Causing, intending, and assisting death. *Journal of Clinical Ethics* (1993), **4**, 112–17.

Callahan, D. *The Troubled Dream of Life: Living with Mortality*. New York: Simon & Schuster, 1993.

Cassel, C. K. and Meier, D. Sounding board: Morals and Moralism in the debate over euthanasia and assisted suicide. *New England Journal of Medicine* (1990), **323**, 750–2.

Foley, K. M. The relationship of pain and symptom management to patient requests for physician-assisted suicide. *Journal of Pain Symptom Management*, (1991), **6**, 289–97.

Gaylin, W., Kass, L. R., Pellegrino, E. D. and Siegler, M. Doctors must not kill. *Journal of the American Medical Association* (1988), **259**, 2139–40.

Quill, T. E., Cassel, C. K. and Meier, D. E. Care of the hopelessly ill: Proposed clinical criteria for physician-assisted suicide. *New England Journal of Medicine* (1992), **327**, 1380–4.

Quill, T. E. Death and dignity: A case of individualized decision making. *New England Journal of Medicine* (1991), **324**, 691–4.

Quill, T. E. *Death and Dignity: Making Choices and Taking Charge.* New York: W. W. Norton, 1993.

Various authors. Dying well? A colloquy on euthanasia and assisted suicide. *Hastings Center Report* (1992), **22**, 6–55.

# "I will give no deadly drug": Why doctors must not kill†

LEON R. KASS

EDITORS' SUMMARY

Leon R. Kass, MD, presents a case against physician euthanasia and physician-assisted suicide. He contends that there are two streams of arguments in favor of euthanasia, one from the standpoint of patient autonomy and the other from the standpoint of physician benevolence or love. The first he reduces to absurdity, arguing that the patient cannot control all aspects of professional ethics. Otherwise patients could request and receive anything they want from doctors. Second, physician benevolence does not adequately circumscribe duties to patients, for otherwise physicians could do many untoward things, like having sex with patients, if they were done out of "love" or "compassion." Further, because of special vulnerabilities of patients, there are limits the profession puts on itself regarding such things as sexual relations with patients and killing patients. Kass's argument hinges on a philosophy of medicine in which there are inherent moral standards in the profession. These standards protect the patient and society from the power that is conferred on physicians through the patient's illness and need for help.

† **Editor's note:** The following article was presented as the American College of Surgeons Lecture on Ethics and Philosophy, which was inaugurated on October 23, 1991, in Chicago, IL. A slightly different version of this article was previously published as "Neither For Love Nor Money: Why Doctors Must Not Kill," in *The Public Interest*, Winter, 1989.

Reproduced with the author's permission from *American College of Surgeons Bulletin*, (1992), 77(3), 6–17.

Is the profession of medicine ethically neutral? If so, whence shall we derive the moral norms or principles to govern its practices? If not, how are the norms of professional conduct related to the rest of what makes medicine a profession?

These difficult questions, now much discussed, are in fact very old, indeed as old as the beginnings of Western medicine. According to an ancient Greek myth, the goddess Athena procured two powerful drugs in the form of blood taken from the Gorgon Medusa, the blood drawn from her left side providing protection against death, that from her right side a deadly poison. According to one version of the myth, Athena gave Asklepios, the revered founder of medicine, vials of both drugs; according to the other version, she gave him only the life-preserving drug, reserving the power of destruction for herself. There is force in both accounts: the first attests to the moral neutrality of medical means, and of technical power generally; the second shows that wisdom would constitute medicine an unqualifiedly benevolent – that is, intrinsically ethical – art.

Today, we doubt that medicine is an intrinsically ethical activity, but we are quite certain that it can both help and harm. In fact, today, help and harm flow from the same vial. The same respirator that brings a man back from the edge of the grave also senselessly prolongs the life of an irreversibly comatose young woman. The same morphine that reverses the respiratory distress of pulmonary edema can, in higher doses, arrest respiration altogether. Whether they want to or not, doctors are able to kill – quickly, efficiently, surely. And what is more, it seems that they may soon be licensed and encouraged to do so.

In one year, in Holland, some 5000–10 000 patients were intentionally put to death by their physicians, while authorities charged with enforcing the law against homicide agreed not to enforce it. Not satisfied with such hypocrisy, and eager to immunize physicians against possible prosecution, American advocates of active euthanasia are seeking legislative changes in several states that would legalize so-called mercy killing by physicians. And American medicine shows increasing signs that it may be willing to participate. A few years ago the editor of the *Journal of the American Medical Association* published an outrageous (and perhaps fictitious) case of mercy killing, precisely to stir professional and public discussion of direct medical killing – perhaps, some then said, as a trial balloon.[1] Since then, we have had Dr Kevorkian's suicide-machine and Dr Quill's published account of suicide-assistance, the latter especially drawing considerable medical support. So-called active euthanasia practiced by physicians seems to be an idea whose time has come. But, in my view, it is a bad idea whose time must not come – not now, not ever. This Chapter is, in part, an effort to support this conclusion. But it is also an attempt to explore the ethical character of the medical profession, using the question of killing by doctors as a probe. Accordingly, I will be considering these interrelated questions: What are the norms that all physicians, *as physicians,* should agree to observe, whatever their personal opinions? What is the basis of such a medical ethic? What does it say – and what should we think – about doctors intentionally killing?

## Contemporary ethical approaches

The question about physicians killing is a special case of this general question: may or ought one to kill people who ask to be killed? Among those who answer this general question in the affirmative, two reasons are usually given. First is the reason of *freedom* or *autonomy*. Each person has a right to control his or her body and his or her life, including the end of it. On this view, physicians (or others) are bound to acquiesce in demands not only for termination of treatment but also for intentional killing through poison, because the right to choose – freedom – must be respected, even more than life itself, and even when the physician would never recommend or concur in the choices made. Physicians, as keepers of the vials of life and death, are morally bound actively to dispatch the embodied person, out of deference to autonomous personal choice.

The second reason for killing the patient who asks for death has little to do with choice. Instead, death is to be directly and swiftly given because the patient's life is deemed no longer worth living, according to some substantive or "objective" measure. Unusually great pain or a terminal condition or an irreversible coma or advanced senility or extreme degradation is the disqualifying quality of life that pleads – choice or no choice – for merciful termination. It is not his autonomy but rather the miserable and pitiable condition of his body or mind that justifies doing the patient in. Absent such substantial degradations, requests for assisted death would not be honored. Here the body itself offends and must be plucked out, from compassion or mercy, to be sure. Not the autonomous will of the patient, but the doctor's benevolent and compassionate love for suffering humanity justifies the humane act of mercy killing.

As I have indicated, these two reasons advanced to justify the killing of patients correspond to the two approaches to medical ethics most prominent in the field today: the school of autonomy, and the school of general benevolence and compassion (or love). Despite their differences, they are united in their opposition to the belief that medicine is intrinsically a moral profession, with *its own* imminent principles and standards of conduct that set limits on what physicians may properly do. Each seeks to remedy the ethical defect of a profession seen to be in itself *a*moral, technically competent but morally neutral.

For the first ethical school, morally neutral technique is morally used only when it is used according to the wishes of the patient as client or consumer. The model of the doctor–patient relationship is one of contract: the physician – a highly competent hired syringe, as it were – sells his or her services on demand, restrained only by the law. Here's the deal: for the patient, autonomy and service; for the doctor, money, graced by the pleasure of giving the patient what he wants. If a patient wants to fix her nose or change his gender, determine the sex of unborn children, or take euphoriant drugs just for kicks, the physician can and will go to work – provided that the price is right.†

For the second ethical school, morally neutral technique is morally used only when it is used under the guidance of general benevolence or loving charity. Not the will of the patient, but the humane and compassionate motive of the physician – not as physician but as human being – makes the doctor's actions ethical. Here, too, there can be strange requests and stranger deeds, but if they are done from love, nothing can be wrong – again, provided the law is silent. All acts – including killing the patient – done lovingly are licit, even praiseworthy. Good and humane intentions can sanctify any deed.

In my opinion, each of these approaches should be rejected as a basis for medical ethics. For one thing, neither can make sense of some specific duties and restraints long thought absolutely inviolate under the traditional medical ethic – for example, the proscription against having sex with patients. Must we now say that sex with patients is permissible if the patient wants it and the price is right, or, alternatively, if the doctor is gentle and loving and has a good bedside manner? Or do we glimpse in this absolute prohibition a deeper understanding of the medical vocation, which the prohibition both embodies and protects? Indeed, as I will now try to show, using the taboo against doctors killing patients, the medical profession has its own intrinsic ethic, which a physician true to his calling will not violate, either for love or for money.

### Profession: Intrinsically ethical

Let me propose a different way of thinking about medicine as a profession. Consider medicine not as a mixed marriage between its own value-neutral technique and some extrinsic moral principles, but as an inherently ethical activity, in which technique and conduct are both ordered in relation to an overarching good, the naturally given end of health. This once traditional view of medicine I have defended at length.[2] Here, I will present the conclusions without the arguments. It will suffice, for present purposes, if I can render this view plausible.

A profession, as etymology suggests, is an activity or occupation to which its practitioner publicly professes – that is, confesses – his devotion. Learning may,

---

† Of course, any physician with personal scruples against one or another of these practices may "write" the relevant exclusions into the service contract he offers his customers.

of course, be required of, and prestige may, of course, be granted to, the professional, but it is the profession's goal that calls, that learning serves, and that prestige honors. Each of the ways of life to which the various professionals profess their devotion must be a way of life worthy of such devotion – and so they all are. The teacher devotes himself to assisting the learning of the young, looking up to truth and wisdom; the lawyer (or the judge) devotes himself to rectifying injustice for his client (or for the parties before the court), looking up to what is lawful and right; the clergyman devotes himself to tending the souls of his parishioners, looking up to the sacred and the divine; and the physician devotes himself to healing the sick, looking up to health and wholeness.

Being a professional is thus more than being a technician. It is rooted in our moral nature; it is a matter not only of the mind and hand but also of the heart, not only of intellect and skill but also of character. For it is only as a being willing and able to devote himself to others and to serve some high good that a person makes a public profession of his way of life.

The good to which the medical profession is devoted is health, a naturally given although precarious standard or norm, characterized by "wholeness" and "well-working," toward which the living body moves on its own. Even the modern physician, despite his great technological prowess, is but an assistant to natural powers of self-healing. But health, though a goal tacitly sought and explicitly desired, is difficult to attain and preserve. It can be ours only provisionally and temporarily, for we are finite and frail. Medicine thus finds itself in between: the physician is called to serve the high and universal goal of health while also ministering to the needs and relieving the sufferings of the frail and particular patient. Moreover, the physician must respond not only to illness but also to its meaning for each individual, who, in addition to his symptoms, may suffer from self-concern – and often fear and shame – about weakness and vulnerability, neediness and dependence, loss of self-esteem, and the fragility of all that matters to him. Thus, the inner meaning of the art of medicine is derived from the pursuit of health and the care for the ill and suffering, guided by the self-conscious awareness, shared (even if only tacitly) by physician and patient alike, of the delicate and dialectical tension between wholeness and necessary decay.

When the activity of healing the sick is thus understood, we can discern certain virtues requisite for practicing medicine – among them, moderation and self-restraint, gravity, patience, sympathy, discretion, and prudence. We can also discern specific positive duties, addressed mainly to the patient's vulnerability and self-concern – including the demands for truthfulness, patient instruction, and encouragement. And, arguably, we can infer the importance of certain negative duties, formulable as absolute and unexceptionable rules. Among these, I submit, is this rule: *Doctors must not kill*. The rest of this article attempts to defend this rule and to show its relation to the medical ethic, itself understood as growing out of the inner meaning of the medical vocation.

I confine my discussion solely to the question of direct, intentional killing of patients *by physicians* – so-called mercy killing. Though I confess myself opposed to such killing even by nonphysicians,[3] I am not arguing here against euthanasia per se. More importantly, I am not arguing against the cessation of medical treatment when such treatment merely prolongs painful or degraded dying, nor do I oppose the use of certain measures to relieve suffering that have, as an unavoidable consequence, an increased risk of death. Doctors may and must allow to die, even if they must not intentionally kill.

## Bad consequences

Although the bulk of my argument will turn on my understanding of the special meaning of professing the art of healing, I begin with a more familiar mode of ethical analysis: assessing needs and benefits versus dangers and harms. Still the best discussion of this topic is a now-classic essay by Yale Kamisar, written more than 35 years ago.[4] Kamisar makes vivid the difficulties in assuring that the choice for death will be freely made and adequately informed, the problems of physician error and abuse, the troubles for human relationships within families and between doctors and patients, the difficulty of preserving the boundary between voluntary and involuntary euthanasia, and the risks to the whole social order from weakening the absolute prohibition against taking innocent life. These considerations are, in my view, alone sufficient to rebut any attempt to weaken the taboo against medical killing; their relative importance for determining public policy far exceeds their relative importance in this essay. But here they serve also to point us to more profound reasons why doctors must not kill.

There is no question that fortune deals many people a very bad hand, not least at the end of life. All of us, I am sure, know or have known individuals whose last weeks, months, or even years were racked with pain and discomfort, degraded by dependency or loss of self-control, or who lived in such reduced humanity that it cast a deep shadow over their entire lives, especially as remembered by the survivors. All who love them would wish to spare them such an end, and there is no doubt that an earlier death could do it. Against such a clear benefit, attested to by many a poignant and heartrending true story, it is difficult to argue, especially when the arguments are necessarily general and seemingly abstract. Still, in, the aggregate, the adverse consequences – including real suffering – of being governed solely by mercy and compassion may far outweigh the aggregate benefits of relieving agonal or terminal distress.

The first difficulty emerges when we try to gauge the so-called need or demand for medically assisted killing. This question, to be sure, is in part empirical. But evidence can be gathered only if the relevant categories of "euthanizable" people are clearly defined. Such definition is notoriously hard to accomplish – and it is not always honestly attempted. On careful inspection, we discover that if the category is precisely defined, the need for mercy killing seems greatly exaggerated, and if the category is loosely defined, the poisoners will be working overtime.

The category always mentioned first to justify mercy killing is the group of persons suffering from incurable and fatal illnesses, with intractable pain and with little time left to live but still fully aware, who freely request a release from their distress – for example, people rapidly dying from disseminated cancer with bony metastases, unresponsive to chemotherapy. But as experts in pain control tell us, the number of such people with truly untreatable pain is in fact rather low. Adequate analgesia is apparently possible in the vast majority of cases, provided that the physician and patient are willing to use strong enough medicines in adequate doses and with proper timing.†

But, it will be pointed out, full analgesia induces drowsiness and blunts or distorts awareness. How can that be a desired outcome of treatment? Fair enough. But then the rationale for requesting death begins to shift from relieving experienced suffering to ending a life no longer valued by its bearer or, let us be frank, by the onlookers. If this becomes a sufficient basis to warrant mercy killing, the category of euthanizable people cannot be limited to individuals with incurable or fatal painful illnesses with little time to live. Now persons in all sorts of greatly reduced conditions – from persistent vegetative state to quadriplegia, from severe depression to the condition that now most horrifies, Alzheimer's disease – might have equal claim to have their suffering mercifully halted. The trouble, of course, is that most of these people can no longer request for themselves the dose of poison. Moreover, it will be difficult – if not impossible – to develop the requisite calculus of degradation or to define the threshold necessary for ending life.

In view of the obvious difficulty in describing precisely and "objectively" what categories and degrees of pain, suffering, or bodily or mental impairment could justify mercy killing, advocates repair (at least for the time being) to the principle of volition: the request for assistance in death is to be honored because it is freely made by the one whose life it is, and who, for one reason or another, cannot commit suicide alone. But this too is fraught with difficulty: how free or informed is a choice made under debilitated conditions? Can consent long in advance be sufficiently informed about all the particular circumstances that it is meant pro-

---

† The inexplicable failure of many physicians to provide the proper and available relief of pain is surely part of the reason why some people now insist that physicians (instead) should give them death.

spectively to cover? And, in any case, are not such choices easily and subtly manipulated, especially in the vulnerable? Kamisar is very perceptive on this subject:

> Is this the kind of choice . . . that we want to offer a gravely ill person? Will we not sweep up, in the process, some who are not really tired of life, but think others are tired of them; some who do not really want to die, but who feel that they should not live on, because to do so when there looms the legal alternative of euthanasia is to do a selfish or cowardly act? Will not some feel an obligation to have themselves "eliminated" in order that funds allocated for their terminal care might be better used by their families or, financial worries aside, in order to relieve their families of the emotional strain involved?[4]

Even were these problems soluble, the insistence on voluntariness as the justifying principle cannot be sustained. The enactment of a law – like the proposed Washington initiative – legalizing mercy killing on voluntary request will certainly be challenged in the courts under the equal protection clause of the Fourteenth Amendment. The law, after all, will not legalize assistance to suicides in general, but only mercy killing. The change will almost certainly occur not as an exception to the criminal law proscribing *homicide*, but as a new "treatment option," as part of a right to "a humane and dignified death." Why, it will be argued, should the comatose or the demented be denied such a right or such a "treatment" just because they cannot claim it for themselves? This line of reasoning has already been the route by which substituted judgment and proxy consent have been allowed by the courts in termination of treatment cases. When proxies give their consent, they will do so on the basis not of autonomy but of a substantive judgment – namely, that for these or those reasons, the life in question is not worth living.

Precisely because most of the cases that are candidates for mercy killing are of this sort, the line between voluntary and involuntary euthanasia cannot hold, and will be effaced by the intermediate case of the mentally impaired or comatose who are declared no longer willing to live because someone else wills that result for them. In fact, the more honest advocates of euthanasia openly admit that it is these non-voluntary cases which they especially hope to dispatch, and that their plea for *voluntary* euthanasia is just a first step. It is easy to see the trains of abuses that are likely to follow the most innocent cases, especially because the innocent cases cannot be precisely and neatly defined away from the rest.†

---

† This is no mere scare-mongering. Recent reports on the practice of euthanasia in Holland provide ample proof. In a study of 300 physicians, published in 1989, Professor F. C. B.

Abuses aside, legalized mercy killing by doctors will almost certainly damage the doctor–patient relationship. The patient's trust in the doctor's whole-hearted devotion to the patient's best interests will be hard to sustain once doctors are licensed to kill. Imagine the scene: you are old, poor, in failing health, and alone in the world; you are brought to the city hospital with fractured ribs and pneumonia. The nurse or intern enters late at night with a syringe full of yellow stuff for your intravenous drip. How soundly will you sleep? It will not matter that your doctor has never yet put anyone to death; that he is legally entitled to do so will make a world of difference.

And it will make a world of psychic difference too for conscientious physicians. How easily will they be able to care whole-heartedly for patients when it is always possible to think of killing them as a "therapeutic option"? Shall it be penicillin and a respirator one more time, or, perhaps, this time just an overdose of morphine? Physicians get tired of treating patients who are hard to cure, who resist their best efforts, who are on their way down – "gorks," "gomers," and "vegetables" are only some of the less than affectionate names they receive from the house officers. Won't it be tempting to think that death is the best "treatment" for the little old lady "dumped" again on the emergency room by the nearby nursing home?

Even the most humane and conscientious physician psychologically needs protection against himself and weaknesses, if he is to care fully for those who entrust themselves to him. A physician-friend who worked many years in a hospice caring for dying patients explained it to me most convincingly: "Only because I knew that I could not and would not kill my patients was I able to enter most fully and intimately into caring for them as they lay dying." The psychological burden of the license to kill (not to speak of the brutalization of the physician-killers) could very well be an intolerably high price to pay for physician-assisted euthanasia.

The point, however, is not merely psychological: it is also moral and essential. My friend's horror at the thought that he might be tempted to kill his patients,

van Wijmen (a supporter of euthanasia) found that more than 40% admitted to having performed *involuntary* euthanasia at least once, and more than 10% five times or more.

The Dutch Government's Report (September 1991), despite its reassuring conclusions, provides even more alarming data: in addition to 2300 cases (per year) of voluntary euthanasia and 400 cases of physician-assisted suicide, there were over 1000 cases of active *involuntary* euthanasia, without the patients' knowledge or consent, including over 100 cases in which the patients were mentally competent. In addition, of 8100 cases in which overdoses of morphine were given with the intent to terminate life, 61% were without patient knowledge or consent. "Low quality of life," "the family couldn't take it any more," and "little hope of improvement" were reasons that physicians gave for killing patients without request.

Finally, an excellent book, *Regulating Death: Euthanasia and the Case of the Netherlands*, by Carlos Gomez, MD, shows how the Dutch practice departs extensively from the guidelines set down by the Dutch Medical Society, and makes clear why the practice of euthanasia is, necessarily, virtually un-regulable.

were he not enjoined from doing so, embodies a deep understanding of the medical ethic and its intrinsic limits. We move from assessing consequences to looking at medicine itself.

## Medicine's outer limits

Every activity can be distinguished, more or less easily, from other activities. Sometimes the boundaries are indistinct: it is not always easy, especially today, to distinguish some music from noise or some art from smut or some teaching from indoctrination. Medicine and healing are no different: it is sometimes hard to determine the boundaries, both with regard to ends and means. Is all cosmetic surgery healing? Are placebos – or food and water – drugs?

There is, of course, a temptation to finesse these questions of definition or to deny the existence of boundaries altogether: medicine *is* whatever doctors *do*, and doctors do whatever doctors can. Technique and power alone define the art. Put this way, we see the need for limits: technique and power are ethically neutral, notoriously so, usable for both good and ill. The need for finding or setting limits to the use of powers is especially important when the powers are dangerous: it matters more that we know the proper limits on the use of medical power – or military power – than, say, the proper limits on the use of paint brush or violin.

The beginning of ethics regarding the use of power generally lies in nay-saying. The wise setting of limits on the use of power is based on discerning the excesses to which the power, unrestrained, is prone. Applied to the professions, this principle would establish strict outer boundaries – indeed, inviolable taboos – against those "occupational hazards" to which each profession is especially prone. Within these outer limits, no fixed rules of conduct apply; instead, prudence – the wise judgment of the man-on-the-spot – finds and adopts the best course of action in the light of the circumstances. But the outer limits themselves are fixed, firm, and nonnegotiable.

What are those limits for medicine? At least three are set forth in the venerable Hippocratic Oath: no breach of confidentiality, no sexual relations with patients, and no dispensing of deadly drugs.† These unqualified, self-imposed restrictions are readily understood in terms of the temptations to which the physician is most vulnerable, temptations in each case regarding an area of vulnerability and exposure that the practice of medicine requires of patients. Patients necessarily divulge and reveal private and intimate details of their personal lives; patients necessarily expose their naked bodies to the physician's objectifying gaze and investigating

---

† See, in Reference no. 2, my essay on the Hippocratic Oath, especially pages 232–40. See also the chapter, "Professing ethically: The place of ethics in defining medicine," especially pages 217–23.

hands; patients necessarily expose and entrust the care of their very lives to the physician's skill, technique, and judgment. The exposure is, in all cases, one-sided and asymmetric: the doctor does not reveal his intimacies, display his nakedness, offer up his embodied life to the patient. Mindful of the meaning of such non-mutual exposure, the physician voluntarily sets limits on his own conduct, pledging not to take advantage of or to violate the patient's intimacies, naked sexuality, or life itself.

The prohibition against killing patients, the first negative promise of self-restraint sworn to in the Hippocratic Oath, stands as medicine's first and most-abiding taboo: "I will neither give a deadly drug to anybody if asked for it, nor will I make a suggestion to this effect . . . In purity and holiness I will guard my life and my art." In forswearing the giving of poison, the physician recognizes and restrains a god-like power he wields over patients, mindful that his drugs can both cure and kill. But in forswearing the giving of poison, *when asked for it*, the Hippocratic physician rejects the view that the patient's choice for death can make killing him, *or assisting his suicide*, right. For the physician, at least, human life in living bodies commands respect and reverence, by its very nature. As its respectability does not depend upon human agreement or patient consent, revocation of one's consent to live does not deprive one's living body of respectability. The deepest ethical principle restraining the physician's power is not the autonomy or freedom of the patient; neither is it his own compassion or good intention. Rather, it is the dignity and mysterious power of human life itself, and, therefore, also what the Oath calls the purity and holiness of the life and art to which he has sworn devotion. A person can choose to be a physician, but he can not simply choose what physicianship means.

## The central core

The central meaning of physicianship derives not from medicine's powers but from its goal, not from its means but from its end: to benefit the sick by the activity of healing. The physician as physician serves only the sick. He does not serve the relatives or the hospital or the national debt inflated due to Medicare costs. Thus, he will never sacrifice the well-being of the sick to the convenience or pocketbook or feelings of the relatives or society. Moreover, the physician serves the sick not because they have rights or wants or claims, but because they are sick. The healer works with and for those who need to be healed, in order to help make them whole.

Despite enormous changes in medical technique and institutional practice, despite enormous changes in nosology and therapeutics, the center of medicine has not changed: it is as true today as it was in the days of Hippocrates that the ill desire to be whole; that wholeness means a certain well-working of the enlivened body and its unimpaired powers to sense, think, feel, desire, move, and maintain itself; and that the relationship between the healer and the ill is constituted, essentially even if only tacitly, around the desire of both to promote the wholeness of the one who is ailing.

The wholeness and well-working of a human being is, of course, a rather complicated matter, much more so than for our animal friends and relations. Health and fitness seem to mean different things to different people, or even to the same person at different times of life. Yet not everything is relative and contextual; beneath the variable and cultural lies the constant and organic, the well-regulated, properly balanced, and fully empowered human being. Indeed, only the existence of this natural and universal subject makes possible the study of medicine.

But human wholeness goes beyond the kind of somatic wholeness abstractly and reductively studied by the modern medical sciences. Whether or not doctors are sufficiently prepared by their training to recognize it, those who seek medical help in search of wholeness are not to *themselves* just bodies or organic machines. Each person intuitively knows himself to be a center of thoughts and desires, deeds and speeches, loves and hates, pleasures and pains, but a center whose workings are none other than the workings of his enlivened and mindful body. The patient presents himself to the physician, tacitly to be sure, as a psychophysical unity, as a *one*, not just a body, but also not just as a separate disembodied person that simply *has* or *owns* a body. The person and the body are self-identical. True, sickness may be experienced largely as belonging to the body as something other; but the healing one wants is the wholeness of one's entire embodied being. Not the wholeness of *soma*, not the wholeness of *psyche*, but the wholeness of *anthropos* as a (puzzling) concretion of *soma-psyche* is the benefit sought by the sick. This human wholeness is what medicine is finally all about.

Can wholeness and healing, thus understood, ever be compatible with intentionally killing the patient? Can one benefit the patient as a whole by making him dead? There is, of course, a logical difficulty: how can any good exist for a being that is not? "Better off dead" is logical nonsense – unless, of course, death is not death indeed but instead a gateway to a new and better life beyond. But the error is more than logical: in fact, to intend and to act for someone's good requires their continued existence to receive the benefit.

To be sure, certain attempts to benefit may in fact turn out, unintentionally, to be lethal. Giving adequate morphine to control pain might induce respiratory depression, leading to death. But the intent to relieve the pain of the living presupposes that the living still live to be relieved. This must be the starting point in discussing all medical benefits: no benefit without a beneficiary.

Against this view, someone will surely bring forth the hard cases: patients so ill-served by their bodies that they can no longer bear to live, bodies riddled with cancer and racked with pain, against which their "owners" protest in horror and from which they insist on being released. Cannot the person "in the body" speak up against the rest, and request death for "personal" reasons?

However sympathetically we listen to such requests, we must see them as incoherent. Such person–body dualism cannot be sustained. "Personhood" is manifest on earth only in living bodies; our highest mental functions are held up by, and are inseparable from, lowly metabolism, respiration, circulation, excretion. There may be blood without consciousness, but there is never consciousness without blood. Thus, one who calls for death in the service of personhood is like a tree seeking to cut its roots for the sake of growing its highest fruit. No physician, devoted to the benefit of the sick, can serve the patient as person by denying and thwarting his personal embodiment. The boundary condition, "no deadly drugs," flows directly from the center, "make whole."

To say it plainly, to bring nothingness is incompatible with serving wholeness: one cannot heal, or comfort, by making nil. The healer cannot annihilate if he is truly to be a healer. The physician–euthanizer is a deadly self-contradiction.

## When medicine fails

But we must acknowledge a difficulty. The central goal of medicine – health – is, in each case, a perishable good: inevitably, patients get irreversibly sick, patients degenerate, patients die. Healing the sick is in principle a project that must at some point fail. And here is where all the trouble begins: How does one deal with "medical failure?" What does one seek when restoration of wholeness – or "much" wholeness – is by and large out of the question?

Contrary to the propaganda of the euthanasia movement, there is, in fact, much that can be done. Indeed, by recognizing finitude yet knowing that we will not kill, we are empowered to focus on easing and enhancing the lives of those who are dying. First of all, medicine can follow the lead of the hospice movement and, abandoning decades of shameful mismanagement, provide truly adequate (and now technically feasible) relief of pain and discomfort. Second, physicians (and patients and families) can continue to learn how to withhold or withdraw those technical interventions that are, in truth, merely burdensome or degrading medical additions to the unhappy end of a life – including, frequently, hospitalization itself. Ceasing treating and allowing death to occur when (and if) it will seem to be quite compatible with the respect life itself commands for itself. For life can be revered not only in its preservation, but also in the manner in which we allow a given life to reach its terminus. To repeat: doctors may and must allow to die, even if they must not intentionally kill.

Ceasing medical intervention, allowing nature to take its course, differs fundamentally from mercy killing. For one thing, death does not necessarily follow the discontinuance of treatment. Karen Ann Quinlan lived roughly 10 years after the court allowed the "life-sustaining" respirator to be removed. Not the physician, but the underlying fatal illness becomes the true cause of death. More important morally, in ceasing treatment the physician need not *intend* the death of the patient, even when the death follows as a result of his omission. His intention should be to avoid useless and degrading medical *additions* to the already sad end of a life. In contrast, in active, direct mercy killing the physician must, necessarily and indubitably, intend *primarily* that the patient be made dead. And he must knowingly and indubitably cast himself in the role of the agent of death. This remains true even if he is merely an assistant in suicide. A physician who provides the pills or lets the patient plunge the syringe after he leaves the room is morally no different from one who does the deed himself. "I will neither give a deadly drug to anybody if asked for it, nor will I make a suggestion to this effect."

## Being humane and being human

Once we refuse the technical fix, physicians and the rest of us can also rise to the occasion: we can learn to act humanly in the presence of finitude. Far more than adequate morphine and the removal of burdensome chemotherapy, the dying need our presence and our encouragement. Dying people are all too easily reduced ahead of time to "thinghood" by those who cannot bear to deal with the suffering or disability of those they love. Withdrawal of contact, affection, and care is the greatest single cause of the dehumanization of dying. Not the alleged humaneness of an elixir of death, but the humanness of connected living-while-dying is what medicine, and the rest of us, most owes the dying. The treatment of choice is company and care.

The euthanasia movement would have us believe that the physician's refusal to assist in suicide or perform euthanasia constitutes an affront to human dignity. Yet one of their favorite arguments seems to me rather to prove the reverse. Why, it is argued, do we put animals out of their misery but insist on compelling fellow human beings to suffer to the bitter end? Why, if it is not a contradiction for the veterinarian, does the medical ethic absolutely rule out mercy killing? Is this not simply inhumane?

Perhaps inhumane, but not thereby inhuman. On the contrary, it is precisely because animals are not human that we must treat them (merely) humanely. We put dumb animals to sleep because they do not know that they are dying, because they can make nothing of their misery or mortality, and, therefore, because they cannot live deliberately (humanly) in the face of their own suffering or dying. They cannot live out a fitting end. Compassion for their weakness and dumbness is our only appropriate emotion, and given our responsibility for their care and well-being, we do the only humane thing we can.

But when a conscious human being asks us for death, by that very action he displays the presence of something that precludes our regarding him as a dumb animal. Humanity is owed humanity, not humaneness. Humanity is owed the bolstering of the human, even or especially in its dying moments, in resistance to the temptation to ignore its presence in the sight of suffering.

What humanity needs most in the face of evils is courage, the ability to stand against fear and pain and thoughts of nothingness. The deaths we most admire are those of people who, knowing that they are dying, face the fact frontally and act accordingly: they set their affairs in order, they arrange what could be final meetings with their loved ones, and yet, with strength of soul and a small reservoir of hope, they continue to live and work and love as much as they can for as long as they can. Because such conclusions of life require courage, they call for our encouragement, and for the many small speeches and deeds that shore up the human spirit against despair and defeat.

Many doctors are, in fact, rather poor at this sort of encouragement. They tend to regard every dying or incurable patient as a failure, as if an earlier diagnosis or a more vigorous intervention might have avoided what is, in truth, an inevitable collapse. The enormous successes of medicine these past 50 years have made both doctors and laymen less prepared than ever to accept the fact of finitude. Physicians today are not likely to be agents of encouragement once their technique begins to fail.

It is, of course, partly for these reasons that doctors will be pressed to kill – and many of them will, alas, be willing. Having adopted a largely technical approach to healing, having medicalized so much of the end of life, doctors are being asked, often with thinly veiled anger, to provide a technical final solution for the evil of human finitude and for their own technical failure: if you cannot cure me, kill me. The last gasp of autonomy or cry for dignity is asserted against a medicalization and institutionalization of the end of life that robs the old and the incurable of most of their autonomy and dignity: intubated and electrified, with bizarre mechanical companions, once proud and independent people find themselves cast in the roles of passive, obedient, highly disciplined children. People who care for autonomy and dignity should try to reverse this dehumanization of the last stages of life, instead of giving dehumanization its final triumph by welcoming the desperate goodbye-to-all-that contained in one final plea for poison.

The present crisis that leads some to press for active euthanasia is really an opportunity to learn the limits of the medicalization of life and death and to recover an appreciation of living with and against mortality. It is an opportunity for physicians to recover an understanding that there remains a residual human wholeness, however precarious, that can be cared for even in the face of incurable and terminal illness. Should doctors cave in, should doctors become technical dispensers of death, they will not only be abandoning their posts, their patients, and their duty to care; they will set the worst sort of example for the community at large – teaching technicism and so-called humaneness where encouragement and humanity are both required and sorely lacking. On the other hand, should physicians hold fast, should doctors learn that finitude is no disgrace and that human wholeness can be cared for to the very end, medicine may serve not only the good of its patients, but also, by example, the failing moral health of modern times.

## References

1. "It's over, Debbie." *JAMA*, 259:272, 1988. In response, see also Gaylin W, Kass LR, Pellegrino ED, and Siegler M: Doctors must not kill. *JAMA* 259: 2139–2140, 1988.

2. Kass LR: *Toward a More Natural Science: Biology and Human Affairs*. New York: The Free Press, 1985; paperback, 1988. (See, especially, chapters 6–9.)

3. Kass LR: Death with dignity and the sanctity of life. *Commentary*, March, 1990.

4. Kamisar Y: Some nonreligious views against proposed "mercy killing" legislation. *Minn. Law Rev.*, 42:969–1042, 1958. [Reprinted, with a new preface by Kamisar, in "The slide toward mercy killing," *Child and Family Reprint Booklet series*, 1987.]

# Voluntary euthanasia and other medical end-of-life decisions: Doctors should be permitted to give death a helping hand

HELGA KUHSE

EDITORS' SUMMARY

Helga Kuhse, PhD, argues that voluntary euthanasia and physician-assisted suicide are moral and ought to be offered by an enlightened public policy. She shows how, in terms of the act, there is no moral difference between killing and letting die. She further argues that intention might differ, but no one can measure or assess the mental states of persons. Intention cannot be regulated. Instead she proposes public criteria concerning consent, which can be regulated. No harms can be foreseen, in her view, either to doctors or to patients, from permitting a small number, who so choose, to die slightly earlier than they would normally, with a relief of pain and suffering.

## The ubiquity of medical end-of-life decisions

Death is no longer the natural event it once was. It is very often the result of a medical decision. A recent large-scale Dutch study suggests that some 38% of all non-acute deaths are the result of medical end-of-life decisions. These end-of-life decisions may involve the withholding or withdrawing of life-sustaining treatment, the administration of potentially life-shortening palliative care, and euthanasia or medically assisted suicide. This means that the contemporary question is not so much *whether* physicians ought to be allowed to give death a helping hand, but rather *when* and *why* and *under what circumstances* they should do so.

Many people believe that it is sometimes morally proper for physicians to withhold or withdraw life-sustaining treatment from their patients, or to administer potentially life-shortening palliative care, but that it is always morally wrong to practice voluntary euthanasia, that is to deliberately end the patient's life, at the patient's request. I call this view "the conventional view" and argue that there are good reasons why we should reject it.

## Two cases and the conventional view

In its 1980 *Declaration on Euthanasia*, the Roman Catholic Church reconfirmed its strong opposition to euthanasia of any kind:

> Nothing and no one can in any way permit the killing of an innocent
> human being, whether a foetus or an embryo, an infant or an adult,
> an old person or one suffering from an incurable disease, or a person
> who is dying.

Despite its firm opposition to euthanasia, the *Declaration* does, however, sanction the administration of potentially life-shortening palliative care and the discontinuation of burdensome treatment.

A similar view underpins the 1987 Policy Statement of the World Medical Association, which reads:

> Euthanasia, that is the act of deliberately ending the life of a patient
> . . . is unethical. This does not prevent the physician from respecting
> the desire of a patient to allow the natural process of death to follow
> its course in the terminal phase of sickness.

These two statements capture the "conventional view." They hold that the intentional or deliberate termination of an innocent human being's life is always morally wrong, but at the same time express the belief that "allowing to die" is sometimes morally proper.

But when does a doctor practice euthanasia, that is when does he or she intentionally or deliberately terminate a patient's life, and when does he or she merely allow a patient to die? Let us consider two cases, the cases of Mrs N and Carla. (Both these cases are recounted in the book *Willing to Listen – Wanting to Die*, see Suggestions for further reading, p. 258. The first case is described by Nicholas Tonti-Filippini; the second by Pieter Admiraal.)

### Mrs N

Mrs N was 45 years old, married and the mother of two teenage children. She was suffering from motor neurone disease, a progressive and terminal condition that eventually renders the patient totally unable to move, to speak, or to breathe on her own.

When Mrs N was admitted to the hospital, she had breathing difficulties and was attached to a respirator. Her condition gradually worsened and hopes of her being able to leave the hospital, even with ventilatory assistance, diminished. She could still communicate with some difficulty, and the nurses reported that Mrs N frequently expressed the wish to have the ventilator withdrawn, even though she knew that this would lead to her death.

The hospital's ethics consultant and a psychiatrist were asked to assess the circumstances of the case. The ethics consultant came to the conclusion that Mrs N's wish to have the treatment withdrawn was not unreasonable under the circumstances, and that her decisions should be respected; and the psychiatrist found that Mrs N was competent to make that decision.

Six days after Mrs N's doctor had told her that he was prepared to discontinue treatment, the ventilator was turned off. To ensure that Mrs N's death from respiratory failure would be as comfortable as possible, the ventilator was not turned off at once, but was gradually turned down, while the oxygen level was increased. Mrs N remained conscious for six hours and, in the seventh hour, died from the effects of carbon dioxide retention, with her husband by her side.

According to the conventional view, Mrs N had been allowed to die. Even though her physician had performed an action – turning off her respirator – he had not, apparently, infringed the rule against the intentional termination of life. According to the conventional view, he had performed a morally good or licit act.

Now contrast the case of Mrs N with that of Carla.

### Carla

Carla was 47 years old. She was married and the mother of four teenage children.

She was suffering from ovarian cancer, which had spread. When her pain could no longer be controlled at home, she was admitted to the local hospital.

She was seen by the palliative care specialist, who told her that he would almost certainly be able to relieve her pain. This proved correct. A needle placed under her skin delivered a continuous infusion of morphine, and after two days Carla's pain was well controlled.

The palliative care specialist visited Carla frequently. During one of those visits and in the presence of her husband, Carla raised the question of euthanasia: would she be helped to die when the stage had been reached where she found her condition unbearable? The palliative care specialist explained to Carla that doctors at the hospital were always willing to discuss the question of euthanasia with terminally ill patients, but that the final decision had to be agreed to by a health care team consisting of two doctors, a nurse, and a minister of religion.

Carla's condition deteriorated quickly. She was vomiting constantly, lost a lot of weight and became extremely weak. At the same time, the tumour continued to grow and was soon obstructing the blood flow in her legs, causing a very painful swelling.

Throughout her illness, Carla remained completely alert. She was a courageous woman and sustained by a strong Catholic faith. Four days before her death, a Roman Catholic Priest administered the Last Sacraments. Soon after, and not unexpectedly, she raised the question of euthanasia.

A family meeting was called and by now all members of the family could accept Carla's decision. Similarly the nurses, who had cared for Carla on a daily basis: they knew her circumstances well and respected Carla's wish to die.

Then there was the gynecologist, the Roman Catholic priest, and the palliative care specialist. They too agreed that the patient's wish should be respected. She had received the best of palliative care, but there was nothing more that could be done to alleviate her suffering.

In consultation with Carla and her family, it was arranged that the palliative care specialist would end Carla's life on the following day. When the time came, Carla once again reaffirmed her desire to die. Her husband, two of Carla's sisters, her priest, and two of the children were by her side when the doctor administered the drugs that would end her life. Carla fell asleep with a smile on her face and died eight minutes later.

According to the conventional view, her doctor had performed an immoral act because he had intentionally terminated the life of his patient.

Should we accept the conventional view? To answer that question, we need to ask: what makes medical end-of-life decisions right or wrong.

## What makes medical end-of-life decisions wrong?

Carla's doctor terminated Carla's life by administering a lethal drug. Mrs N's doctor terminated his patient's life by turning off a respirator. Both doctors acted upon the request of a competent and terminally ill patient, both knew that their patient would die as a consequence of what they did, and both, we can safely assume, acted from the highest of motives: respect for a suffering patient's wish to die with dignity. And yet, according to the conventional view, Carla's doctor acted wrongly: he performed an inherently immoral act, but Mrs N's doctor performed a morally good act. Why? What could be the morally relevant feature by which we might be able to distinguish the two cases?

### *Causing death and allowing nature to take its course*

We can distinguish the two actions by their causal features. On the one hand, Carla died from the administration of a lethal drug. The drug caused her death, and the doctor who administered it therefore "killed" Carla. Mrs N's death, on the other hand, was not caused by a lethal substance. Rather, she died from respiratory failure, from her disease. Because of that, we can say that her doctor "allowed" her to die.

Does this difference between killing and allowing to die make a *moral* difference? Some people think that it does because when a doctor discontinues life-sustaining treatment he or she merely allows nature to take its course. But is this right? Take Mrs N's case. It is true. Mrs N died from "natural causes" – respiratory failure; but that this respiratory failure led to the patient's death was not a natural event. It was the result of a medical end-of-life decision and a direct consequence of what the doctor did: turning off the respirator. This means that from the *moral*, rather than from the merely causal, point of view the actions of the two doctors are indistinguishable. Both doctors are equally responsible for the deaths of their patient.

### *What's in a word?*

Some opponents of voluntary euthanasia rely heavily on the negative connotations of the word "killing." There is no doubt "killing" ordinarily describes a very bad act; the term "allowing to die," however, has a positive appeal – especially when used in the context of terminal illness, where death is often a welcome friend. We must, however, be careful that our moral judgments are not based simply on the emotional ring of words.

Killing a person is the worst thing we can do to someone who is enjoying life and wants to go on living. But is killing another person *always* a bad thing? In other words, if we welcome death as a friend at the end of a nontreatment decision, why can't we invite it along as Carla and her doctor did?

And what about allowing or letting people die? Surely that's not always a good thing. Letting healthy people die when we could easily save them can be just as bad, and sometimes worse, than killing them. It would be worse if instead of killing them swiftly and painlessly, we deliberately allowed people to die slowly and painfully.

The last point has obvious implications for the practice of medicine and the duties and responsibilities of doctors. Now, Mrs N may not have asked for voluntary euthanasia. But what if she had expressed the wish to be helped to die in the way Carla did? "Being allowed to die" can be a drawn-out, distressing and undignified process – a contradiction of everything the person ever strived for in life. Dying may take hours, days, weeks or even months – depending on the patient's medical condition. It took Mrs N nearly seven hours to die after the respirator was switched off. For some people this will already be far too long. They fail to see why they must – in the name of a moral view that they do not share – die slowly, perhaps gasping for breath and consciously experiencing the effects of respiratory failure, when they could so easily be helped to die. In a case such as this, killing a person would be better – not worse – than allowing him or her to die.

The above issues need discussing. Such discussions are, however, often sadly lacking in the writings of some of the most vocal and zealous defenders of the conventional view. Instead of presenting readers with rational arguments for their view, these writers will often rely on little more than the emotional and chilling ring of words. Doctors who practice euthanasia will often be described as "killers," or as being "licensed to kill"; patients are "being put down" or "culled;" and the very idea of "the physician-euthanizer" is labelled a "deadly self-contradiction." In case this is not yet enough to convince their readers of the evils of voluntary euthanasia, there might also be scenarios in which doctors figure as "highly competent hired syringe[s]" or "poisoners" who "will be working overtime" once the prohibition on the intentional termination of life is lifted.

This is not the way to argue about the morality of voluntary euthanasia. We can and must do better than that.

### Intending death and foreseeing death

Now, a supporter of the conventional view might say that I have so far missed the point. What distinguishes the case of Mrs N from that of Carla is that the doctor in the second case, but not the doctor in the first case, *intended* his patient's death. He must have *wanted* Carla to die – why else would he have administered a lethal drug? Mrs N's doctor need not have intended his patient's death – he did not *want* her dead. He merely wanted to discontinue burdensome treatment, thereby allowing her to die with dignity.

What is wrong, then, with voluntary euthanasia according to the conventional view is that the doctor who practices it must intend the patient's death, whereas a doctor who allows a patient to die does not. There are two problems with this distinction.

First, intentions – as understood by defenders of the conventional view – are ultimately mental states, that is, they are private and not immediately visible to an outside observer. When I stretch my legs while you are walking towards me, am I merely intending to make myself more comfortable, or do I want you to trip? Similar questions can be raised with regard to medical end-of-life decisions. When a doctor administers life-shortening palliative care or turns off a respirator, does he or she intend to relieve pain and discomfort, or does he or she intend the patient's death? Can one actually perform an action, such as turning off a respirator knowing that the patient will die, without also intending the patient's death?

The opaqueness of intentions has implications for public policy, a point we shall return to in a later section. Here we are focusing on the moral nature of actions. This brings me to my second point: should we accept the view that the moral goodness or badness of actions depends on what a person directly intends? And should we accept the view that a doctor who deliberately ends a patient's life is a bad doctor? I can see no reason for accepting either one of those propositions. Here is why.

### Autonomy and well-being

I take the view that decision-making in the practice of medicine should – other things being equal – be based on the patient's best interests. For the purposes of this chapter, we are concerned with the interests of patients who can make decisions for themselves, who are competent. Like all patients, competent patients have an interest in well-being; that is, in freedom from pain, in having their functioning restored, and so on. But a competent patient's interests reach far beyond simple well-being. A competent patient has an overriding interest in autonomy or self-determination; that is, to make decisions about the kind of life he or she wants to live and what kind of a person he or she wants to be.

Respect for the patient's autonomy and the patient's own understanding of well-being now guides decision-making when refusal of treatment is at issue and is recognized in law. The case of Mrs N is an example of this. She was able to choose between a longer or a shorter life by accepting, or refusing, life-sustaining treatment. Respect for autonomy should also, in my view, guide the administration of potentially life-shortening palliative care, and of euthanasia. A doctor acting in accordance with this principle does not, when practicing voluntary euthanasia, perform a morally bad action, and this does not make him or her a bad doctor. The physician is morally in the same position as one who, at the patient's request, discontinues life support: he or she is a good doctor performing a morally good action.

In other words, what I am suggesting is that we stop asking whether a doctor "causes" death or "merely allows it" to happen, and whether he or she "directly intends" it, or "merely foresees" it. These distinctions do not, in my view, have inherent moral significance. What we should be striving for instead is to ensure that patients die a dignified death. A dignified death is the kind of death a person wants to die, a death that respects the patient's values and beliefs, his or her own evaluation of bearable or unbearable suffering, and which fulfils, rather than contradicts, that person's life-story. Many patients' needs will be satisfied by being allowed to die and or being the recipient of good palliative care. But being allowed to die will not meet the needs of all patients. Sadly, pain cannot always be relieved, nor is it always possible to control the many distressing symptoms that may be associated with the dying process. In circumstances such as these, some patients will ask their doctor for direct help in dying. Many doctors would like to assist their patients in such circumstances because they believe that voluntary euthanasia or medically assisted suicide is not only compatible with good medical practice, but actually required by it. Both the patient and the doctor may regard it as a moral act, and view the law that prevents them from carrying out this act as unjust and immoral.

## Public policy

### *Moral plurality*

This leads us into questions of public policy. People who approach ethics from different moral, cultural or religious perspectives will often arrive at different answers to morally controversial questions. These answers have their source in particular value systems and can therefore not be shown to be true or false, in the ordinary sense of those terms.

What should be an appropriate social response? Given that there is fundamental disagreement about the morality of a practice, how should modern pluralist societies such as the USA or Australia respond to it? Should they allow or prohibit the practice, and on what grounds?

## *Personal liberty*

It is now widely accepted that personal autonomy or liberty is a very important value and that it is inappropriate for the state either to adopt a paternalistic stance towards its mature citizens, or to restrict their freedom through the enforcement of a particular moral point of view. Only if one person's actions cause harm to others is it considered legitimate for the state to step in, and to bring in laws that restrict individual liberty. As the British philosopher John Stuart Mill put it in his famous essay *On Liberty* (1859):

> The only purpose for which power can rightfully be exercised over
> any member of a civilised community, against his will, is to prevent
> harm to others ... Over himself over his own body and mind, the
> individual is sovereign.

The argument from liberty or autonomy suggests that people should, under the appropriate circumstances, be free to commit suicide, and that those who are terminally or incurably ill should be able to enlist the help of willing doctors to end their lives. But voluntary euthanasia and assisted suicide are almost everywhere prohibited by law. While some countries and states allow assisted suicide, voluntary euthanasia remains a crime in all countries, with the exception of the Northern Territory of Australia, where laws allowing voluntary euthanasia were enacted in 1995.

Carla died in the Netherlands where procedures are in place that allow physicians to practice voluntary euthanasia by gaining immunity from prosecution. Mrs N died in Victoria, one of the Australian states where doctors are permitted to allow their patients to die but are prohibited, by law, from practicing voluntary euthanasia, as in most other countries.

There is considerable public and professional support for a liberal approach towards voluntary euthanasia. Australian surveys suggest that 75% of nurses, 60% of doctors and 78% of the general population want voluntary euthanasia legalized, and that every third doctor who has been asked by a patient to do so has, in fact, practiced it, even though it is a criminal offense. Surveys in other countries have shown similar results. Why, then, has the law not been changed?

## *Bad consequences*

The law must not be changed, it is often said, because a public policy that allows voluntary euthanasia will have bad overall consequences.

Many claims about bad consequences rest on the assumption that voluntary euthanasia is so inherently different from other medical end-of-life decisions that it raises entirely new issues – issues that are not already raised by the other end-of-life decisions. But this assumption is largely wrong. Voluntary euthanasia is very much like other medical end-of-life decisions.

It is thus sometimes claimed that patients cannot rationally or autonomously choose euthanasia, because they might be depressed or their minds clouded by medication. Similar arguments are raised against doctors practicing it – sometimes coupled with the claim that doctors have a duty to benefit their patients, not to end their existence. After all, how can a doctor benefit a patient when the act of euthanasia ensures that the patient is no longer around to receive the benefits?

If opponents of voluntary euthanasia really believed this kind of argument, then they would also have to hold that no patient can ever autonomously refuse life-sustaining treatment or choose life-shortening palliative care, and that no doctor must ever cooperate with a patient in life-shortening decision.

But this is not what most opponents of voluntary euthanasia believe. Rather, they seem to assume that a patient can rationally refuse treatment (and that doctors ought to cooperate with this decision), but that the patient cannot rationally or autonomously choose voluntary euthanasia. This is inconsistent. The question is whether a patient can rationally choose an earlier death over a later one, and that choice is made in either case. Hence, if a patient can rationally opt for an earlier death by refusing treatment or by accepting life-shortening palliative care, he or she must also be able rationally to opt for an earlier death by euthanasia – and a doctor must be able to benefit a patient by euthanasia, just as he or she can by implementing the other end-of-life decisions which also lead to the patient's non-existence.

Another frequent objection is that it would be impossible to frame laws and provide safeguards against abuse. What this objection overlooks is that the opportunity for abuse already exists. If doctors wanted to abuse the powers they already have, they could do so by simply allowing their patients to die or by administering life-shortening palliative care. What is more, detection would be less likely than it would be in the case of unconsented-to euthanasia. There would be no non-therapeutic toxic substance in the patient's bloodstream, the patient having died from "natural causes" or from the administration of drugs used in palliative care.

Would the doctor–patient relationship be threatened if physicians were permitted to practice voluntary euthanasia, as it is sometimes claimed (see Leon Kass this volume)? I cannot see why. On the contrary, many patients would derive considerable peace of mind from the knowledge that their doctors would be willing to give death a helping hand – if this should ever become necessary, which it might not.

More recently, a new claim has been put forward by people opposed to voluntary euthanasia. They argue that voluntary euthanasia should not be legalized because the experience from the Netherlands (where doctors have been able to practice voluntary euthanasia for the last decade or so) shows that it will lead to abuse – and they base their findings on a recent large-scale study, mentioned at the very beginning of this chapter. The argument is that the study shows that the introduction of voluntary euthanasia has led to abuse – it is claimed that physicians did not always obtain their patients' consent when they withdrew or withheld treatment, administered life-shortening palliative care, or administered euthanasia.

But how can a single study – so far the only one of its kind anywhere in the world – possibly show that the introduction of voluntary euthanasia *has led* to abuse? To demonstrate that, at least two studies would be needed – one conducted before the practice of voluntary euthanasia was introduced and one conducted some time after it. Only then would we be able to compare figures and be able to say that there is more or less abuse.

Also, we do not know whether there is more or less abuse of the practice in the Netherlands than in countries such as the USA or Australia, where voluntary euthanasia is unlawful. It may be less, or more. We simply do not know.

## Conclusion

The best way of preventing abuse is to make medical end-of-life decisions transparent. This means focusing on consent, rather than on the notions of causation and intention that underpin the conventional view. As I noted above, often only the physician will know what is directly intended when he or she makes an end-of-life decision. According to the above-mentioned Dutch study, in the Netherlands a nontreatment decision precedes 17.5% of all deaths; another 17.5% of all deaths are preceded by the administration of potentially life-shortening palliative care, and 3% of all deaths result from voluntary euthanasia or medically assisted suicide. In other developed countries, figures are likely to be similar. The breakdown shows that even in a country where voluntary euthanasia may be openly practiced by doctors, it is a relatively infrequent end-of-life decision. This suggests that the *scope* for abuse – in terms of the sheer numbers of patients affected – is much greater in the context of nontreatment and palliative care decisions than it is in the voluntary euthanasia context. Why, then, are those who fear abuse almost exclusively and unreasonably focusing their attention on voluntary euthanasia? Is it because they are subscribing to the private view that voluntary euthanasia – as a case of the intentional termination of life – is morally wrong whereas allowing to die is not?

To sum up, then: doctors should be permitted to practice voluntary euthanasia because it respects the patient's autonomy and his or her own evaluation of suffering, because it is compatible with sound medical practice (and perhaps required by it), and because there are no good reasons to think that voluntary euthanasia will have bad consequences in practice.

### Suggestions for further reading

Charlesworth, M. *Bioethics in a Liberal Society*. Cambridge: Cambridge University Press, 1993.

Humphry, D. and Wickett, A. *The Right to Die – Understanding Euthanasia*. New York: Harper & Row, 1986.

Kuhse, H. *The Sanctity-of-Life Doctrine in Medicine – A Critique*. Oxford: Oxford University Press, 1989.

Kuhse, H. (ed.) *Willing to Listen – Wanting to Die*. Melbourne: Penguin Books, 1994.

Rachels, J. *The End of Life: Euthanasia and Morality*. Oxford: Oxford University Press, 1987.

Rollin, B. *Last Wish*, New York: Simon & Schuster, 1985.

Sacred Congregation for the Doctrine of the Faith *Declaration on Euthanasia*. Vatican City: Vatican, 1980.

# Humans as research subjects

HERMAN WIGODSKY and SUE KEIR HOPPE

EDITORS' SUMMARY

Herman Wigodsky, MD, PhD, and Sue Keir Hoppe, PhD sketch the evolution of ethical principals governing research on human subjects. They trace the origin of the Nuremburg Code regarding such research to the practices of famous researchers in one of their laboratories, as well as to early guidelines established, ironically, in Germany itself. The guidelines were further enhanced through the Declaration of Helsinki and through the development of Institutional Review Boards in the USA and comparable research ethics committees in other countries. Problems still exist because of variability among committees and countries and disagreement on ethical issues surrounding research on genetics, mental illness, AIDS, and other incapacitating and stigmatizing disorders. The authors assert that government regulation of human subject research can be effective if flexibility of implementation and consideration of local conditions and mores are maintained.

The use of humans in biomedical research probably dates back to the beginnings of medicine itself, but societal interest in the ethics of using humans as subjects in research was not a major concern until the latter part of the nineteenth century. In 1865, Claude Bernard, a professor of physiology at the Sorbonne in Paris, wrote a classic treatise on the science of clinical experimentation. Bernard stated that physicians should never perform on a human an experiment that might be harmful to any extent, even though the resulting knowledge might be advantageous to science or to the health of others. His treatise outlined the ways in which clinical experiments should be organized properly to answer the questions posed – what is termed the "scientific method."

With the advent of the germ theory of disease in the nineteenth century, bio-medical experimentation increased. Still, there was little general social concern about the ethics of such experimentation and, in fact, raising ethical concerns was sometimes dangerous. In 1901, a Russian, Smidovich, writing of necessity under the pseudonym Vikenty Veressayev, published a book describing atrocious human experiments on the transmissibility of syphilis and gonorrhea. He portrayed experimenters as "bizarre disciples of science" and scientific "zealots" and appealed to society as the protector of human subjects.

During the first half of the twentieth century, medical schools in the USA depended largely upon charity hospitals to provide patients for teaching students. The use of charity patients for experimentation came to be justified as a means of patients repaying society for free health care by loaning their bodies for the advancement of science and, thus, the improvement of humankind. There were few ethical restrictions on research at that time; most medical researchers were regarded as persons of good will who had benefit to the patient, as the researchers saw it, as the principal goal.

In 1939, one of us, H. W., began to conduct experiments on the effects of simulated high altitudes on respiratory processes in human volunteers under the supervision of Andrew Ivy in the physiology laboratories at Northwestern University Medical School. The altitude experiments were some of many different studies involving human subjects that were conducted in Ivy's laboratories over the years. There were no formal criteria for the protection of human research subjects, but Ivy insisted that no experiment be undertaken in his laboratories unless the experimenter was one of the research subjects (generally the first). Whenever feasible and/or appropriate, no experiments using humans were undertaken in Ivy's laboratories until exhaustive studies had been done in animals. A complete explanation of each experiment – the hypothesis being tested, results of animal studies, procedures to be performed, and possible risks – was given to each volunteer, usually a faculty or staff member, medical student, or patient.

Ivy's attention to research ethics was commendable for the times, but we point out that he also conducted experiments on patients in a mental hospital to study typhoid fever transmission. Having discovered that the typhoid germ grew best in media containing bile, Ivy hypothesized that typhoid was carried by persons in whom the germ resided chronically in the gall bladder. Since the disease broke out periodically in mental hospitals, psychiatric patients were an ideal study population. Although participation of the patients was "voluntary," there is some question as to whether such experiments would be allowed at the present time, based on issues of the competence of the patients and their ability to give "true" informed consent.

The advent of the Second World War and the identification of numerous pandemic diseases and war-related illnesses led to the focusing of military and civilian research efforts on these conditions. Of serious diseases, malaria was prominent and the search for anti-malarial drugs engaged wide attention. By the end of the war, information had been collected on hundreds of other drugs and chemicals, and was used later in the search for cancer treatments. Examples of nonnoxious, but important, physiological studies were those on night blindness and color vision deficiencies.

During the Second World War, H. W. served as Executive Officer of the Research Division of the Office of the Air Surgeon of the Air Force and was engaged in the identification and exploration of man–machine and man–environment relationships. Through interfaces with various operational groups such as Armament, Signal Corps, and Air Intelligence, H. W. became aware of research being conducted in Germany that had death as an end-point. Ivy, his mentor from Northwestern, was then Civilian Director of the Naval Medical Research Institute. H. W. told Ivy of the research atrocities in Germany about which he had heard. Ivy subsequently gathered additional information through Navy Intelligence, confirming the reports.

The full panoply of the Nazi experiments, conducted in the name of scientific research, gradually unfolded as the war came to an end. In 1946, the USA conducted the Second War Crimes Trial of some of the Nazi physicians involved in the atrocities. The case, United States v. Karl Brandt [Hitler's personal physician] *et al.*, charged 24 defendants. Ivy, as an internationally known scientist and human researcher, was a logical choice for nomination by the American Medical Association to the Secretary of War as a witness for the USA in the trial.

It is interesting to note that in 1900, prior to the Nazi experiments, the Prussian Minister of Religious, Educational, and Medical Affairs ordered clinic directors to limit participation in medical experiments to competent adults who had consented to take part after being fully apprised of possible adverse consequences. Each clinic director could independently authorize experiments but had to keep a record of compliance with the Minister's directive. In 1931, as the result of accusations in the Press of abuse of human research subjects, the German Minister of the Interior published regulations on the use of new therapies and human experiments that remained binding during the Third Reich. However, the declaration by the Germans that Jews, other nationalities, socialists, gypsies, etc., were "nonpersons" made it possible for some otherwise scientifically sound experimenters to conduct questionable research on members of these groups. As "nonpersons," the individuals were not covered by the German ethical directives of 1900 and 1931.

At the Second War Crimes Trial of the Nazi physicians, eight defendants were acquitted, seven were condemned to death, and nine served 10 to 20 years in prison. There was evidence of criminality against other experimenters but not all could be tried, for various technical reasons or excuses. Some were aided (illegally) by the United States military to come to the USA, where their "knowledge" was to be utilized in military laboratories. It is reported that one such person was the instigator of infamous experiments with lysergic acid diethylamide (LSD), a hallucinogen with many adverse effects, in studies for the Central Intelligence Agency. It was reported that some individuals were slipped LSD without their knowledge or consent.

George Annas has written that the "tribunal's focus was on the criminal nature of the Nazi experiments," but the "judges were also grappling with much broader ethical concerns regarding medical research. The trial court sought a historical framework of medical standards from which to judge the Nazi physicians and attempted to elucidate the scope of medical experimentation undertaken by the Nazis, and other physicians and scientists, during the Second World War. Finally, the court attempted to establish a set of principles of human experimentation that could serve as a code of research ethics." Andrew Ivy and Leo Alexander, a Boston neurologist/psychiatrist who were also witnesses at the trial, were instrumental in codifying these research ethics, by organizing their testimony as a series of principles that could be utilized by the judges in their final decision. The principles appear to have been derived from many sources, but our earlier description of the procedures used in Ivy's laboratories lends credence to the belief that what came to be known as the Nuremberg Code is in many respects a reflection of this great researcher's personal standards.

The Nuremberg Code consists of ten principles. In abbreviated fashion these are:

1. The voluntary consent of a human subject to experimentation is absolutely essential. The subject should have legal capacity to consent and sufficient knowledge and comprehension of the nature of the experiment and risks involved to make an informed and enlightened decision.
2. The experiment should be designed to yield fruitful results for the good of society, unobtainable by other methods or means of study and not random and unnecessary in nature.
3. Anticipated results of the experiment should justify its performance.
4. The experiment should be conducted so as to avoid all unnecessary physical and mental suffering and injury.
5. No experiment should be conducted where there is *a priori* reason to believe that death or disability will occur.

6. The degree of risk to be taken should never exceed the humanitarian importance of the problem to be solved by the experiment.
7. The experimental subject should be protected against even remote possibilities of injury, disability, or death.
8. The experiment should be conducted only by scientifically qualified persons.
9. During the course of an experiment, a human subject should be free to end the experiment.
10. During the course of an experiment, the scientist in charge must be prepared to terminate the experiment at any stage, if there is reason to believe that continuation of the experiment is likely to result in injury, disability, or death to the subject.

The importance of the Nuremburg Code cannot be overemphasized, both in terms of its influence on research ethics in general and on the development of subsequent research guidelines and regulations. From time to time attempts have been made to resurrect some of the infamous Nazi experiments, such as those on high altitude or exposure to low temperatures. Proponents of such attempts appear to be unaware of the vast data that had already been acquired in civilian and military studies in the USA, Canada, and Great Britain, under conditions that were protective of the human research subjects involved and, in stark contrast to the Nazis, did not use death as an end-point. The attempts to resurrect Nazi data raise significant ethical concerns that go beyond their use. Should data obtained in an unethical study be published or used in any form? Is it permissible to use such data if no relevant data are available from ethically conducted studies? It is our position that data obtained in an unethical study should not be published and should be ignored. The issue, however, remains controversial.

In 1954, the World Medical Association (WMA) adopted a set of ethical principles for research and experimentation, intended to be professional guidelines designed by physicians for physicians. The WMA principles went beyond the Nuremburg Code in that they provided for surrogate consent in the event a research subject was too ill to consent for him- or herself. Then in 1964, the WMA issued the Declaration of Helsinki, a major contribution of which was the distinction between clinical research occurring in conjunction with medical care, and clinical research with no therapeutic potential. The 1975 revision of the Declaration contained the first reference to the need for committee review of the ethics of proposed human subject research.

The Declaration of Helsinki, which was most recently revised in 1989, is one of the most widely referenced sets of international guidelines. However, there are many other on-going efforts to address international, shared concerns about research ethics. For example, the Council for International Organizations of Medical Sciences (CIOMS) and the World Health Organization (WHO) have convened conferences and proffered international ethical guidelines on a number of topics since 1982. The sponsorship of CIOMS and WHO in the development of research ethics guidelines assures wide international attention to such issues, but neither the Nuremberg Code, the Declaration of Helsinki, nor the CIOMS/WHO guidelines have binding legal authority.

In the USA, the post-Second World War increase in research funds available through Congress to the National Institutes of Health (NIH) was accompanied by a growing interest in ethical problems surrounding research. The NIH opened a Clinical Center in 1953, enabling NIH scientists to recruit patients with specific diseases and normal volunteers for study. The first federal document, entitled "Group consideration for clinical research procedures deviating from accepted medical practice or involving unusual hazard," was issued, requiring committee review of research in the Clinical Center. In 1966, the Surgeon General of the United States Public Health Service (PHS) ordered that no extramural research or research training grant involving humans be awarded by the PHS unless a committee of the grantee institution had reviewed the ethics of the proposed work. The committee was to assess the adequacy of measures to protect the rights and welfare of potential research subjects, including procedures for obtaining informed consent, and the potential risks and benefits of the research. Subsequent revisions of the Surgeon General's policy statement allowed grantee institutions to give "assurance" of prior committee review on an institution-wide basis, rather than for individual projects and required committees to consider local laws and norms in their deliberations. The Institutional Relations Branch of the NIH Division of Research Grants was established to carry out the Surgeon General's order. The Branch later was upgraded and the name was changed to, and remains, the Office for Protection from Research Risks (OPRR).

Also in 1966, ethicist Henry Beecher published a seminal article in the *New England Journal of Medicine* in which he "exposed" 22 examples, selected from the published scientific literature, of "unethical or questionably unethical" research procedures. The article received wide attention from the media. Revelations of unethical biomedical research escalated over the ensuing years, including, for example, the so-called Tuskegee Syphilis Study in which treatment was withheld from poor African-American men who did not know they were part of a long-term research project on the "natural course" of the disease conducted by the PHS. While unethical biomedical research has been of predominant social concern, behavioral research has been suspect as well. A classic example of such research is experiments that were done at Yale University on "conditions of obedience and disobedience," in which students were led to believe that they were administering potentially harmful electrical shocks to others when they failed to learn. Although the students were "debriefed" (told that they had been deceived and no real shocks were delivered), some argue that the psychological trauma of realizing that one's behavior would have seriously hurt another person was not erased by the debriefing procedure.

In 1971, the United States Food and Drug Administration (FDA) promulgated regulations requiring peer review of all new and unapproved drugs and devices used in institutions. In 1974, the United States Department of Health, Education and Welfare (DHEW) issued the first federal regulations on Institutional Review Boards (IRBs), committees to be established in institutions to carry out the peer review required by NIH in the 1966 Surgeon General's order. In the same year, Congress passed the National Research Act, which mandated IRBs and called for the creation of the National Commission for the Protection of Human Subjects of Biomedical and Behavioral Research.

The charge to the National Commission was to identify problems in carrying out government obligations for the ethical conduct of research utilizing human research subjects and to suggest means for solving the problems. The National Commission deliberated for four years and in 1978 issued the *Belmont Report*, named after a conference center near Baltimore at which the Commission met. The *Belmont Report* is a succinct summary of three basic principles that should be used to evaluate the ethicality of human subject research. The principles include:

1. *Respect for persons*, which includes requirements to treat individuals as autonomous and to protect those who are not autonomous. An autonomous person is one "capable of deliberation about personal goals and of acting under the direction of such deliberation." The process of informed consent derives from this principle and includes issues of full disclosure of information about a study to a research subject by an investigator, the investigator's certainty that the information has been understood by the subject, and that a subject's participation is voluntary.

2. *Beneficence*, a concept based on the Hippocratic axiom "do no harm." Beneficence involves maximizing benefits and minimizing harms. This ethical standard requires that risks to research subjects are reasonable in relation to anticipated benefits.

3. *Justice*, which implies fairness in the distribution of both benefits and burdens of research – especially that the "burdens" or risks of research are not borne only or mainly by disadvantaged groups (e.g. the poor, ethnic or racial minorities) when potential advantages of research stand to benefit all. This principle involves equitable subject selection.

In 1981, major revisions of FDA and DHEW regulations on the protection of human subjects in research were made. The revisions built upon the ethical principles that had evolved since the Second World War and set out fairly specific procedures to be followed by IRBs. For example, the regulations specified the minimum size and composition of an IRB – five members, including a person who is not affiliated to the institution and one whose concerns are in nonscientific areas. The rules also specified categories of research, primarily based on potential risks to subjects, for which there are different levels of review. Some types of research (e.g. medical record or vital statistics searches without notation of a person's identity) require less scrutiny than others (e.g. first-time use of cancer medicines in humans). Basic elements of informed consent, and conditions under which consent is to be obtained, were outlined in detail. Finally, categories of special, "vulnerable" populations such as children and prisoners were defined and additional precautions pertaining to their involvement in research were stipulated. The regulations have been revised twice – once to incorporate changes after experience in implementing them and then to expand their authority beyond the NIH to other governmental entities as well.

Despite the relative specificity of the current regulations governing human subject research, many aspects of the system of IRB review have not been, and perhaps cannot be, codified. The work of the National Commission on vulnerable populations was never completely implemented; in particular, regulations pertaining to the "institutionalized mentally infirm" were not finalized because agreement on them could not be reached. Thus, IRBs wrestle with proposed research on persons with schizophrenia and other mental disorders, especially when the research involves a "wash-out" period preceding the trial of a new medicine (i.e. an effective medicine is stopped so that the influence of the new one can be evaluated). Scientific breakthroughs and novel technologies, as well as changing social norms and values, also challenge IRBs to make judgments about issues not specifically addressed in the regulations. Many of the studies undertaken as part of the Human Genome Project, designed to locate on the DNA all genes in the body, have little individual benefit to subjects who take part, but social and economic risks such as the ability of subjects, or members of their families, to obtain or retain health insurance are great. Similar privacy and confidentiality issues exist in research into the acquired immunodeficiency syndrome (AIDS), ranging from basic epidemiologic studies of transmission to clinical trials of AIDS vaccines and cures. While genetic and AIDS research are relatively new, studies of human reproduction have traditionally presented IRBs with difficult ethical decisions. When *in vitro* fertilization techniques were being pioneered, there were questions about whether the techniques should be performed only with married couples, in the interest of a resulting child, and whether "ownership" of frozen embryos was pertinent when time between fertilization and implantation was sufficient to include the possibility of a couple's divorce.

Based on the IRBs with which we are familiar, there is little question that the establishment of the review system has done a great deal of good in protecting the rights and welfare of human research subjects. Although our evidence is mostly anecdotal, sensitivity on the part of investigators to issues of protection of human research subjects has increased as a result of the IRB review process. Improved quality of proposed research studies also has been attributed to the IRB system.

However, there is substantial variability in the ways in which IRBs (and similar ethics committees in other countries) carry out the regulations and attend to the ethical principles outlined in the *Belmont Report*. Generally, IRB deliberations focus on informed consent and risk–benefit assessment, with relatively less attention directed to equitable subject selection. The United States NIH Revitalization Act of 1993 was an attempt to redress these priorities by mandating the inclusion of previously underrepresented women and minorities as research subjects. IRBs also devote more attention to the initial review of proposed studies and less to evaluating studies in progress. An on-going evaluation of the IRB system, initiated and funded by the PHS, will yield systematic information that can be used to describe the range of variability and factors that are associated with the relative effectiveness of IRBs in protecting human research subjects. Some of the factors include the institutional setting(s) in which an IRB is located (e.g. university, medical school, hospital, pharmaceutical company), financial and intellectual support given an IRB by the host institution(s), and the amount of research conducted in the institution(s). To some extent, the variability that occurs among IRBs is by design, in that IRBs were created as agents or committees of their own institutions, not as arms of the federal government. While inconsistency among IRB decisions can be frustrating to investigators and to sponsors of research who decry "ethical relativism," the dictum to consider local community and institutional norms and practices is a strength of the system.

The regulations governing human subject research technically pertain only to studies supported by federal monies, but most IRBs in prominent universities and research centers apply the regulations to all research, irrespective of funding source. Moreover, DHHS funds now support research in over 50 countries outside the USA. To qualify for this support, grantees must agree to be guided by United States regulations for the protection of human research subjects or to have in place regulations which parallel these. In both these ways, DHHS has exerted enormous influence, nationally and world wide, in the conduct of human research.

Increased scrutiny of research on humans has brought attention to two other ethical problems in research – possible conflict of interest between investigators and subjects and so-called scientific misconduct, including but not limited to, deviation from an approved research protocol (e.g. enrolling non-eligible subjects), falsification of research data, and plagiarism. In 1992, the Office of Research Integrity (ORI) was created as an independent office within DHHS, replacing the Office of Scientific Integrity at NIH and the Office of Scientific Integrity Review at DHHS. Regulations and enforcement actions not withstanding, some investigators still act in an unethical manner. The impact of recent media revelations, such as the reported falsification of records in a large, multicenter study comparing lumpectomy to mastectomy in preventing breast cancer recurrence, further erode public confidence in the scientific community. As ethicist Judith Swazey has noted, early instances of scientific misconduct were viewed as "idiosyncratic events, perpetrated by individuals with various types of behavioral pathology and/or stress reactions" and whistleblowers as "more suspect and deviant than those accused of misconduct." Now, misconduct cases are seen as involving "certain patterned phenomena" that necessitate government regulation of procedures for investigating and dealing with allegations of misconduct.

The patterned phenomena to which Swazey refers relate to profound changes that have taken place in our perceptions of the meaning of science in general and biomedical, behavioral, and medical research in particular. In earlier times, scientists were held in awe and, generally, were affiliated with universities or specialized centers such as the Rockefeller Institute. Their goals were the generation of knowledge for knowledge's sake. As late as the 1930s, scientists who were employed by industry generally were looked upon with considerable suspicion that their research was not truly unbiased.

However, the appearance of "technocracy" began to legitimize the division of scientists and their work into basic and applied. With the advent of the Second World War, many scientists undertook targeted or applied research without criticism. After the war, there was an explosion of the student population in universities and increasing concern about the governance of universities and their principles. Research funding shifted dramatically to government support and scientists soon seemed to be more loyal to their funding sources (e.g. the NIH) than to their own universities. Scientists also became more mobile as they were able to move their research funds with them when they changed institutions. Dramatic advances in technology were followed by increased interest in patentability of scientific techniques and applied end-products of basic research. The 1990 passage of legislation in the United States permitting institutions to patent federally sponsored research caught the eye of institutional administrations always hungry for new sources of revenue. This challenged further the principles, morality, and ethics of the universities and their faculties.

These changes were accompanied by the expansion of existing ethical concerns and the appearance of new ones. The enormous growth in the numbers of scientists meant that there were more opportunities in the field for persons with marginal sensitivity to research ethics. As we have noted, concern on the part of Congress and other governmental agencies about ethical problems in federally supported research led to the establishment of federal standards of conduct in research. Despite the fact that we can identify many of the factors that have invited increasing governmental regulation of human subject research, we have not solved all of the ethical problems involved and new ones appear constantly. The problems are social in nature and government intervention to protect human subjects of research will likely continue, a prediction reminiscent of Smidovich's appeal for help from society mentioned at the start of this chapter.

## Suggestions for further reading

Annas, G. J. and Elias, S. (eds.) *Gene Mapping: Using Law and Ethics as Guides.* New York: Oxford University Press, 1992.

Annas, G. J. and Grodin, M. A. (eds.) *The Nazi Doctors and the Nuremberg Code: Human Rights in Human Experimentation.* New York: Oxford University Press, 1992.

Barber, B., Lally, J. J., Markarushka, J. L. and Sullivan, D. *Research on Human Subjects: Problems of Social Control in Medical Experimentation.* New York: Russell Sage Foundation, 1973.

Beauchamp, T. L. & Childress, J. F. (eds.) *Principles of Biomedical Ethics.* New York: Oxford University Press, 1989.

Bernard, C. *An Introduction to the Study of Experimental Medicine.* New York: Dover, 1957.

Faden, R. R. and Beauchamp, T. L. (1986). *A History and Theory of Informed Consent.* New York: Oxford University Press, 1986.

Gray, B. *Human Subjects in Medical Experimentation.* New York: John Wiley and Sons, 1975.

Greenwald, R. A., Ryan, M. K. and Mulvahill, J. E. (eds.) *Human Subjects Research: A Handbook for Institutional Review Boards.* New York: Plenum Press, 1982.

Jones, J. H. *Bad Blood: The Tuskegee Syphilis Experiment.* New York: Free Press, 1981.

Levine, R. J. *Ethics and Regulation of Clinical Research*, 2nd edn. New Haven, CN: Yale University Press, 1988.

McNeil, P. M. *The Ethics and Politics of Human Experimentation.* Cambridge: Cambridge University Press, 1993.

Mendelsohn, E., Swazey, J. P. and Taviss, I. (eds.) *Human Aspects of Biomedical Innovation.* Cambridge, MA: Harvard University Press, 1971.

# Research involving children as subjects†

ROBERT J. LEVINE

EDITORS' SUMMARY

Robert J. Levine, MD, analyzes the reasoning behind excluding or involving children as research subjects, reasoning that would be applied analogously to any vulnerable population unable to fully consent. On the one hand, if we exclude children from research, we may actually put them more at risk in the future when they are treated with drugs under less controlled circumstances. On the other hand, if we include them then we seem to violate the stringent standards of informed consent, perhaps subjecting them to unanticipated side-effects. Some people argue that children are incapable of giving consent and should never be treated. But a compromise has been reached in the last 20 years by which consent for minimal risk procedures may be given by surrogates, sometimes with the assent of the minor. This model may be applicable to research on other vulnerable populations.

In 1971 the *Lancet*, a leading British medical journal, published a scathing critique of a series of experiments conducted at the Willowbrook State School, an institution for "mentally defective persons" in New York City. The first group of experiments were done to study the nature of the disease, infectious hepatitis. Later experiments were performed to test the effects of gammaglobulin and other

† Portions of this article are excerpted or adapted from Levine, R. *Ethics and Regulation of Clinical Research*, 1986, with the permission of the publisher.

substances in preventing or reducing the severity of the disease. The research subjects, who were all mentally retarded children, were deliberately infected with the hepatitis virus; early subjects were fed extracts of stool from infected individuals and later subjects received injections of more purified virus preparations. The researchers defended their experiments by pointing out that the vast majority of children admitted to Willowbrook acquired hepatitis spontaneously and, perhaps, it would be better for them to be infected under carefully controlled conditions in a program set up to provide the best available treatment.

An additional criticism was leveled against the recruitment policies. During the course of these studies, Willowbrook closed its doors to new inmates owing to overcrowded conditions. However, the hepatitis program, because it occupied its own space in the institution, was able to continue to admit new patients as each new study group began. Thus, in some cases parents found that they were unable to admit their children to this institution unless they agreed to their participation in these studies.

In response to this as well as to other exposés of apparently unethical research practices, the United States Congress in 1974 established the National Commission for the Protection of Human Subjects of Biomedical and Behavioral Research. The National Commission was directed to identify the basic ethical principles that should underlie the conduct of biomedical and behavioral research involving human subjects and to recommend guidelines that should be followed to assure that such research would in the future be carried out according to the requirements of the basic ethical principles.

The National Commission identified three basic ethical principles. The first of these, *respect for persons*, according to the National Commission, requires that individuals should be treated as autonomous agents; this means others are not to touch them or to encroach upon their privacy without their permission. In the context of research, this creates a general requirement for informed consent. Respect for persons further requires that persons with diminished autonomy and thus in need of protection are entitled to such protection.

The second principle, *beneficence*, according to the National Commission consists of two general rules: (a) do no harm and (b) maximize possible benefits and minimize possible harms. In the context of research, this principle is satisfied by refraining from the deliberate infliction of injury on research subjects and by assuring that the risks of research are in a reasonable relation to the expected benefits.

The third principle, *justice*, requires a fair or equitable sharing of burdens and benefits. As interpreted by the National Commission, justice requires special protection for those who are weaker, more vulnerable, or less advantaged than others. They are to be protected from bearing more than their fair share of the burdens of participation in research. Further, they are to be assured equitable access to the benefits of research.

In retrospect, the Willowbrook studies appear to have violated each of the three basic ethical principles. Not only was there no informed consent, the parents were under great pressure to enroll their children in the research project. Otherwise, they would either have had to find other generally less satisfactory accommodations for their retarded children or wait a long time for an opening at Willowbrook that was not reserved for the research program. The children were exposed to a high probability of contracting a serious disease – one for which there was no effective treatment at the time. And finally, because mentally retarded children must be considered a highly vulnerable population, they should have been regarded as among the least eligible candidates for participating in research of this type.

With regard to developing guidelines to assure that research involving children would be conducted in accordance with the basic ethical principles, the National Commission was presented with one fundamental question: should we do research involving children as subjects? If we do, then we shall be doing research on individuals without their informed consent, an apparent violation of the principle of respect for persons unless one can assure that they will be protected sufficiently from harm. If we do not, then we shall be depriving children, as a class of persons, of the benefits of research, an apparent violation of the principle of justice. This chapter will first provide an account of the issues and arguments that the National Commission considered as it was deciding how to interpret the principles of respect for persons and justice as they applied to children as research subjects. This will be followed by a review of the major features of United States federal policy regarding research involving children, a policy that was established in response to the recommendations of the National Commission.

## Respect for persons

The central problem presented by proposals to do research involving children as subjects is that children, as a class of persons, lack the legal capacity to consent. A leading international code of research ethics, the Declaration of Helsinki, promulgated by the World Medical Association in 1964, recognizes the necessity of involving those who are legally incompetent as research subjects. In such cases it calls for the consent by the parent, guardian, or legally authorized representative; such consent is commonly referred to as "proxy consent." Much of the debate about the ethics and law of research involving children focused on the nature of the procedures for which a proxy may consent. These debates reflect the requirement of the principle of respect for persons that individuals who are not capable of protecting themselves through informed consent are entitled to protection from harm. As we shall see, the controversy centered on how much risk of harm could be authorized by parents or guardians.

An extreme position in this debate is presented by the theologian, Paul Ramsey, who bases his argument on a strict interpretation of the ethical principle of respect for persons, i.e. that we must let persons alone unless they consent to be touched. Consequently, he argues that the use of a nonconsenting subject such as a child is wrong whether or not there is any risk simply because it involves an "unconsented touching." As he puts it, one can be "wronged without being harmed." "Wrongful touching" can be made right only when it is for the good of the individual. Therefore, proxy consent may be given for nonconsenting subjects only when research activity includes therapeutic interventions related to the subject's own recovery. Ramsey acknowledges that, in some cases, there may be powerful moral reasons for involving children in research having no therapeutic components; however, "it is better to leave [this] research imperative in incorrigible conflict with the principle that protects the individual human person from being used for research purposes without . . . consent." He continues that it would be immoral either to do or not to do the research, but he maintains that one should "sin bravely" in the face of this dilemma by sinning on the side of avoiding harm rather than attempting to promote the general welfare.

When the principle of respect of persons is interpreted strictly, however, the unconsented touching of a competent adult is wrong even if it benefits that person. In that case, why should potential benefit justify such touching of a child? Theological ethicist, Richard McCormick, proposes that the validity of such interventions rests on the presumption that the child, if capable, would consent to therapy. This presumption, in turn, comes from a person's obligation to seek therapy, an obligation which people possess simply as human beings. Because people have an obligation to seek their own well-being, we presume that they *would* consent if they *could* consent and thus presume also that proxy consent for therapeutic interventions will not violate respect for them as persons.

By analogy, argues McCormick, people have other obligations as members of a moral community to which one would presume their consent; McCormick calls this "their correctly construed consent." One such obligation is to contribute to the general welfare when to do so requires little or no sacrifice. Therefore, McCormick concludes that nonconsenting subjects may be used in research not directly related to their own benefit so long as the research fulfills an important social need and involves *no discernible risk*. In McCormick's view, respecting persons includes recognizing that they are members of a moral community with its associated obligations.

Ramsey counters this argument by claiming that children are not adults with a full range of duties and obligations. Therefore, they have no obligation to contribute to the general welfare and respect for them requires that they be protected from harm and from unconsented touching.

Philosopher-ethicist Benjamin Freedman bases his argument on the same premise as Ramsey's, i.e. a child is not a moral being in the same sense as an adult. However, his analysis yields the same conclusion as McCormick's. Exactly because children are not autonomous, they have no right to be left alone. Instead, they have a right to be taken care of. Thus, the only relevant moral issue is the risk involved in the research; the child must be protected from harm. Therefore, Freedman agrees with McCormick that children may be used in research unrelated to their therapy provided it presents to them no discernible risk.

Philosopher-ethicist Terrence Ackerman next entered the debate claiming that we tend to fool ourselves with procedures designed to show respect for the child's very limited autonomy. He claims that the child tends to follow the course of action that is favored by the adults who are responsible for the child's well-being. He further contends that, in general, this is as it ought to be. "Once we recognize our duty to guide the child and his inclination to be guided the task becomes that of guiding him in ways which will achieve his well-being and contribute to his becoming the right kind of person."

Willard Gaylin, a psychiatrist and ethicist, tells the story of a man who acted in accord with Ackerman's position. After directing his 10-year-old son to cooperate with having his blood drawn for research purposes, he explained that his direction arose from his moral obligation to teach his child that there are certain things one does to help others even if they do cause a bit of pain: "This is my child. I was less concerned with the research involved than with the kind of boy I was raising. I'll be damned if I was going to allow my child, because of some idiotic concept of children's rights, to assume that he was entitled to be a selfish, narcissistic little bastard."

As we shall see, the policy recommended by the National Commission reflected its conclusion that, because infants and very young children have no autonomy, there is no obligation to respond to them through the usual devices of informed consent. Rather, respect for infants and very small children requires that we protect them from harm. "Minimal risk" was identified as a threshold standard in that research presenting more than minimal risk requires special justification; "no discernible risk" was rejected as too restrictive. The National Commission also recommended procedures for respecting the developing autonomy of older children and adolescents by engaging their "assent."

## Justice

As a consequence of the uncertainties about the ethical propriety of and legal authority to do research on children, there was in the 1960s and 1970s a great reluctance in the USA to do studies to determine the safety and effectiveness of drugs in children. As a result, pediatrician Harry Shirkey, proclaimed in 1968 that "infants and children are becoming the therapeutic orphans of our expanding pharmacopoeia." Since 1962, nearly all drugs had been required by the United States Food and Drug Administration (FDA) to carry on their labels one of the familiar "orphaning" clauses: for example, "not to be used in children," "is not recommended for use in infants and young children, since few studies have been carried out in this group," "clinical studies have been insufficient to establish any recommendations for use in infants and children," and "should not be given to children." By 1975 over 80% of all drugs prescribed for children included such orphaning clauses on their labels.

The therapeutic orphan phenomenon is not limited to children. Very similar conditions exist in the use of drugs for pregnant women and in young women generally. Pregnant women were usually excluded from drug trials and from many other types of research because the unborn child is seen by some as a nonconsenting subject who might be peculiarly vulnerable to the effects of drugs. Women who are biologically capable of becoming pregnant were commonly excluded from many types of research, including drug trials. Given suitable plans for contraception, it would be reasonably safe to include them in most studies; however, many investigators did not wish to assume the burden of discussing plans for contraception with prospective subjects. Moreover, many investigators feared the potential legal consequences of a failure to prevent pregnancy during a study. (Recently the United States government has required inclusion of women who could become pregnant as research subjects in most types of medical research, including drug trials.)

The therapeutic orphan phenomenon represents a serious injustice. If we consider the availability of drugs proved safe and effective through the devices of modern clinical research a benefit, then it is unjust to deprive classes of persons such as children and women of this benefit. This injustice is compounded as follows. If we were to do clinical trials in children as we now do in adults, the first administration of various drugs to children would be done under conditions much more controlled and much more carefully monitored than is customary in the practice of medicine. It is likely that adverse drug reactions that are peculiar to children would be detected much earlier than they are now; consequently, either we could discontinue administration of the drugs to children or we could issue appropriate warnings to physicians who are prescribing the drugs. The prevailing practice in the USA is to ignore the orphaning clauses on the package labels. Consequently, there is a tendency to distribute unsystematically the unknown risks of drugs in children and pregnant women, thus maximizing the frequency of their occurrence and minimizing the probability of their early detection. Parenthetically, it should be noted that most drugs that prove safe and effective in adults do not produce unexpected adverse reactions in children; however, when they do, the numbers of harmed children tend to be much higher than they would be if the drugs had been studied systematically before they were introduced into the practice of medicine.

The public policy based on the recommendations of the National Commission has largely relieved the problems associated with the therapeutic orphan phenomenon in children. In particular, the recommendation that risks "presented by an intervention that holds out the prospect of direct benefit for the individual subject" may be considered differently from risks presented by procedures designed to serve solely the interests of research has greatly facilitated the ethical conduct of clinical trials in children.

## Public policy

The National Commission concluded that children, because they have limited capacity to consent, are vulnerable or disadvantaged in ways that are morally relevant to their involvement as subjects of research. Therefore, the National Commission interpreted the principle of justice as requiring that we facilitate activities that are designed to yield direct benefit to the children-subjects and that we encourage research designed to develop knowledge that will be of benefit to children in general. However, we should generally refrain from involving children in research that is irrelevant to their conditions as individuals or at least as a class of persons. The principle of respect for persons was interpreted as requiring that we show respect for a child's capacity for self-determination to the extent that it exists. Although they are legally incapable of consent, many can register knowledgeable agreements (assents) or deliberate objections, terms that I shall discuss shortly. To the extent that the capacity for self-determination is limited, respect is shown by protection from harm. Accordingly, the National Commission recommended that the authority accorded to children or their legally authorized representatives to accept risk be strictly limited; any proposal to exceed the threshold of "minimal risk" requires special justification.

Let us now consider in more detail some of the features of the National Commission's recommendations.

All plans to do research involving human subjects, including children must be reviewed and approved by an Institutional Review Board (IRB), a multidisciplinary committee that has the responsibility of protecting the rights and welfare of the research subjects. For research involving children, "the IRB is required to determine, . . . where appropriate, that studies have been conducted first on animals and adult humans, then on older children, prior to involving infants." This requirement reflects the interpretation of the requirement of the principle of justice that vulnerable persons are to be afforded special protection from the burdens of research. Adults are perceived as less vulnerable than older children who, in turn, are less vulnerable than infants. Investigators in the USA who propose to do research on children without having first done such research on animals, adults, or both are obliged to persuade the IRB that this is necessary. The strongest justification would be that the disorder or function to be studied has no parallel in animals or adults. In such cases, when the research presents any risk of physical or psychological harm or significant discomfort, researchers are expected to initiate their work on older children who are capable of assent, before they involve infants.

## Assent and permission

The National Commission abandoned the use of the word "consent" except in situations in which an individual can provide "legally effective consent" on his or her own behalf. As a corollary to this, the phrase "proxy consent" was discarded.

Adequate provisions must be made for "soliciting the assent of children (when capable) and the permission of their parents or guardians." The transactions involved in obtaining assent and parental permission are essentially the same as those for informed consent.

According to the National Commission, a child with normal cognitive development becomes capable of meaningful assent at about the age of 7 years, although some may be younger and some older. Mere failure to object is not to be construed as assent. Researchers are also expected to respond to a child's "deliberate objection." Some children who are incapable of meaningful assent are able to communicate clearly their disapproval or refusal of a proposed procedure. A 4-year-old may protest, "No, I don't want to be stuck with a needle." By contrast, an infant who might cry or withdraw in response to almost any stimulus is not capable of deliberate objection. A child's deliberate objection usually should be regarded as a veto to his or her involvement in research. Of course, there are exceptions to this general rule; most importantly, parents and guardians may overrule a young child's objection to interventions and procedures that hold out the prospect of direct benefit to the child.

### *Permission*

Unless otherwise specified, the involvement of children as research subjects must be authorized by the permission of a parent or guardian. If more than minimal risk is presented by a nonbeneficial intervention or procedure, permission must be obtained from both parents unless the IRB agrees that this is an unreasonable requirement. It would be unreasonable if, for example, one of the parents is deceased, unknown, incompetent or not reasonably available. In other classes of research, the IRB may decide that the permission of one parent is sufficient.

The National Commission indicated that the parental or guardian permission should reflect the collective judgment of the family that an infant or child may participate in research. In research projects for which the permission of one parent or guardian is sufficient, for example research in which the risks or discomforts are related to a therapeutic intervention, it may be assumed that the person giving formal permission is reflecting a family consensus.

The requirement for parental or guardian permission assumes that the child is living in a reasonably normal family setting and that a normal loving relationship exists between the child and the parents. If there is cause to suspect that no such loving relationship exists, the IRB may require different procedures.

## Requirements varying with the degree and nature of risk

Research that presents to children no more than minimal risk may be conducted with no requirements other than those already mentioned. According to the definition in the federal regulations: "*Minimal risk* means that the probability and magnitude of harm or discomfort anticipated in the research are not greater in and of themselves than those ordinarily encountered in daily life or during the performance of routine physical or psychological examinations or tests." Examples of procedures presenting no more than minimal risk are: routine immunizations, modest changes in diet or schedule, physical examinations, obtaining blood and urine specimens, and developmental assessments. Many routine tools of behavioral research such as most questionnaires, observational techniques, noninvasive physiological monitoring and puzzles also present no more than minimal risk. Questions about some topics, however, may generate such anxiety or stress as to involve more than minimal risk. Research in which information is gathered that could be harmful if disclosed should not be considered of minimal risk unless there are secure measures in place to maintain confidentiality.

### Minor increments above minimal risk

When minor increments above minimal risk are presented by procedures that do not hold out any expectation of direct health-related benefit for the child, some special justifications are required. The IRB must agree that such interventions or procedures present experiences to the subjects that are "reasonably commensurate" with those inherent in their actual or expected medical, psychological, or social situations, and are likely to yield generalizable knowledge about the subject's disorder or condition. The requirement that experiences be reasonably commensurate with those inherent in their actual or expected situations calls for some clarification. First, it means that the procedures to be followed are those that they will ordinarily experience by virtue of their having or being treated for their disorder or condition. Thus, it might be appropriate to invite a child with leukemia who has had several bone marrow examinations to consider having another for research purposes. One could not, however, justify extending a similar invitation to a normal child. This requirement makes it very difficult to develop normal control data for examinations and other procedures that present more than minimal risk.

The requirement for commensurability reflects the National Commission's judgment that children who have had a procedure performed upon them would be more capable than are those who are not so experienced to base their assents on familiarity with the procedure and its attendant discomforts; thus, their decisions to participate will be more informed.

The IRB must further agree that the anticipated knowledge is of "vital importance" for understanding or amelioration of the subject's disorder or condition. This is a much stronger requirement than that for minimal risk research for justification in terms of developing knowledge that will be of use to the class of children of whom the particular subject is a representative. This requirement thus establishes a higher standard for assessing the importance of the knowledge to be gained. It also strengthens the general requirement to use children as subjects only in research that is relevant to their own disorder or condition.

This policy creates a requirement for the IRB to make two difficult judgments for this class of research. What is "vital importance"? What is the upper limit of "minor" in assessing an increment above minimal risk? Those who developed this policy showed wisdom by resisting demands to define the boundaries of these terms. IRBs and researchers were thus challenged to explore these concepts and to develop functionally relevant definitions as they consider problems presented by particular proposals to do research. As they share and debate the fruits of their explorations with one another, we can expect a gradual refinement of our understanding of these concepts and how to use them.

Among the procedures that have been approved by IRBs as presenting minor increments above minimal risk are bone marrow aspirations in children with leukemia and single additional spinal taps in adolescents who have already had at least one for a neurological disorder. IRBs have rejected proposals to do left heart catheterizations on children who have no known heart disease but who are at risk for the development of heart diseases in the future.

Proposals to do research presenting "more than minor increments" above minimal risk must be justified according to even more stringent standards. Approval of such proposals in the USA may be granted only by the Secretary of the Federal

Department of Health and Human Services after consultation with a panel of experts in pertinent disciplines.

### Interventions presenting the prospect of direct benefit

Research protocols that present more than minimal risk of physical or psychological harm to children but in which the risk is "presented by an intervention that holds out the prospect of direct benefit for the individual subject" may be considered differently. In such cases, the IRB must agree that such risk is justified by the anticipated benefits to the subjects and that the relation of the anticipated benefit to such risk is at least as favorable to the subjects as that presented by available alternative approaches.

This policy calls for an analysis of the various components of the research protocol. Procedures that are designed solely to benefit society or the class of children for which the particular child-subject is representative are considered as the research component. Judgments about the justification of risks imposed by such procedures are to be made as already discussed. For example, if minor increments above minimal risk are presented by nonbeneficial procedures, the special protections already discussed must be observed.

The therapeutic components of the research protocol, however, are to be considered precisely as they are in the practice of medicine. Risks are justified by anticipated benefits to the individual subjects and, further, by the assent when appropriate of the child and the permission of the parents or guardians.

In general, a child's objection should be binding unless the intervention holds out a prospect of direct benefit that is important to the health or well-being of the child and is available only in the context of research designed to evaluate its safety and effectiveness.

The general presumption is that parents may make decisions to override the objections of school-age children in such cases. However, in some circumstances the objections of teenagers to decisions on their behalf by parents may prevail. In the practical world of decision-making about who can authorize a therapeutic procedure, whether it be investigational or accepted, it rarely suffices to point to the law and thereby identify the person who has the legal right to make the decision. Many factors must be taken into account in reaching judgments about the capability of various persons to participate in and, in the event of irreconcilable disputes, to prevail in such choices. In general, these judgments become more complicated as the child gets older or as the stakes get higher. The necessary considerations have all of the richness and complexity of the same considerations encountered in the course of the practice of medicine. This is because this class of research activities resembles medical practice at least as much as it does the conduct of other classes of research involving human subjects.

In some jurisdictions of the USA adolescents are permitted to obtain medical treatment or advice for certain diseases or conditions without the permission or knowledge of their parents. Such diseases and conditions include, but are not limited to, sexually transmitted diseases and birth control. Adolescents are generally permitted to participate without parental permission in research designed to advance knowledge of these conditions even when the research has no therapeutic components. Mature minors are also permitted to authorize their own participation in many types of research that present minimal risk (particularly social and behavioral research), even though it has no relation to any disorder or condition they might have.

## Epilogue

Early on in this chapter I observed that the fundamental question presented to the National Commission in the 1970s was: "Should we do research involving children as subjects?" Then I offered two succinct statements of the meanings of answering "Yes" or "No." Now let us reflect further on these two statements.

"If we do, then we shall be doing research on individuals without their informed consent, an apparent violation of the principle of respect for persons unless one can assure that they will be protected sufficiently from harm." Now, 20 years later, we have a much richer understanding of the meaning of "sufficient protection from harm." Armed with this understanding we have been much more confident about setting the limits on what types of research can be authorized by parents and guardians on behalf of their children, particularly when the children can participate in the decision-making process by giving their assents. In short, we have refined our capacity to show respect for children.

"If we do not, then we shall be depriving children as a class of persons of the benefits of research." In the early 1970s, the prevailing perception was that participation in research was a burden. Persons or groups of persons who could not protect themselves from the burdens of research participation through the devices of informed consent were labeled vulnerable and therefore in need of special protections. Now we recognize that some of our past policies and practices designed to protect vulnerable persons from unwanted or unwarranted burdens actually deprived them of important benefits.

The development of United States federal policy on research involving children reflects these and other refinements in our understanding of how we ought to treat children as individuals and as a class of persons.

## Suggestions for further reading

Freedman, B. *Moral Responsibility and the Professions.* New York: Haven Publishing, 1982.

Gaylin, W. and Macklin, R. (eds.) *Who Speaks for the Child?* New York: Plenum Press, 1982.

Grodin, M. A. and Glantz, L. H. (eds.) *Children as Research Subjects: Science, Ethics and Law.* New York: Oxford University Press, 1994.

Levine, R. J. *Ethics and Regulation of Clinical Research*, 2nd edn. Baltimore, MD: Urban & Schwarzenberg, 1986.

McCormick, R. A. *Moral Theology Challenges for the Future: Essays of Richard A. McCormick.* New York: Paulist Press, 1990.

Melton, G. B., Koocher, G. P. and Saks, M. J. (eds.) *Children's Competence to Consent.* New York: Plenum Press, 1983.

Ramsey, P. *The Patient as Person.* New Haven, CN: Yale University Press, 1970.

# Future challenges of medical research review boards

CHARLES R. MacKAY

EDITORS' SUMMARY

Charles R. MacKay, PhD, probes the principles, structure, and procedures governing American Institutional Review Boards (IRBs) that examine all medical research that might involve risk to human subjects. Procedures and structures like this have been adopted by many advanced countries. After detailing the ethical principles of respect for persons, beneficence, and justice, he explores challenges for these principles and the doctrine of informed consent for research on AIDS, increased research on women and minorities, genetics, Alzheimer's disease and mental disorders, and human embryos. The waiver of the need for consent is a difficult and complex topic that must be judged differently for each new challenge. As medicine and science advance, the structure for research review must also be adjusted.

For a decade from 1974 to 1983, leaders in the research field engaged in sustained efforts to establish effective national standards to protect individuals who take part as subjects of biomedical and behavioral research. In America, they were given that charge by the United States Congress after dozens of hearings had produced findings of repeated failures to safeguard the rights and well-being of citizens in research activities conducted and supported by the United States government. What emerged was a comprehensive system, based on widely shared ethical standards incorporated into a detailed set of federal regulations, administered by the

National Institutes of Health (NIH), intended to govern the conduct of research. Agreement was reached among scientific, academic and government leaders that the new standards would effectively forestall reoccurrence of the notorious episodes that prompted the need for reforms. In fact, the standards were to gain world-wide acceptance as the model adopted by other nations and international bodies. But some recent disclosures cast a shadow over this optimism.

A reporter's efforts to uncover secret research led to government acknowledgment that hundreds of radiation experiments had been carried out for over two decades under United States government sponsorship at major government and university research centers prior to 1974. Anticipating intense public concern, President William J. Clinton established in December 1993 a presidential level commission to investigate the research and to determine whether or not unethical research may be continuing. One is led inescapably to ask: "Why didn't the earlier intensive review unearth this dubiously ethical research?" "How does it affect the credibility of the extensive reforms now in place?" or "Is this no more than a delayed aftershock?"

Another unfolding saga involves cover-ups of irregularities in decades-long government-supported research of the world's largest program of breast cancer studies. The irregularities and cover-ups occurred in the face of explicit measures to prevent them. Even when confronted with the facts, the researchers stubbornly sloughed off the gravity of the lapses. No doubt, the largest casualty is the reliability of information on breast cancer treatment. But evidence of seriously flawed procedures for obtaining informed consent and frequent deviation from the safety limits of the protocols leads directly to questions about the adequacy of protections for research subjects. The researchers' "stone-walling" appears to mock the trust placed in them and to belittle the values on which the system rests.

Still other unnerving incidents lead to questions about the candor of researchers. When federal agencies entrusted with protection of research subjects squabble about where to attribute responsibility for multiple deaths in an experimental study of liver disease, public confidence in the government's system to protect individual research subjects is eroded. How reliable is the system to protect those who volunteer to take part in experimental drug trials? Did the decades-long efforts of reform really achieve their aims? Or are these atavistic behaviors isolated events?

## The protection system

Over the last decade, developments have placed enormous pressures on the system relied upon to protect research subjects. First, the workload has grown exponentially. The threefold increase in volume of protocols involving human subjects has more than offset the major provisions made originally to allow opportunity to subject research with significant risks to intensive review while handling minimal or no risk research in a streamlined fashion.

Second, new ethical questions of historic proportions have emerged. Sweeping changes in the directions and imperatives of research have occurred as researchers turned to deal with: the AIDS epidemic; human gene therapy and the revolution in molecular biology and DNA technology that made possible the Human Genome Project; initiatives to include the elderly, women and minorities in research; more rapid transfer of research findings to clinical settings through clinical outreach programs; research involving *in vitro* fertilization and the human embryo; research in the field of emergency medicine.

Third, marked shifts in the patterns of funding for research and its organization have altered traditional academic research. The system devised in the 1960s and 1970s was designed around the paradigm of researchers working on their own projects within an academic environment. Since that time, the portion of total biomedical research funding by NIH has shrunk from just under 50% to 30%. As competition for funds intensified, researchers have turned to sources and opportunities that do not fit the academic model. They have been forced to seek alternative sources because NIH was making proportionately fewer awards. The growth of industry-sponsored research and the potential of biotechnology has increased the industry funding to be now 50% of total national health research and development. There is thus more research outside the direct sphere of NIH influence and oversight. Indeed, most United States scientists today work for corporations. These new sources of funding carry more potential for introducing tangible and commercial interests, possibly more influential than academic advancement and scientific prominence. More and more research is organized on a collaborative, multicenter basis, reaching into clinical settings where the influences of the closed academic setting are not powerful. Government oversight is thus more remote and attenuated. Yet the agenda for protecting research subjects is tougher than ever. And it is caught up in the squeeze on the academic research environment and the clamor for return on investment.

Thus, as the research enterprise extends the frontiers of human capabilities, it challenges many long-held notions and forces us to redefine values. From the ability to sustain human life with only marginally human traits to the capacity to manipulate the molecular vectors of human traits, the questions of when to allow death and how to create life artificially are literally unprecedented. Innovations such as organ transplants, the artificial heart, test-tube babies, genetic therapy, and genetic engineering are also unprecedented. The questions these advances thrust on us have never been faced before. Transplanting a baboon heart into a tiny infant, artificially fertilizing eggs retrieved from the tissue of an aborted fetus, tethering near lifeless bodies to mechanical life support devices – the new technologies force us to redefine our notions of life and death. They require us to come to grips with the questions: "If these research notions are to be redefined, who is to do it?" "Is it in the hands of the one who controls the technology, or in the hands of those whom it is intended to benefit?"

Thus, it is no coincidence that leadership in research has been paralleled by developments in research ethics. Research ethics instructs us that researchers may not legitimately pursue their interests at the expense of the interests of the individuals who take part as subjects of research. It reminds us that those individuals alone rightfully define their own interests. Research ethics has developed a framework that allows us to analyze the implications of advances in research and their application to clinical medicine.

## An ethical framework for research

This ethical framework is a charter, an amendment to the social contract that bestows on researchers their authorization to conduct research involving other fellow human beings as subjects. Whether clinicians or scientists, it is not one's inherent right to embark on research efforts outside the bounds of the bedrock values of our society. Thus, the framework spells out explicitly the precise steps researchers are to follow in protecting subjects. This charter is underscored by the Nuremberg Code, the Helsinki Code, the *Belmont Report* and federal or other government regulations that prescribe the terms under which research may be carried out. In grappling with the advances of biomedical and behavioral research, our moral sensibility has become more acute. Lack of moral clarity is no longer a mitigating but is, instead, an aggravating factor. The right of researchers to test the limits is one that needs to be circumscribed by valid ethical concerns of our society and culture. The researchers who take this mission upon themselves must accept the responsibilities that go with it. An overview of the ethical framework will give us the background for understanding the imperatives of research involving human subjects.

Key events in the early days of the biological revolution shaped scholarly and public debate about the complex questions arising from and in research in the life sciences. In the early 1960s, nations were startled by the reports that a new tranquillizer, thalidomide, caused grotesque deformities in infants born to women who took the drug during pregnancy. Moreover, the women were not told that the drug was experimental or possessed any possible adverse consequences. Later another drug, DES (diethylstilbestrol), was found to cause cancer to develop in the maturing reproductive systems in adolescent children of women who had used the drug to prevent miscarriages. The dramatic impact of these reports left the public fearful of hidden hazards in research medicine and wary that patients might be unsuspecting research subjects.

Other revelations of unethical research reinforced the concern that citizens, particularly those whose abilities to understand or whose social position left them open to exploitation, might be unwitting research subjects. In New York, senile Jewish residents were injected with live cancer cells at the direction of a researcher who though stripped of his medical license, was later given an award for his distinguished contributions to cancer research. A prestigious and later acclaimed researcher at the Willowbrook State School, an institution for retarded children, required parents to have their retarded children infected with the hepatitis virus as a condition for their entry into the institution (see Levine, this volume). The most flagrant episode was the disclosure that the United States Public Health Service, whose ethical standards were extolled as the acme, had for 40 years experimented on hundreds of black men from rural Alabama without their knowledge in a syphylis study, denying them access to medical treatment of known effectiveness. The nation and the Congress were outraged that government physicians had abused their position of trust and continued an experiment decades after treatments were available (see Levine, this volume).

Thus, it was against a background of notorious abuses that the American Congress mandated a review of the ethics of research. What emerged from the decade of intense high-level study and deliberations were the elements of a "system" to guard against a repeat of similar abuses, while allowing urgently needed research to continue. The system encompasses two major realities: the conceptual ethical framework of principles guiding the activities of researchers and a highly developed infrastructure which operationalizes the ethical framework. This section briefly reviews the bedrock values of society articulated in that framework and describe how the infrastructure operates to fulfill the goals of the framework.

The keystone of the system is the body known as the Institutional Review Board, or IRB. (In other countries it is called the ethics committee.) The purpose of this committee is to ensure that knowledgeable and disinterested individuals independently review any research that involves human subjects. Its composition reflects both scientific expertise and sufficient understanding of the ethical dimensions of research. To ensure independence, at least one member must be unaffiliated with the organization conducting the research. The committee must include a nonscientist. Many research organizations have more than one IRB to handle the review load.

The IRB's responsibility is to determine that no individual faces unwarranted or undisclosed risks and voluntarily agrees to take part in the research. The IRB has full authority to require safeguards that in its judgment are needed to protect the rights and welfare of the individual subjects, both in the actual research activity and in the procedures for recruiting and selecting subjects, including procedures for informed consent. The IRB has great authority over research under its jurisdiction. Its review and approval is necessary before any research involving human subjects may be conducted. IRBs review the research periodically throughout the life of the project to respond to any emergent problems or new information that might affect the rights and well-being of the subjects. It has authority to require modifications to protect subjects or suspend approval if hazards occur or researchers deviate from the terms of approval. Neither the researcher nor the organization may reverse any modification the IRB has required, or overturn its determinations.

The fairly detailed records of the NIH allow some estimation of the investment of time and resources devoted to ensuring the protection of human subjects. Records do not provide exact numbers of individuals who are involved as subjects of research, though rough estimates indicate that at any given time in the USA they number over 1.5 million. Nationally, personnel directly involved in IRB review of the research under the NIH's purview devote at least 70 000 person-hours each month, solely to review activities. Taken together with estimates of staff and administrative costs of review, this translates into a figure of about $160 000 000 annually for review costs alone, but does not reflect any calculation of time of researchers and their staff in planning and carrying out aspects of the research related to protecting humans, such as obtaining consent or increased monitoring for safety. Nor does it reflect so-called opportunity costs, those costs associated with the need for increased time in process-related activities, some of which may be mainly bureaucratic functions of routine record-keeping and filing of reports.

The operations of the sprawling infrastructure are grounded on a coherent ethical framework that integrates the administrative tiers of responsibility – federal, institutional, IRB, and researchers. There are three levels to the ethical framework for protecting human research subjects: principle, procedure, and practice.

### *Level of principle*

By now the principles of the protection of research subjects are widely known and applied in other areas. The *Belmont Report* (1978), which synthesized the years of deliberations of the United States National Commission for the Protection of Human Subjects, laid out the basic principles that guide researchers and society in assessing the involvement of individuals as subjects of research in morally legitimate ways: respect for persons, beneficence and justice. The articulation of principles represents a marked advance over earlier ethical codes and precepts, which remain limited in their prescriptive formulation. The three principles highlighted by the National Commission sum up the basic ethical traditions of Western culture.

### *Respect for persons*

American society places a high value on the rights and dignity of the individual. In some contexts, individual rights are regarded as inalienable and United States civil and legal institutions are mobilized to secure and protect them. In the context of research, the notion of "respect for persons" means that to ensure full self-determination the informed consent of the research participant must precede involvement in the research. The *Belmont Report* specifies the distinct components of informed consent: information, comprehension, and "voluntariness." Only after providing the prospective subject with an understandable, complete description of the procedures, the risks and benefits, and the alternatives is it morally legitimate to involve individuals in research. Further, the individual must comprehend what is involved in taking part in the research. "Voluntariness" means that individuals must not be induced or coerced to take part in the research under some physical or moral threat, for example the denial of medical treatment or the potential of less than adequate care, if they refuse or withdraw. They must also be free to discontinue their participation at any time, with no loss of benefits.

### *Beneficence*

"Beneficence" has two dimensions: the avoidance of harms, and the maximization of benefits. This means that there must be a systematic assessment of the research to identify potential risks, the safeguards against those risks and the likelihood that the research will actually lead to results that benefit individuals and society. Poorly designed research may not only cause immediate harm to individuals; if its results are equivocal, it will be necessary to repeat the research, thus placing other individuals at the same risk or inconvenience.

### *Justice*

The principle of "justice" has several meanings in the context of research, but the meaning of equity or fairness in the selection of the subjects is at the root of all. This means not only that benefits and burdens of a particular research project should be fairly apportioned, but that, to the fullest extent possible, individuals should have access to research that may benefit them individually or as a class. Increasingly, as research is seen as an avenue to improve both understanding and treatment of illness, there has been strong pressure to initiate research directed at groups and conditions that have been neglected. Thus, justice has both a positive thrust – addressing needs – and a negative sense of avoiding exploitation of some groups, particularly those who may be vulnerable owing to their social or health status. Quite recently, the federal agencies that oversee clinical research have adopted, owing to Congressional prompting, a more proactive stance regarding the participation of women and minorities in clinical research to better address the need of these long underserved groups in the population.

## *Level of procedure*

These principles are transcribed at the level of procedure in the formal Federal Regulations for the Protection of Human Subjects. The regulations spell out the basic elements and procedures for obtaining and documenting informed consent as the essential standard for respecting individual autonomy. They provide for a process designed to communicate sufficient information, to promote comprehension and ensure voluntariness – the essential components for informed consent. Not only do the regulations enumerate specific items of information that must be conveyed to the subject. The regulations go a step further in authorizing the IRBs to "require that information, in addition to that specifically mentioned in [the regulations] be given to the subjects, when in the IRB's judgment the information would meaningfully add to the protection of the rights and welfare of subjects."

The principle of beneficence is embodied in procedures in two ways: requirements for review, and criteria for approval. Authorities, qualifications, and composition of the reviewing body, the IRB, are intended to ensure maximum independence and disinterestedness as well as to promote full discussion and consideration. Their disapprovals cannot be reversed by institutional or governmental officials. In turn, federal regulatory offices independently evaluate the qualifications, composition, and independence of each IRB. Records of deliberations and decisions must reflect dissent and are open to review by federal officials.

More substantively the deliberations are guided by clear instructions on the criteria to be followed in determining whether to approve the research. These criteria provide for assessment of risks and benefits as well as several other factors that directly and indirectly impact on the overall reasonableness of the risks. Besides assuring themselves of the adequacy of circumstances and process for obtaining consent, the IRB is required to review the way subjects are selected and recruited, particularly when the subjects are vulnerable to coercion or undue influence because of illness or other disadvantage, to guard against pressures or exploitation. Now, in addition, IRBs must also consider whether the recruitment of subjects ensures adequate representation of those who can benefit from the research. Special emphasis is placed on ensuring that "there are adequate provisions to protect the privacy of subjects and . . . the confidentiality of data." These factors evidence how questions of justice and fairness are intertwined with the evaluation of risks and benefits. Individuals vary in their assessments of risks and benefits and their willingness to take chances. But some situations per se are unacceptable, either because of inherent hazards, or because the totality of factors indicates a disproportionately serious adverse impact for some individuals or class of individuals. Risks acceptable for many can become too great for some because the preponderance of the factors involved shifts the balance.

## Level of practice

On the level of practice, the system has functioned remarkably well by several measures. As the complexity and volume of research has increased, the process of review and the built-in safeguards have adjusted to ensure that important and needed research moves ahead without diminution of protection for subjects. The system has, in fact, effected a lasting and measurable change in the attitudes and behaviors of researchers and particularly in the generation of researchers who have entered the field since it has been in place. Though recent empirical data on the performance of the system are not a refined measure, they suggest some ways in which the system appears to accomplish its goals effectively.

As part of its operations in support of biomedical and behavioral research the NIH requires all proposed projects to undergo rigorous scientific review. Three times each year over 150 panels of scientific experts meet to review proposals. About 10 000–12 000 of the projects propose to involve human subjects. Before the project is accepted for scientific review. The NIH requires that documentation for each project from both the local IRB and the institution that the research subjects will be adequately protected. Thus, the research plan must describe the adequacy of the safeguards and the reasonableness of the risks for subjects, as well as other ethical considerations, such as recruitment and consent procedures.

Even so, the expert scientists are instructed to assess these procedures independently. Whenever the reviews identify problems, NIH officials insert a block in the computer system that generates award notices until the problem is resolved. After scientific review, the NIH advisory boards and councils, which comprise both scientists and members of the public, also review the research that is to be actually funded. Senior NIH officials managing the scientific program and those in charge of protecting human subjects must concur that the matter is cleared up before the computer block can be lifted and an award made.

These additional levels of review complement local review by concentrating a depth and breadth of expertise to identify in advance of the research possible hazards that might not be detected at the local level. Not only are these added measures of protection. The national-level review addresses in part another concern frequently raised: local IRBs lack sufficient expertise to assess highly technical and innovative procedures that characterize so much of today's research. Expert national review is already an element built into the system. Prospective review eliminates foreseeable problems and potential hazards; the rare problems are isolated and not systemic. The overall net effect of the review system has been to restore public confidence in the value of research by demonstrating that compromising the rights and well-being of individuals is not an inevitable social cost of research. A major study by the NIH is underway to assess IRB performance and to yield information on steps that might further strengthen it.

Earlier descriptions of the enormous increase in the volume of human subjects research, the changing patterns of research funding and organization and the sweeping change in directions and imperatives of research indicate the real need to revisit the system in some depth. The study will employ several methods: survey, on-site examination of IRB records and written materials, and an independent panel of reviewers to examine selected cases. Overall, more than 250 separate measures of performance have been identified to develop findings on the adequacy of performance at four levels: outcome, output, process, and resources. At the level of outcome, for example, the study will examine the extent to which risk–benefit assessment and the informed consent process are operating effectively. On the level of output, the study will provide data on IRB workload and the time and effort absorbed in the types of modification IRBs make, often skewed in the direction of bureaucratic changes with little impact on consent or research procedures. On the level of process, the study will yield systematic information on actual operations and procedures of IRBs, which will be invaluable in contributing to improved operation. Finally, the study will develop information on actual costs and effort involved in carrying out IRB functions. Thus, the study will provide empirical data on how well the system is functioning and the basis for any recommended modifications.

## Future challenges

The IRB system came into existence in response to concerns that research was harmful. However, the stunning successes of research and the effectiveness of the IRB system itself have created a different set of expectations. Increasingly, research is seen as beneficial. How are IRBs adjusting to this new climate? How is this more benign view of research going to affect the way IRBs operate? What are the ethical challenges to be addressed over the next few years?

### *Acquired immunodeficiency syndrome (AIDS)*

Some point to the AIDS story as a turning point in attitudes toward research. Those at risk for the grim, rapid course of AIDS had no alternative but research.

Experimental interventions at least had some prospect of benefit, whatever the risks. This called for a different IRB response. It was argued that those facing the agonizing consequences of HIV infection had prior moral claim to assess the risks of experimental drugs. In this situation, the role of the IRB was to ensure that they had correct understanding of the risks and no illusions about the results. Although the combination of a major health threat and an activist group in command of its destiny might warrant a less "protectionist" IRB stance, not all at risk of developing AIDS fit this picture. Infants, drug abusers, adolescents, and teenage mothers represent an ever-increasing proportion of AIDS cases. In these populations, IRBs cannot limit their role to enhancing autonomy. It is already difficult enough to try to ensure it. Yet the principles of beneficence and justice strongly urge offering the experimental alternative to those who are desperately ill.

This is not a simple decision. Scientific methodology requires randomization to ensure the highest level of reliability in the results. This rules out the operation of even factors unconsciously biasing the choice of individuals to receive the intervention and puts the intervention to rigorous test. But it means that not all who take part in the research will obtain whatever benefit the experimental intervention offers. This is considered ethical because the benefit is not yet proven and its risks may not be fully known, so those who do not receive it are also spared the risks. In the case of competent adults the IRB has reasonable grounds to conclude that they have sufficient self-determination to process the information and voluntarily agree. There is no such basis among the other populations. Thus, although there are compassionate grounds to include individuals whose capacity to comprehend is doubtful or absent, it is not justifiable to include them because they do not understand the research. The IRB encounters the inescapable tension of its role when the benefits of research are available only in the restricted form of randomized protocols.

### Mental and cognitive disorders

These ethical tensions are not unique to AIDS. Research is urgently needed on Alzheimer's disease and other dementing disorders, especially as we experience greater longevity and an increasingly older population, because these currently untreatable and progressive diseases will become more common. Also, research on drugs for mental diseases requires exquisite sensitivity to protecting subjects from harm. Yet some avenues of research are blocked because of the impossibility of ensuring the informed consent of individuals whose mental faculties are inherently impaired. This problem restricts some research undertaken to understand and treat many forms of mental illness. In these areas, when the interventions hold no clear benefit for the individual subjects, as in randomized trials or in studies that involve invasive procedures of greater than minimal risk, only voluntarily consenting competent individuals may take part in the research. The imperatives of beneficence and justice – extending the benefits of research to those conditions and individuals with greatest need – are to be reconciled with the principle of respect for persons.

## *Women and minorities*

One especially thorny nest of questions that IRBs face immediately with no thoughtful national guidance is the fallout of recent major policy changes related to the inclusion of women and minorities in clinical research. Two recent federal initiatives have brought about this situation. One is the new policy of the United States Food and Drug Administration to eliminate long-standing restrictions on the involvement of women of child-bearing potential in clinical studies of new drugs, devices and biological compounds. The other is the requirement of the United States NIH Revitalization Act of 1993 imposing strict standards on involvement of women and minorities in clinical research conducted or supported by the NIH. Although both steps are welcome and long-overdue rectifications of discredited practices, they challenge IRBs in several ways.

For decades research has been haunted by the specter of thalidomide and diethylstilbestrol (DES) (see Levine, this volume). But in time it became clear that these were often pretexts for skirting potential liability questions and for simplifying the tasks of researchers who were reluctant to deal with the complications that female physiological and metabolic processes introduced into research. The result was a strong indication of a systematic bias excluding women and their health concerns from research. Women certainly and minorities – for other reasons – have been underrepresented in research. This effectively denies them the opportunity of participating in research, often directly beneficial to individuals, and more broadly leads to neglect of questions directly related to their health status and effective treatment. But, as noted above, participation in research is not an unequivocal benefit.

Accordingly, researchers and IRBs must now give consideration to measures to safeguard women and minorities against the risks of research. Women are still susceptible to the mutagenic and, if pregnant, teratogenic side-effects of many interventions. We have virtually no data and precious little experience on the myriad social and cultural factors that have the potential to heighten not only psychosocial risks, but also medical risks of many minority groups. For example, health and dietary practices among many groups may introduce complications of protocol compliance: underlying genetic and familial characteristics not well described in medical literature suggest a special level of vigilance in monitoring minorities. It is not encouraging that the many pages of policy and guidance coming out of NIH in implementing these requirements make no mention of concrete measures, such as requirements for pregnancy testing, or general exhortations to be alert to special needs of these populations for protection. Unfortunately, the measures used to safeguard these research subjects may amount to little more than incorporating into the informed consent an equivalent of the warning label on cigarettes or alcoholic beverages, the *caveat emptor* approach. This essentially minimalist approach, besides failing to address substantive questions, contributes to devaluing the importance of informed consent.

## *Informed consent problematics*

Despite enormous achievements of scholarship and policy development on its importance, informed consent still serves as a "lightning rod" for expressions of dissatisfaction with the operations of the system for protection of research subjects. Both detractors and supporters of the system pounce on informed consent with equal vengeance. For example, in open national meetings of serious scholars interested in safeguarding individual privacy, informed consent was denounced as meaningless, in a nihilistic diatribe. On the other side, those who embrace the primacy of autonomy and informed consent are sorely vexed by the way the legalistic consent form has usurped the opportunity for communication, dialogue, understanding and shared decision-making that are the heart of the process. While the breadth of these issues calls for a separate treatment, several questions regarding informed consent loom as immediate challenges to IRBs and researchers.

As noted earlier, the requirement for informed consent prior to participation in research makes research involving cognitively impaired persons difficult and even impossible. A similar situation faces researchers and IRBs in the case of research on emergency medical interventions, a pressing research priority. Victims of cardiovascular attacks or of severe trauma often suffer complete or partial loss of consciousness, thus rendering them incapable of informed consent. The law provides for interventions in such life-threatening situations, based on the premise that individuals would have consented to the interventions as in their own best interests had they been capable. Thus, experimental interventions of direct benefit to individuals are acceptable without consent, following this line of thought.

However, the technique of randomization requires that the individual first give consent to the research intervention. It clearly is no longer arguable that the experimental procedure is in the best interests of the individual. The aim is rather to find out the answer to a research question, even though the researcher continues to exercise due diligence for the welfare of the research subject. One cannot presume that individuals have expectations of being randomized when they are carried into an emergency room.

A way around this predicament may be found in an interpretation of NIH regulations. IRBs are authorized to waive the consent requirements if certain conditions can be met. This authority is different from what exists in the FDA rules to permit use of experimental drugs or devices in medical emergencies in clinical practice when no equivalent alternative exists. The NIH rules apply in situations of "minimal risk"; that is, when the seriousness and likelihood of the risks are no greater than those faced in ordinary life. It has been argued that the risks facing an individual needing life-saving intervention meet that test, when the procedures to which random assignment is made are at least equivalent to alternative available interventions for the life-threatening condition. Accordingly, the IRB could determine to waive the requirement for informed consent, if the other conditions are met.

Could an equivalent case be made for the research on dementing disorders or on mental illness? Probably not, because, although the risks of ordinary life for those afflicted by those conditions are extremely serious, they are not faced with a life-and-death situation. Many of the interventions proposed exceed either in known risks of invasiveness or unknown potential side-effects what would be the ordinary physical condition of the individuals without the procedures. Thus, it is hard to argue in this fashion that proposed research risks would fall within the range of minimal; that is, equivalent to the necessarily aggressive life-saving intervention in the emergency or acute setting.

### Genetics

The approach to resolving another impasse on consent in genetic studies may also lie in the waiver provision. In one scenario, existing biological specimens collected under a variety of conditions over time now loom as critical to advancing our understanding the genetic basis for health and disease owing to the advent of powerful technologies of DNA analysis. However, it is impractical and even impossible to recontact the individuals to obtain their consent. Some have argued that no DNA or genetic research carries minimal risk because of potential harms to the individuals from both discovering or disclosing things about them that are not only private but closely guarded secrets integral to their self-identity. While it is important not to downplay risks associated with genetic research, these claims seem to reflect an extreme position. Evidence of harms generally ensuing from DNA analysis has not been systematically collected and analyzed.

Without disputing the claims, still it is possible to hypothesize situations where the possibility of the harms could become vanishingly small. Thus, protection of the information might be so strong that there would be far less likelihood of the harm from mishandling of the information than from a hypothetical deliberate action of retrieving saliva or a fallen strand of hair for analysis. Another argument against use of the waiver provision is based on the individual's inherent right to the information in the DNA as intensely intimate. In this line of argument a consent waiver would sacrifice a fundamental right, thus failing to meet a condition of the waiver. This view holds that the genetic information encoded in nucleic acid is literally information, a result of too much metaphorical license in usage of the word "information," to say nothing of an almost paranoiac extreme view of personal privacy. Clearly a significant number of very technical and quite costly intermediate steps are required to transform DNA sequences into meaningful information. Again the burden of proof should fall on those who contend there is a definable interest that is sacrificed.

Other ethical issues raised by genetic studies present greater difficulty because paradigms have not yet emerged to guide us. In some areas, even when there is not consistency among researchers and IRBs about what is the best balance between protecting the subjects and accomplishing the aims of the research, there is, nonetheless, agreement about the items of information to be disclosed to subjects as part of the consent process. Thus, it is common to see information relative to: whether specimens collected will undergo DNA analysis; access of other researchers to the specimens or to the research information and the conditions of such access, including any specific consent from the subject; whether subjects may be recontacted, by whom and under what circumstances; possible implications for employment and insurance of the findings, and other items. Other areas are more problematic and so how they are handled and what is disclosed are more likely to vary, depending on the style and convictions of the individual researchers. Unlike other areas of research there was no settled corpus of experience for IRBs to turn to, which could furnish precedent and basis for working out detailed norms that appeared successful over time in protecting the interests of subjects without impeding research.

Though some direction and guidance for IRBs has been developed, there remains an enormous shortfall in addressing their information needs and questions. For example, the great diversity in the field itself translates into major uncertainties for IRBs in handling of information related to early or presymptomatic testing. This is further complicated by the rapid strides in the field that identify the genes or gene markers for many diseases and conditions when there is no specific intervention to treat or cure them.

This situation causes IRBs some difficulty in determining when genetic information is really beneficial to subjects: sometimes early information is too uncertain to be meaningful; sometimes information is reliable but no therapy exists; sometimes the information seems harmful in itself, as in presymptomatic testing for which no treatment exists; sometimes the information disrupts family comity, or in the case of children, renders them liable to be stigmatized. The large number of genetic diseases and the variability of their expression contributes to the uncertainties that plague IRBs.

### The human embryo

A new and ethically controversial application of recent findings in genetics has emerged in research on *in vitro* fertilization and the human embryo. It is now possible to biopsy a single cell from the 4- to 8-cell stage embryo, freeze the embryo for possible later implantation and subject the biopsied cell to a battery of genetic tests, such as those for sickle cell anemia, cystic fibrosis, Huntington's disease, Down syndrome and other genetic conditions. If the results show presence of the genetic or chromosomal defect, the frozen embryo simply would not be implanted. The risks and inconveniences associated with harvesting ova make it unlikely that this procedure would be elected by any but the couples already planning *in vitro* fertilization to treat fertility problems. However, in the USA this means millions of couples might be candidates for this approach. This clinical application is based on techniques, which, if indications are correct, will come into wider use when NIH begins to fund human embryo research within the near future.

In February 1994, a 19-member task force began deliberating and receiving extensive public comment on the ethical issues involved in human embryo research. As the panel's views began to crystallize, it was clear that its recommendation would be that human embryo research could be done ethically and it began to outline the conditions and restrictions that might attach to such research. The potential for increased understanding of human reproduction, early development and disease are simply too compelling in the panel's view to forego after they reached a determination that early stage *in vitro* laboratory embryos were not developed to the stage that warranted protections given to the fetus or infants. In its report in 1995, limits on laboratory testing were proposed: prohibiting research beyond 14 days; not using ova, sperm or embryos whose donors had not explicitly consented to research use; special protections in the case of ova or embryos destined for transfer; prohibition of all forms of commercialization, commodification or payment in connection with gamete or embryo donation. The panel's final recommendations require approval at several levels of government. Already there is mounting opposition from some groups and in the Congress to the prospect of human embryos research. Here the lessons of earlier government inaction after recommendation of advisory groups is instructive. In the cases of both *in vitro* fertilization and fetal tissue transplant, failure to take action to implement the recommendations led to blocking of government funding in these two important research areas. The result has been not only to slow scientific understanding of these fields, but also to drive them outside the sphere of academic and peer-reviewed research. The resultant uneven progress has not served the public interest.

## Suggestions for further reading

Annas, G. J. and Grodin, M. A. *The Nazi Doctors and the Nuremberg Code: Human Rights in Human Experimentation.* New York: Oxford University Press, 1992.

Fox, R. C. and Swazey, J. P. *Spare Parts: Organ Replacement in American Society.* New York: Oxford University Press, 1992.

Holtzman, N. A. *Proceed with Caution: Predicting Genetic Risks in the Recombinant DNA Era.* Baltimore, MD: Johns Hopkins University Press, 1989.

Jones, J. H. *Bad Blood: The Tuskegee Syphilis Experiment,* 2nd edn. New York: The Free Press, 1992.

Rothman, D. J. *Strangers at the Bedside: A History of How Law and Bioethics Transformed Medical Decision Making.* New York: Basic Books, 1991.

Veatch, R. M. *The Patient as Partner: A Theory of Human Experimentation Ethics.* Bloomington, IN: Indiana University Press, 1987.

# Animals in research

## FRANKLIN M. LOEW

### EDITORS' SUMMARY

Franklin M. Loew, DVM, PhD shows how the arguments today for and against the use of animals in research are essentially like those anti-vivisectionist arguments of a hundred years ago. Those who argue for using animals cite the benefit to human beings. These are utilitarian arguments. Those arguing against using animals in research and testing argue, either from the rights of animals or from utility, that the benefits to humans do not outweigh the burdens to animals from suffering and confinement. Loew includes the recent efforts to find alternatives to using animals, as well as the extensive lobbying and media campaigns of both sides of the issue.

## History

The animal research controversy has a long history and it seems to follow a 50-year cycle of waxing and waning. From 1850 to 1900, the controversy about using animals this way grew and required the serious attention of leaders of society toward the end of the century. From 1900 to 1950, the issue gradually disappeared from view as a significant societal problem. Then, from 1950 onward, it began to develop once more and by the 1980s, again demanded the serious attention of politicians, scientists and the public. It is not clear whether the issue will begin to fade away again in the twenty-first century or if its new intellectual under-pinnings will sustain it. However, the issues and arguments put forward in the nineteenth century about vivisection as it was called are, with one exception, exactly the same as those we are dealing with today. They still remain largely unresolved. The one exception is the use of alternatives to animals, which holds out the promise, in the view of its proponents, of having the fruits of research without having to bear the cost in animal pain, distress and death.

## Animal numbers

The statistics on laboratory animal numbers in the USA are crude and relatively unreliable. In Europe, Britain has kept figures on laboratory animal use for over 100 years and most countries in the European Union are now required to collect and report accurate statistics on animal use. These figures indicate that animal use has been falling in Europe since the late 1970s and early 1980s. In Switzerland and Great Britain, for example, animal use has fallen by 50% from 1980 and 1975, respectively (to around 1 million animals in Switzerland and 3 million animals in Great Britain in 1992). For other countries, laboratory animal use has fallen by 20–40%.

In the USA, the data on laboratory animal numbers are not as good. However, annual surveys were conducted in the 1960s by the National Academy of Science's Institute for Laboratory Animal Resources (ILAR) and, from 1972, by the United States Department of Agriculture (USDA), which has kept statistics on dogs, cats, primates, rabbits, hamsters and guinea-pigs but not rats and mice. It is possible to track the number of these animals used annually from around 1960 and the data show that numbers peaked in the late 1960s, fell rapidly in the early 1970s, remained stable for the next ten years and then began to fall slowly from around 1985.

Since 1968, the number has fallen by over 50%. However, rats and mice usually account for 80–85% of the laboratory animal total and it could be argued that the ILAR/USDA data do not reflect trends in mouse and rat numbers (ILAR does record mouse and rat use but the data exist only for the 1960s, 1971 and 1978 – a 40% decline was recorded between 1968 and 1978). Other data, from the Department of Defense (DoD) and corporate laboratory records, indicate that mouse and rat use fell from 25% (DoD) to 70% (Hoffman-La Roche) during the 1980s.

## How much animal pain and distress?

Public opinion polls and reaction to media stories indicate that, when the public becomes concerned, it is primarily concerned with laboratory animal pain and distress. Even the (painless) killing of laboratory animals is also perceived to carry a cost (particularly by those who work in research laboratories). However, we have very few data on the extent of animal pain and distress in research. The USDA requires registered laboratories to report their animal use (not including rats and mice) in three categories – research causing no pain/distress (category C), research causing pain/distress that is relieved by drugs (category D), and research causing pain and distress that is not relieved by drugs (category E). However, the USDA has never provided guidelines to help institutions decide how to classify their research (for example, if drugs are given to relieve pain for some, but not all of the time, should it be placed in category D or E?).

Nevertheless, the USDA returns indicate that 5–6% of all animal research is placed in category E, but there are very large differences among institutions and states. For example, Kansas reports that over 40% of its animals are used in category E research while many other states that use large numbers of animals report that less than 1% of all their research falls into category E. Some corporations that do toxicity testing (where pain-relieving drugs are usually not used) report no animal use in category E. Many non-profit institutions are very reluctant to place animal research in category E because they believe they will be targeted by animal activists if they do. Thus, it is very probable that the USDA statistics underreport laboratory animal pain and distress, however mild some of it may be.

The only country that has collected systematic data on animal pain and distress is the Netherlands. Their 1990 *Annual Report on Animal Experimentation* notes that 53% of the animals experience minor discomfort, 23% moderate discomfort, and 24% severe discomfort. About one-fifth of the animals in this last category were given medication to alleviate pain. Examples of procedures that would place animals in the "severe" category are prolonged deprivation of food or water, some experimental infections, tumor research and $LD_{50}$ testing (finding the dose at which 50% of the animals die). Laboratory animal research causes less pain and distress than is implied by animal protection literature but more animal pain and distress than is claimed by research advocates.

## Regulatory structures

Prior to 1970, animal research was largely unregulated in the USA. In 1966, the American Laboratory Animal Welfare Act was passed to regulate dog and cat dealers but research institutions were not included. In theory at least, many institutions had animal care committees on their books at this time but, if they functioned at all, they were mainly concerned with allocating space for research animals and setting the rates for maintaining animals in the facilities.

In 1970, the Laboratory Animal Welfare Act became the Animal Welfare Act and all institutions registered under the Act were required to follow regulations that governed the care of dogs, cats, primates, rabbits, hamsters and guinea-pigs.

How those animals were actually used remained outside the scope of the Act. Nonetheless, in response to rising public criticism, institutions began to address the question of how research animals should be used in experiments in addition to the routine care and housing they should receive. In 1981, the University of Southern California reworked its animal care committee and started to oversee how animals were used at the university. A local animal activist was even appointed to sit on the committee. Other institutions began to follow this lead.

In 1985, the United States Public Health Service (PHS) revised its animal use policy and required all institutions receiving its funds (mainly from the National Institutes of Health (NIH)) to establish animal care and use committees to review and approve animal research protocols. The new policy was based on the model of the Institutional Review Boards (IRBs) established in the 1970s to review research using human subjects. The new animal research committees began to grapple more and more with how animals should be used. Then, at the end of 1985, major amendments to the Animal Welfare Act were passed that required all registered research institutions (not just those receiving PHS funding) to establish Institutional Animal Care and Use Committees (IACUCs). The IACUC was required to review and approve animal research protocols prior to any animal research being conducted and to pay particular attention to reducing pain and distress in research animals. In addition, the amendments required institutions to address the psychological well-being of primates and the exercise and socialization needs of laboratory dogs.

Today, those using laboratory animals in the USA have to conform to a wide range of housing and care standards and also have to address a variety of issues dealing with how the animals are used. In particular, if the animals are likely to experience pain and distress (even if alleviated by anesthetics or analgesics) the investigator has to demonstrate that he or she has looked for alternatives. IACUCs also pay much greater attention to the need to prevent pain and distress. However, there are still tensions about any interference with how animals are used and the boundaries of IACUC power to prevent particular research projects.

## Justifying animal research

Animal research is almost always justified in terms of its great utility in improving human and animal health, while the costs of such research in terms of animal harm and distress are considered to be small by comparison. Sometimes, it is also argued that animal research has played an important role in the development of basic knowledge about biology.

Although some critics argue that animal research has played no role in the advance of medical knowledge, such arguments are plainly wrong. There are many examples where animal research and testing have played an important part in the development of new knowledge or insights that have led to improvements in medical therapy. Some animal research projects have proved to be more important than others, but experience indicates that it is not possible to predict which research is likely to be more important than other research in building our understanding of human and animal biology and disease.

In the past 10 to 15 years, research advocates have begun to draw on more emotional arguments to prove that animal research is necessary, rather than simply listing the medical advances that are based on animal research. Patients who have benefited from modern medical technology have come forward as spokespersons to endorse the importance of animal research. This approach has been developed to counter the strong emotional arguments of the critics of animal research.

## Criticizing animal research

The critics of animal research have always employed emotion-laden images to protest the use of laboratory animals but, in the past 20 years, they have also developed a range of reason-based arguments that are grounded either in moral philosophy or that employ methods of argument and citation used in scientific discourse. By adopting a scientific style of argument, animal research critics are tapping into the authority and credibility that science enjoys in modern society.

Animal research is criticized on moral grounds either because animals are argued to have inherent moral rights that would prevent their use in research (rights-based arguments) or because animal research causes more animal harm and distress than an equivalent degree of benefits for humans and animals (utilitarian or consequentialist argument). The rights-based arguments do not necessarily hold that animals and humans have the same rights. The utilitarian argument is very similar to that used to justify animal research. The difference between the research advocates and the utilitarian critics is that the critics argue that animal research causes considerable animal pain and distress for little or no real benefit for the most part, or for benefits that could be gained from using alternatives.

The critics have also put forward a range of technical arguments claiming that animal research is either not necessary or not as important as implied by the research advocates. These arguments may be summarized as:

1. Better use of preventive medicine will eliminate the need for animal research.
2. Public health and epidemiological research is far more important than animal research in improving public health.
3. Clinical research has provided the key insights in advances in medical

treatment and animal research has merely been employed to dramatize clinical findings.

4. The development of alternatives eliminates the need to use animals.

The importance of preventive medicine, and of public health, epidemiological and clinical research is not in question in this debate. However, research advocates do not accept that the above approaches are either being ignored or that they obviate the need for animal-based research. In addition, alternatives have not advanced to the point where they could replace all animal use.

There is plenty of room for legitimate and even interesting and constructive argument in debating the relative importance of animal research and its cost–benefit characteristics. Unfortunately, arguments are usually presented in relatively absolute terms and the research establishment has shown little interest in debating the technical merits of animal research with their critics for fear that it may give the critics what is perceived to be undeserved legitimacy.

## Animal testing

Laboratory animal use in testing is different from animal research because the main aim of testing is either to establish whether a product is safe for use (e.g. vaccines and biologicals) or to determine the level and type of toxicity associated with a new product (e.g. new drug testing). No hypothesis is being developed or tested in routine animal testing.

Animal testing accounts for between 10% and 20% of all laboratory animal use. Most test regimens for the toxicity or hazard (identifying safety) estimates of a chemical or product employ animals at some point. Such tests have been developed over the past 60 years because of a perceived public health need and because the common laboratory animals, being mammals like humans, are viewed as being sufficiently like humans to provide useful data comparable to responses in human beings.

In the past 20 years, criticisms of such tests have grown and have stimulated a widespread reevaluation of the need for and role of, animal testing. In addition, animal protection criticism and the rapid advance of biological technology have spurred interest in toxicity testing that does not use whole animals (animal organs, animal or human cells, and computer modeling are some of the possible alternatives that are being explored).

In Europe, Japan and the USA, there are numerous projects to develop and validate alternatives for animal testing. Regulatory authorities are working to harmonize testing requirements and support the validation of alternatives. Industrial and academic toxicologists have largely accepted the need to develop, validate and implement alternatives. However, establishing hazard or safety is not easy and data from a laboratory mammal still provide a level of regulatory history and confidence that is not yet seen with the new alternative tests. As experience with the new alternative tests grows, and as knowledge about toxic mechanisms continues to increase rapidly, so the need to perform animal tests will decline. However, animal testing will not disappear in the foreseeable future.

## Animal use in education

Animals have traditionally been used in educational exercises to teach manual skills or to demonstrate known principles of biology or methods of research. Animal protection advocates oppose most use of animals in education because, they argue, the skills, principles and methods can now be taught just as effectively using models, computers or some other teaching aids. Research advocates resist this criticism because they see educational exercises using live and dead animals as essential in stimulating interest in biology and in teaching future members of the public biological literacy and the importance of biological and medical research.

Currently, the debate over animal use in schools focuses on dissection and a student's right to opt out of the laboratory without penalty. Several states have passed laws that specifically permit a student the right to choose. Research advocates are concerned about this because they perceive that, if students are allowed to opt out of dissection, it challenges the school's authority to teach what it considers necessary and how it should be taught and it also might lead to declining standards of biological literacy.

Ironically, the country's medical and veterinary schools are now allowing their students to opt out of animal laboratories if they so choose. Thirty-four of the 126 medical schools have no animal laboratories and 61 of the remainder allow students to opt out of animal laboratories. More and more veterinary schools are allowing students to opt out of the surgery on purchased laboratory dogs and are teaching surgery skills via other means (e.g. student spay/neuter clinics on animals from a local humane society).

There are very few empirical data that either support or refute the contentions of either side. This is an issue where the firmness of the conclusion is inversely proportional to the amount of hard evidence supporting it. The evidence that is available supports the contention that factual (declarative) knowledge can be learned just as effectively from books, lectures and videotape but that problem-solving skills (procedural knowledge) is much more effectively learned by performing laboratory exercises. In addition, unpublished research suggests that factual knowledge and formation of values are unrelated.

## Alternatives

The concept of alternatives developed from a 1959 book that suggested that researchers should seek to *Replace* animal use where possible, *Reduce* animal use where possible, and *Refine* animal research techniques so as to reduce animal pain and distress as much as possible. These three Rs now constitute what most people identify as Alternatives although there is a tendency for both sides to focus on replacement and ignore reduction and refinement.

As mentioned above, animal use in American laboratories has dropped by up to 50% in the past 20 years and it is generally considered that part of this fall resulted from the promotion and adoption of the idea of alternatives but nobody knows how much. In addition, more attention is being given to reducing animal pain and distress in research.

In the USA, there is a certain amount of "schizophrenia" about the concept of alternatives. While corporate toxicologists and regulatory scientists have mostly accepted the term "alternative" and are comfortable working to develop and implement alternatives, academic scientists and their main funding source, the NIH, and many research advocacy organizations reject the use of the term, preferring to use "adjunct" and "complementary methods." Those who reject the term tend to see it as a Trojan Horse planted by the animal protection movement that will lead to great harm for medical research if allowed to gain a foothold.

It appears as though most of the public who pay attention to this issue use the term "alternative" and so do legislative bodies. The United States Congress recently mandated the NIH to develop a plan for promoting and implementing alternatives but, to date, only the National Institute of Environmental Health Sciences (which happens to be heavily involved in developing new toxicity testing methods) has publicly embraced the term.

## Role of the media

Scientific organizations have often suggested that animal activists have skillfully manipulated the media (thereby gaining an unfair advantage) because of the images of animals under experiment that they have provided or because animal activists have particular public relations skills. It is true that animal images have a particular pull on the public (equal to human infants) but there is no evidence that animal protection organizations have any greater public relations skills than the scientific organizations who defend the use of animals.

Throughout the 1970s and the early 1980s, the general media's coverage of animal protection issues was largely favorable to the animal groups. However, this began to change around 1985/86. More articles critical of the tactics and claims of the animal groups appeared. The change was not the result of a reappraisal by journalists but by more proactive and aggressive tactics by research advocates who decided that the animal rights threat warranted significant attention. Once they set their mind to it, the scientific organizations and specialized groups formed to defend animal research could call on significant resources, including funding, sophisticated public relations skills and experience, many excellent contacts with the media, and high profile and respected spokespersons. Given the skills and contacts available to the research and testing community, it could be considered something of a surprise that the animal protection community can point to any positive stories in the past five years, let alone still to be holding their own.

It has sometimes been argued that the media converted the animal research controversy from a nonissue into a major story. However, it is not clear that the media have such power. In the 1930s and 1940s, the powerful Hearst newspaper chain adopted the anti-vivisection cause and yet, after two decades of campaigning against animal research, the public still favored animal research by an overwhelming margin. The media does not convert nonissues into major stories. Instead, skilled journalists have sensitive news antennae that sense the moods and concerns of the public before others do and then develop stories that address those concerns. Thus, journalists do not make a public issue so much as articulate it when public concerns reach a certain level.

## Tactics and strategies

### Animal protection

The animal protection groups have traditionally relied on "public education" and new legislation to change animal research practices. The public education initiatives were designed to inform the public about the "horrors" of animal research and then legislative initiatives would be introduced to eliminate the problems and to regulate any remaining use of animals. With the growth of the movement, other tactics were developed and implemented.

High profile campaigns succeeded against narrowly defined targets that were chosen to provide maximum advantage for the critics (e.g. cat sex experiments at the American Museum of Natural History, pig "torture" experiments sponsored by Amnesty International, and eye irritancy and lethal dose ($LD_{50}$) testing by the cosmetic industry). Campaigns with more diffuse goals (e.g. the national ProPets campaign against the laboratory use of unclaimed pound dogs and cats) generally did not have the same success.

People for the Ethical Treatment of Animals (PETA) used undercover investigations and material stolen by the Animal Liberation Front (ALF) to expose research practices. They were particularly successful with two early cases – the exposés of the Institute for Behavioral Research (the Silver Spring Monkey saga[1]) involving an undercover investigation (or infiltration depending on one's point of view), and the University of Pennsylvania head trauma laboratory, involving videotapes of the experiments on baboons stolen by the ALF and later edited into a half-hour exposé.

The use of stockholder resolutions as a way of bargaining with public corporations began in the 1980s and is now a common tactic.

Animal protection organizations composed of, and aimed at, specific professions were established (e.g. Physicians Committee for Reform of Medicine, Association of Veterinarians for Animal Rights, and Psychologists for the Ethical Treatment of Animals). These groups provided a source of expertise and credibility to the animal movement and also served as something of a counterbalance to the existing professional societies that supported animal research, like the American Medical Association.

The animal protection "movement" also continued its legislative lobbying and public education but, with even more members and more money, it was able to do both more effectively. Many of the organizations hired Washington lobbyists to represent their interests. Fund raising and public education mailings were distributed to a million or more constituents as opposed to 100 000. Both of these actions increased the political impact of the animal protection groups on Capitol Hill in Washington.

While the initial undercover investigations and break-ins by the ALF were aimed specifically at exposing conditions in animal research laboratories (i.e. the liberation of information), there were also cases of vandalism (up to and including arson) and anonymous threats were issued against research scientists and their families. These tactics of intimidation led some research advocates to categorize animal rights (and some animal welfare) organizations as violent, anti-science groups and even as supporting terrorism. Such categorization began to have an impact, and many of the establishment animal protection groups publically criticized acts of vandalism and intimidation as being counter to animal rights philosophy (i.e. no harm to any sentient being, including humans). The boundaries of legitimate protest and civil disobedience in animal protection campaigns remain to be defined and articulated.

### *Research advocacy*

Research advocacy and professional scientific and health organizations tended to ignore the animal protection movement until the early 1980s. A new United States research advocacy organization, the (now National) Association for Biomedical Research, was started in 1979 because existing organizations were perceived to be unable to deal with the expanding animal protection movement. In July 1985, Margaret Heckler, then Secretary of the Department of Health and Human Services, suspended a grant to the University of Pennsylvania head trauma laboratory because of violations of animal care and use policies. This was a wake-up call for the research community, who began to develop programs to counter the animal rights movement.

The Association for Biomedical Research (which had many corporate members) and the National Society for Medical Research (which had many university and medical school members) combined forces to form the National Association for Biomedical Research. Many states either established state-based societies for medical research or revived organizations that were active in the early 1900s but had then gradually fallen into a dormant state.

These groups developed a range of tactics and approaches. They monitored state and federal legislatures and lobbied against animal protection legislative initiatives. In Congress, they introduced and eventually saw passed and signed into law an act making theft and destruction of property at a research facility a federal crime and subject to the jurisdiction of the Federal Bureau of Investigation (FBI). They developed numerous brochures and other materials for the public, including a rather successful series of posters. They supported the development of patients' organizations to counter animal protection campaigns and emphasize the importance of animal research to the advancement of medical knowledge. They also developed a variety of curricular and other materials aimed at school teachers and school children that are designed to confirm the importance of animal research and reaffirm how good laboratory animal housing and care are.

While research advocacy organizations like to argue that animal protection groups together have a very large annual budget to devote to campaigns against animal research, the playing field is more level now than it was in the 1970s. While the national animal protection groups in the USA probably together devote around $15 million annually to the animal research issue, they often do not work together or coordinate their activities.

The research advocacy groups together currently devote around $5 million a year to support the need for animal research. However, these funds do not include the activities of the professional scientific and medical societies, of the NIH, or the many corporations that are now actively engaged in the debate. Given the fact that the research establishment also has better access to the sources of power and the policy makers in America, the debate over animal research now would probably favor those who support the need to use animals in the laboratory.

## Conclusion

It is likely that the balance of public opinion will begin to edge back towards greater support for the use of animals if current trends and tactics for and against the use of animals in research remain unchanged.

### NOTE

1. The Silver Spring case involved an animal activist from PETA, Alex Pacheco, volunteering at Dr Edward Taub's laboratory in Silver Spring, MD, for the summer of 1981 to find out first hand what went on in research. At the end of the summer, evidence provided by Pacheco to the Montgomery County Police led to the charging of Dr Taub with cruelty to animals and to the confiscation of 17 monkeys housed in the laboratory. The subsequent cruelty trials, Congressional hearings, NIH investigation and later battles for the custody of the monkeys horrified the scientific community, upset many in Congress and in the general public, and helped to boost PETA from a small grass-roots group into a rapidly growing national organization.

### Suggestions for further reading

Loew, F. M. Animals in research: Public policy determinants. In *Through the Looking Glass*, ed. M. A. Novak and A. J. Petto, 11–19 Washington, DC: American Psychological Association, 1991.

Loew, F. M. Animal agriculture. In *The Genetic Revolution*, ed. B. Davis, pp. 118–31. Johns Hopkins University Press, 1992.

Rollin, B. *The Frankenstein Syndrome*. New York: Cambridge University Press, 1995.

Rowan, A. *Of Mice, Models and Men*. Albany, NY: State University of New York, 1984.

Rowan, A. and Loew, F. M. *The Animal Research Controversy*. Boston, MA: Tufts Center for Animals and Public Policy, 1995.

Tannenbaum, J. *Veterinary Ethics*, 2nd ed. St Louis, MO: Mosby, 1995.

# Taking duties seriously: Medical experimentation, animal rights, and moral incoherence†

DANIEL A. MOROS

EDITORS' SUMMARY

Daniel A. Moros, MD, presents a tightly reasoned account of why ascribing rights to animals is, in his view, morally incoherent. He examines both strong and weak arguments for not permitting the use of animals, from a rights and utilitarian point of view. His aim in examining each argument is to show that there is a significant moral difference between animals and humans. More specifically his aim is to demonstrate how incoherent are the arguments for the claim that humans have duties to animals because animals have rights. Essentially this means that rights demand a corresponding change in human behavior toward animals. He finds no basis for such rights. He does not, however, argue that we should not be kind to animals as a virtue of human society. His argument provides solid support for using animals in medical research and teaching.

Modern medicine is an experimental science built upon, and demanding, research with both animals and humans. Almost all of its researchers, practitioners, and regulatory bodies accept the moral precept that, before human research can be justified, animal models must be appropriately exploited. Hence, a decision, either

---

† An earlier version of this chapter was presented at the Oxford University–Mount Sinai Medical School Conference on Medical Ethics, April 1992.

implicit or explicit, on the moral standing of animals is unavoidable for all involved in the medical enterprise.

Arguments about the moral standing of animals and the propriety of experimenting with animals are typically based on claims about rights or utility. Animal rights supporters assert that differences between humans and animals are not significant enough to exclude animals from the moral community, particularly if we allow that human infants and severely retarded human beings have rights. Utilitarians concerned with the treatment of animals claim that animals experience pain and pleasure, and so have interests, and therefore, must be accorded moral standing. Indeed, two social movements parallel these philosophic positions – an animal rights movement and an animal welfare movement.

Those arguing that animals may be used in research (and teaching) also tend to embrace either a rights or utility perspective. Rights theorists simply deny that animals have moral standing, arguing that they are not part of our moral community. Utilitarians focus on the necessary role of animal experimentation in combating human disease and they emphasize the harm to present and future generations that will be avoided because of animal research.

Moral analysis ultimately confronts us with how our behavior ought to be influenced by our insights. Clearly, proponents of animal rights want to influence what people do rather than judge or alter the behavior of other animals. They want people to stop eating animals, to stop experimenting with animals, to stop using animal skins for clothing. They do not claim that humans have a duty to train carnivores to stop eating other animals in order to protect herbivores from their "natural" predators.

I am interested only in examining arguments which claim that humans have duties to animals because of an animal's rights. Arguments for animal rights that impose no human duties do not ask for any change in our behavior. In what follows I elaborate on inconsistencies in both utilitarian and rights-based claims about human duties toward animals. To reveal the inconsistencies I examine implications for human relationships that logically follow when we elevate the moral standing of animals as compared to human beings.

We demand consistency in our moral reasoning and it is appropriate to challenge individuals if they embrace distinct and incompatible ethical perspectives, one creating duties toward animals and another creating duties to our fellow human beings. Whatever basic principles we embrace must be applied to both the human and nonhuman domains. I argue that in order to be consistent and also make sense of claims for humans having duties to animals (including a duty to leave animals alone), we must weaken many of our basic moral notions of rights, duties, and autonomy to such an extent that the concepts become impotent when applied in ethics, and especially in medical ethics. Thus, rather than occupying the moral "high ground," animal rights advocacy may reflect a lack of moral insight and a failure in public moral discourse.[1]

## Animal rights

Ultimately the claim that animals have rights may take one of two forms. An animal's right to life may be regarded as either (a) equivalent to or (b) weaker than that of a person. I call these two views the "strong and weak positions on animal rights'." In the second case (the weak position) there are circumstances in which it is appropriate to cause the death of an animal but where it would not be correct to cause the death (or even risk the life) of a human being. I try to demonstrate here that the strong position is untenable and that the weak position fails to create powerful duties for humans.

### *The strong position on animal rights*

The animal rights position implies that there can be a natural (biologic) self-sustaining community all of whose members have important rights (e.g. not to have their lives abruptly terminated for the benefit of others) but none of whom have any duties. Yet two aspects of the animal world create particular difficulty for any claim to an animal having rights. First, the biologic (animal) world has been shaped by, and is permeated with, predatory activity. Indeed, the animal world cannot exist without the checks and balances of predation. Therefore, if animal rights exist and are recognizable to thoughtful people by a sort of moral intuition, the natural world must be viewed as inherently immoral (as opposed to merely nonmoral) as such rights are never respected. Second, it seems unreasonable to regard animals as having duties, and more peculiar still to consider that they should be punished for disregarding their duties. Thus, the animal rights position requires that rights be thoroughly disconnected from duties. In fact, if we look closely we see that the claim for animal rights has no moral significance except where animals share the world with human beings. According to the animal rights advocate, humans (but not animals) have duties to respect the rights of animals to life, freedom from human-induced pain and confinement, etc.

A variety of ethical problems emerge from the strong position on animal rights. Foremost is the difficulty in adjudicating competing rights. For example, imagine that you are a superb marksman and that while carrying a rifle you see a wolf attacking a human infant. A strong animal rights position would require that you regard the animal as exercising its rights to hunt and to eat (easily derived from its right to life). Just as clearly, the wolf would have to be regarded as violating another creature's right to life. Any argument that supports shooting the wolf and saving the baby will need to ignore the fact that the baby is human, because if we rank the value of persons above that of other animals we will have abandoned the strong position on animal rights.

Perhaps we can construct an algorithm that makes the aggressor always the wrongful party. Then we would be obliged to shoot the wolf and save the child. However, given this rule, when a farmer's dog chases and catches a squirrel, we become obliged to shoot the dog. And if we do not, then we are again embracing an important distinction between the human baby and all other animals. Furthermore, what are we to do if it is a human hunting an animal for sport? Do we shoot the person, or does the human's pursuit of pleasure also take precedence over the animal's right to life, however much we might disapprove of this human activity.[2]

Preserving the strong position on animal rights by always choosing to sacrifice the aggressor creates additional problems. Among humans, if we know that one group is murdering another, we generally regard ourselves as having some obligation to try to save the innocent. The force of this obligation varies with our proximity to the events and it certainly increases when we have the means to intervene. Now the animal world is filled with hunters – minks, bobcats, owls, etc. They kill the innocent (herbivores). The algorithm suggests that we have an obligation to hunt the hunters, eliminating carnivores whenever possible. Obviously, fulfilling this obligation would lead to an ecological disaster. Alternatively, we can recognize the predator's right to hunt everything including the human baby, or simply value humans above animals, or we may decide that appropriating rights talk to discussions of animal relationships is inappropriate.

Is there another principle for ranking rights that would both preserve the strong position and still direct us to save the human infant? Perhaps we are allowed to exercise our preference where there is no clear priority for one moral choice over another. However, the decision to save the infant would then become merely a matter of preference rather than duty, and a decision not to save the infant would be neither wrong nor punishable. Clearly, all of the above ethical problems vanish if we reject the notion of animal rights. However, they might also be mitigated by embracing a weaker position on animal rights.

### The weak position on animal rights

A fallback position on animal rights would maintain that even for humans the right to life is not absolute, that rights frequently conflict, and that in circumstances where human rights conflict with animal rights, a specific judgment will be necessary. This weaker position maintains that there are circumstances that justify infringing an animal's right to life but not a human's right. Thus, an animal's right creates a duty, which to some degree restricts our behavior. It is, however, clearly less weighty than a person's right to the same thing. My objective in this section is to demonstrate that only the strong position on animal rights confers significant obligations on humans.

What constitutes an adequate reason to violate an animal's right to life? Clearly we allow carnivores to hunt and to kill herbivores. Thus, the right of a carnivore to hunt and eat (a necessary part of its right to life) is at least the equivalent to the right to life of the hunted. So it would seem that a person's significant need for food will warrant killing animals. Further, a person would not be morally required to risk his or her own life and wait until the verge of starvation to act on such a need. If such personal risks were required, the weak and strong positions on animal rights would converge (i.e. without some human deaths there would be no meaning to the term "risk").

As the weak position requires that there not be too great a risk to humans in respecting the rights of animals, it is not yet clear how compelling these animal rights are. If we again look to the predators we see numerous instances of killing that are not individually necessary. We do not feel morally required to act when a well-fed young cat torments a mouse as it practices its skills. We seem to accept the equation that the carnivore's need to master its skills takes precedence over the life of a herbivore. However, since a person's life is more important than that of any other animal (the weak position), it should be acceptable for a physician to practice on animals to master skills, especially when those skills will ultimately be employed to save human lives. Similarly, testing new pharmaceutical agents on animals prior to the introduction of the agents in human trials can also be justified on the grounds that no humans should die in order to safeguard the life of an animal. The weak position on animal rights then is compressed to a claim that animals have some rights, are without duties, and may justifiably be sacrificed for a variety of human purposes.

If the weak position on animal rights is accepted, presumably we should avoid pointlessly harming animals, and all harming would require justification in terms of competing human rights. But there is an odd asymmetry here. Prudence dictates that we should be cautious in enforcing the moral requirement to respect animal rights, because, in the face of uncertainty and ambiguity, we must be certain that we do not exact too great a human price, and infringe upon the rights of humans.

How broadly may human rights be interpreted? What kind of human purpose will justify violating an animal's right to life? Again, we can gain some insight by examining the implications of our tolerance of predation. Clearly the young cat playing with a mouse, or the well-fed wolf of summer that could survive days without a meal, have no overwhelming, immediate need for their next kill, and yet we do not think we are ethically compelled to protect their innocent prey. Thus, an animal's right to life may be trumped by another animal's less pressing need. Indeed, there is no right of any animal against another that compels us to interfere. In our social world, ordinary property rights involving inanimate objects have greater force in compelling our behavior than do any rights of animals in the wild. The weak position on animal rights is ineffectual in that it imposes only relatively "lightweight" duties. It could not support most animal rights claims.

## Rights and duties

The absence (and unreasonableness) of *any* correlative notion of animal duty within either the strong or weak positions on animal rights results in moral claims with little persuasive power. While a discussion of the correlativity of rights and duties is beyond the scope of this chapter, it seems clear that animal rights supporters must either reject correlativity or, alternatively, regard nonhuman animals as having only insignificant duties (whatever they might be) that are never enforced.[3] In either case since, on their account, having rights does not immediately create duties in others, they must now come up with an alternative explanation of why humans have serious duties to animals as well as to one another. The best explanation that I can suggest on their behalf is that, with human understanding and rationality, comes an appreciation that rights (all rights) must be respected. Thus adult humans have duties, but animals that cannot appreciate rights do not have duties.

Such an explanation of duties, rooted in knowledge and understanding, raises another question. When did humans acquire the duty to respect the life of animals? It seems unlikely that this duty existed for humans in earlier hunting societies or that it exists even now in nutritionally marginal peasant societies. Under these circumstances humans would seem the equivalent of typical predators (carnivores), i.e. dependent on the eating of meat and fish and without the knowledge of what could be substituted for meat and fish in order to maintain an adequate diet. It would seem that the alleged duty to avoid meat-eating is a by-product of the growth of knowledge, technology and, ironically, animal experimentation.

The commitment of animal rights advocates requires them to maintain that having duties by dint of being rational nevertheless does not create any significant difference in the moral worth or standing of such individuals. Ethics expert Peter Singer, for example, has drawn an analogy between animals, human infants, and the mentally retarded. He argues that if we recognize the rights of infants and the mentally retarded (perhaps with lesser intellect than some animals), we will find no reasonable grounds to deny rights to animals. However, we can value rationality and still offer reasonable grounds for selectively extending rights to children and the mentally retarded. For example, we can argue that we recognize duties toward infants (i.e. infants have rights) because of their potential for developing into rational beings who will have duties.[4] As regards the mentally retarded, we can argue that they should be treated as if they have the full complement of rights in order to maintain our sensitivity to all humanity because, if we treat the mentally retarded as if they have no rights, we will be all the more likely to ignore the rights of less impaired and fully autonomous human beings. This argument is based on the conditioning of behavior and not the rights of the mentally retarded. Perhaps we grant the retarded many rights because it is risky to draw too fine a distinction about who is and who is not fully human. This would be an argument from prudent moral concerns rather than from a view of the intrinsic worth of the severely mentally retarded. However, without settling on a definitive argument, it seems obvious that there are possible grounds for recognizing a right to life of the profoundly mentally retarded that do not immediately extend comparable rights to animals.[5]

If the primary failure of the animal rights position is its vision of a world in which animals have important-sounding rights and no correlative duties, then an ethically significant dividing line is determined by our notion of duty. Prudential concerns may lead us to extend rights to creatures without similarly extending and enforcing obligations. However, this generosity is possible only within the context of a community of individuals who have both rights and duties.

## Duties to animals: Utilitarian perspectives

Even if the case against the strong and weak positions on animal rights were convincing (I hold it is not), the utilitarian could maintain that "rights talk" fails to capture what is morally significant about our treatment of animals. If we grant that animals can experience pain, utilitarianism can maintain that our duty to minimize pain applies to animals as well as humans. To quote philosopher Jeremy Bentham, "the question is not 'Can they talk?' but 'Can they suffer?'" (*Introduction to the Principles and Morals of Legislation*, 1789)

Ultimately a utilitarian calculation directs our action, i.e. informs us of our duty. As with rights-based arguments reviewed above, the utilitarian case for considerate treatment of animals also may take one of two forms: i.e. (a) humans have duties to animals that can be powerful enough, at times, to override important duties to other humans; and (b) human duties to animals rarely, if ever, take precedence over duties to other humans. I will refer to these views respectively as the strong and weak utilitarian positions on duties to animals. I try to show here that the first position is untenable and that the second imposes little restraint on human conduct.

### The strong utilitarian position on duties to animals

By "simple utilitarianism" I refer to a view that the correctness of an action is judged by the extent to which it maximizes the balance of pleasure over pain (or happiness over unhappiness) and, further, that pleasure and pain are simple, straightforward concepts even if, at times, difficult to assess precisely.[6] A difficulty with any simple utilitarian theory is that a utilitarian calculation requires some underlying assumption about human nature and what is pleasurable and painful for humans. As presumptuous as such assumptions may be in the case of humans, they are hopelessly confusing when extended to animals. We have little sense of animal pleasure other than simply imagining animals to be like people. We have no way of knowing how to equate different animals with one another or with humans. For example, the world will support many more rats than persons or elephants. Is the pleasure of a rat comparable to that of an elephant? If so, on utilitarian grounds we should be moving in the direction of more rats and mice and fewer elephants and human beings.

At this point we confront difficulties analogous to those encountered with the strong position on animal rights. Ultimately the purpose of the utilitarian calculus is to determine correct action – our duty. If animal pleasure and pain are equated with human pleasure and pain, then our duties to animals do not disappear when we merely cease to exploit them for our benefit as pets, dairy cows, etc. The human infant in need has no greater moral claim upon us than any other helpless baby animal. When the well-fed cat toys with a mouse, we have a duty to aid the mouse. And since we have a duty to maximize animal pleasure and minimize pain, we ought to be at war with the predators.

If we reject such a conclusion, then we will have rejected the strong utilitarian position on human duties to animals. If we are to justify our intuition that it is proper to abandon the strong position on duties to animals and still continue to cleave to a utilitarian framework for our ethical reasoning, we then either must, by fiat, simply value humans above animals or reject mere pleasure and pain as the critical parameters for our utilitarian calculation. For the latter approach we will require a modified version of utilitarianism that emphasizes goals and values (necessarily human goals and values, since these are not characteristics of animal life) in order to explain why human needs properly trump animal pain and suffering. We routinely use such a modified utilitarianism in our dealings with one another. Physicians cause pain in order to achieve "more important" goals, which may not be readily characterizable as producing more future pleasure. Parents recognize a duty to discipline children, though they may find this inconvenient and unpleasant. While an explicit, consistent rendition of this more "refined" utilitarianism has proved elusive, some version is implied by our behavior.

## *The weak utilitarian position on duties to animals*

Even if we reject the strong utilitarian position on duties to animals, which equates humans and many or most animals, there may still be a weak position that confers significant duties on human beings. The weak utilitarian position on duties to animals would maintain that because animals suffer pain, they are entitled to some consideration: *no claim is made in terms of animal pleasures or goals.* Two problems arise immediately. First, in actual circumstances, it is impossible to make a utilitarian calculation without some notion of pleasure or of a desirable goal. Since we routinely find it acceptable to cause pain to human beings with the aim of producing a greater or compensating good (e.g. medical treatment may cause pain, athletic competition may cause pain, meeting a deadline for a paper may be unpleasant), there is no obvious reason not to cause animal pain in return for a human benefit. Only the exact limits are unclear. If our calculation considers only animal pain, almost all painless killing of animals or experimentation on anesthetized animals is acceptable. The only open question is how much pain is acceptable? Furthermore, and perhaps more importantly, under circumstances where animal pain is a means toward achieving a significant human goal, it is difficult to see how to factor-in a duty to animals unless, by chance, the human goal itself is first judged to have a negative utility for humans. Thus, even in the case of animal experimentation that causes "unnecessary pain," there may be grounds for criticizing the experimenter, but no actual duty to interfere.

## Human nature versus animal nature

The above criticisms of the weak and strong position on duties to animals slides over one major difficulty. Entry into the moral community is based on some poorly delineated notion of human nature broadly characterized by having purposes and goals – and perhaps the potential for being a rational and autonomous agent. One cannot casually dispense with such concepts, because without them there is no way to regulate our behavior with fellow human beings. However, it is quite possible that the capacity for rational, autonomous action is shared by some other members of the animal kingdom. The inconsistencies demonstrated in the critique of the strong position on animal rights apply only to extending rights to all animals, or even all mammals. We can still argue about particular cases, e.g. whether or not we have duties to dolphins, whales, the great apes, and perhaps some others.[7]

## Kindness to animals: Is this a virtue?

Rejecting animal rights does not mean that people ought not be kind to animals. But a demand for humane treatment of animals must be rooted in a notion of how

people ought to behave, e.g. that they should encourage kindliness and thoughtful-
ness, that they should model such traits (or virtues) for each other and for the next
generation. There is a straightforward utilitarian argument for seeing kindness to
animals as a virtue, i.e. it encourages the habit of considerate treatment of fellow
humans. Furthermore, animals certainly seem to experience pain and presumably
also experience pleasure. We should want people to act in such a way as to favor
minimizing pain unless they have a clear reason for doing otherwise. However,
such a virtue does not impose a duty and must be informed by the realization that
we should always be cautious in restricting the highly valued freedom and rights
of persons in defense of animals.[8]

## Animal rights and medical ethics

It is appropriate to ask how the moral perspectives embedded in the animal rights
positions (particularly the strong position) would function in the context of medical
ethics. Here we are concerned with patient autonomy, patient rights, physician
duties, and the adjudication of conflicting duties. For example, the range of a physician's duties vary according to the autonomy (to some extent linked to the
rationality) of the patient. While the physician must ultimately honor the decisions
of an autonomous individual, caring for patients of diminished competence imposes
additional duties on the physician. While much of the work of modern bioethics has
focused on identifying and correcting an often casual disregard of patient autonomy,
the fact remains that paternalistic intervention is sometimes a duty.

The link between right, duty, rationality, and autonomy is critical. While we
may have difficulty providing precise definitions of these concepts, some core
meaning is readily available. Furthermore, the central issue for applied ethics may
turn out not to be "who is rational or truly autonomous?" but rather "who is not?"
Once we accord high value to autonomy, we must presume that people are in fact
acting autonomously unless there are strong grounds for doubt.

An animal rights supporter might argue that animal behavior raises questions
of rationality and thus prudence dictates that we extend the presumption of autonomy to animals and with it all of the rights that we extend to persons. However,
with such an argument the animal rights supporter must degrade the notion of
"rationality" (along with the notion of "rights") in order to include animals within
the moral community. Having abandoned the traditional criteria of autonomy and
rationality, the animal rights advocate is now left to wrestle with inconsistencies
such as excusing transgressions by persons with diminished rationality (and
accepting increased responsibilities for such persons) while at the same time credit-
ing animals with the appearance of rationality.

A related effort to ground animal rights, adopted by animal rights campaigner Tom Regan in his 1983 book, *The Case for Animal Rights*, is to impoverish our notion of autonomy. In this work Regan argues that there are two useful notions of autonomy. He correctly attributes to the philosopher Kant the view that "individuals are autonomous only if they are capable of acting on reasons that they can will that any other similarly placed individual can act on." Regan acknowledges that it is "highly unlikely that any animal is autonomous in the Kantian sense." However, Regan offers a second concept of autonomy. In this second sense, "individuals are autonomous if they have preferences and have the ability to initiate action with a view to satisfying them."

This view of autonomy as an explicitly stated preference is, however, unacceptable in medicine. If a patient states a preference for treatment X, but is uninformed, the physician has a duty to inform, rather than to simply cooperate. If the patient clearly wants to achieve goal Y, but insists on a treatment that makes the achievement unlikely or impossible, the physician has a duty to clarify the issues and rationalize the decisions. However, the physician must remain alert to the moral requirement to respect autonomy, which may at some point lead to cooperation without further obstruction. Distinguishing between such physician duties requires a notion of autonomy that goes beyond the mere ability to indicate a preference. Human beings have goals and often act for reasons. It is difficult to understand how we can honor human goals (at times rejecting stated preferences) while highly valuing animal preferences, perhaps at the price of sacrificing human goals. If we often view goals as more important than preferences, we must view creatures that have goals and act for reasons as being ethically more significant than creatures that can not be so characterized.

## NOTES

1. Rejecting the concept of animal rights should not be taken as a denial of the value of kindly behavior toward animals. It does imply, however, that the value of kindly behavior toward animals must be derived either on simple utilitarian grounds or from a beneficial effect upon persons and not from fulfilling a moral obligation to respect the rights of animals.

2. Since animal rights can clearly conflict with each other, if we take them seriously we will need a principle for ranking them, and there is no obvious one available.

3. Correlativity is the thesis that every right of an individual is associated with a corresponding duty for some other person. Thus the right of free speech imposes a duty of tolerance on all of us. A right to medical care implies that someone has a duty to provide it.

4. An argument from potential to become an adult human will solve the problem of distinguishing human infants from other animals but has additional implications. For example, if the issue of animal rights is decided on the basis of this claim, it will be

difficult to justify abortion (except in cases such as major central nervous system abnormalities or unlikely survival), because clearly the normal fetus has the potential to become an adult human.

5. There may also be rationales for limiting this right, but this issue is not central to the concerns of the present chapter.

6. With the exception of some animal welfare theorists, few people seriously embrace this simple utilitarian view. Thus, if throwing a single celibate (so we do not need to consider future persons) Christian to the lions creates tremendous pleasure for thousands, most would not support this action as correct. And if a medical research project will benefit thousands, including many of the patient participants, most do not believe that it would be proper for a doctor to conduct such research without the uncoerced agreement of the patient.

7. One can also argue that prudence may support extending the moral domain even further. However, arguments based on prudence must acknowledge scientific perspectives on animal life. Thus, if we have good reason to believe that a sense of purpose comparable to that of a person would be impossible to generate within the context of a rat's brain, then prudence would dictate against extending rights to such creatures and limiting the freedom of human action towards them.

8. Arguments for the preservation of endangered species or environmentally responsible social policies ought not to be based on duties to animals but rather on our duties to our fellow beings and future generations.

## Suggestions for further reading

Carruthers, P. *The Animals Issue: Moral Theory in Practice*. Cambridge: Cambridge University Press, 1992.

Regan, T. *The Case for Animal Rights*. Berkeley, CA: University of California Press, 1983.

Rollin, B. *Animal Rights and Human Mortality*. Buffalo, NY: Prometheus Books, 1992.

Singer, P. *Applied Ethics*. New York, London: Oxford University Press, 1993.

# Animal rights and social practices

TED BENTON†

EDITORS' SUMMARY

Ted Benton, PhD, claims that animal rights activists tend to be practical, while philosophers of animal rights are concerned with abstraction. To be effective, moral theory has to become more sensitive to the many practical contexts in which the issues arise. Benton then examines how utilitarians can trump any claims of animals for individual rights in favor of a general good outcome for all. By contrast, Tom Regan's deontological view of animal's "inherent value" would not permit such overriding of their rights. The problem of this view is that it would be difficult to argue that such inherent value is an absolute right when other beings with rights come into conflict. And what are we to make of forms of human life, such as the severely mentally retarded, and infants? Benton then proposes that there are many other conditions contributing to the value of an individual who is not a moral agent. These then influence our moral judgments when rights conflict. A good example is the difference between a factory farm and pet-keeping. To be effective, action must appeal both to moral theory and to the practical social context of action.

Philosophical discussion about the moral status of animals has emerged only quite recently along with a parallel rise to prominence of social movements that agitate

† The author wishes to acknowledge the support of an ESRC Senior Research Fellowship (No. HS247505494).

and campaign against what they see as abusive treatment of animals in our societies. Typically, social movements focus on specific types of abuse – seal culling, whaling, hunting for sport, the fur trade, safety testing of commercial products, zoos and animal circuses, intensive farming, neglect of and cruelty to pet animals, and, of course, use of animals as experimental subjects in scientific research. By contrast, the philosophical dispute about the moral standing of animals tends to work with highly abstract concepts and principles, descending only at a rather late stage in the argument to a consideration of how these principles might generate decision-making procedures for particular cases.

This gap between our need for ways of thinking about how we should act in particular situations on the one hand, and the abstract concepts of most moral theories on the other, has two unfortunate consequences. The first is that activists, understandably impatient for action to stop manifest abuse, are liable to dismiss the philosophers as an academic irrelevance. The second consequence is that philosophical opponents of the idea of a positive moral standing for animals find it too easy to pick holes in the philosophical case for animal rights without being required to face up seriously to the concrete moral issues at stake.

In what follows, I hope to offer some reasons for thinking that the moral status of animals *is* an important question, one which requires both a philosophical response and concerted social and political action. However, I suggest – though not fully argue for – a shift in the theoretical and philosophical terms in which the "animal question" is posed.

## Animals in philosophy

First, however, it is necessary to devote some space to a brief consideration of the main philosophical positions that have become established contenders in this arena. Interestingly, there has been relatively little philosophical work that argues explicitly against a positive moral standing for animals, though very recently the balance has shifted. Historically, the great weight of philosophical tradition in the West has not assigned much significance to the issue, but, where it has been addressed, the verdict has been against animals. Animals have been supposed to lack rationality, language, autonomy of the will, or, even, consciousness itself. Lacking these morally decisive attributes, animals have, generally, been excluded from moral consideration. Contemporary philosophers on the same side of the argument also tend to pick on one or other of this list of attributes. M. P. T. Leahy, for example, uses the later work of the philosopher Ludwig Wittgenstein to undermine the case for animal rights on the grounds that they are unable to participate in the "language games" through which moral concepts get their meaning. Peter Carruthers derives moral rights and obligations from a notion of contract, and so excludes animals from the human moral community on the basis of their inability to enter into contractual relationships.

This is not the place for a detailed treatment of these issues, but it does seem to me a rather telling point that neither the historical nor the contemporary opponents of a positive moral standing for animals are willing to draw the conclusion that our treatment of animals is therefore a matter of moral indifference – that we may treat them just as we like. These philosophers seem to recognize – even share – the widespread moral intuition of non-philosophers that it *is* wrong to be cruel to nonhuman animals, and that it *does* matter, morally, how we treat them. Generally, this uncomfortable tension is resolved by postulating what Tom Regan calls "indirect duties." That is to say, I owe a duty to be kind to an animal, not to the animal itself, but to some human moral agent: this might be its owner, someone who cares about it, or indeed, myself. But it remains unexplained why, if, indeed, animals have no moral status in themselves, it should matter to other humans, or to myself, how I treat them. Sometimes it is argued that there is some sort of causal connection between being cruel to animals and being cruel to humans but this would be hard to substantiate. Some in the German National Socialist Party leadership, for example, were famously sentimental about animals.

## Utilitarianism

Among the most well-known and influential philosophical theories that have been used in favor of animals is utilitarianism. For utilitarians, the decisive moral attributes are a capacity for pleasure or pain, and, since at least some nonhuman animals are clearly sentient in the required sense, their pleasures and pains should be included alongside those of humans in our moral choices between different possible courses of action.

From the standpoint of animal protection, utilitarian moral theory has the great advantage of focusing moral concern on psychological attributes which (more or less) uncontroversially cross species boundaries. However, utilitarian approaches to the moral status of animals are vulnerable to a series of criticisms that are also leveled at the theory in its application to humans. Two fairly obvious ones are, first, that not all of the morally important things in human life can be reduced to quantities of pain and pleasure, and, second, that some pleasures may be more inherently valuable than others and some pains more worth suffering than others. Interestingly, given the presumably less psychologically and socially complex lives of most nonhuman animals, it could be argued that utilitarianism is on stronger ground dealing with animals than it is with humans (at least with respect to these two lines of criticism).

However, it is a third problem with utilitarianism that is to the fore in the "animals" debate. This is to do with utilitarianism's commitment to "consequentialism" – i.e. to the doctrine that the moral rightness or wrongness of an action (or class of actions) is solely a matter of the good or bad consequences of the action. Because of this feature, it is often argued by opponents of utilitarianism that it can allow, or even recommend, harming some individuals so that greater benefits can be conferred on the majority. So, while utilitarians might insist that animal pain should be taken account of in a dispute over, for example, medical experimentation, the recommendation might still be that the experiment should be conducted because the overall consequence might be an aggregate increase in the balance of pleasure over pain across the whole community of sentient beings. There is, in other words, nothing to rule out a utilitarian approach giving with one hand and taking back with another.

### The rights view

Philosophers who take this to be a decisive objection to utilitarianism have tended to adopt an alternative, "deontological" basis for moral theory. This is not committed to ignoring the consequences of actions, but does insist that the moral character of an action is not *solely* a matter of its consequences: actions may be right or wrong (to some degree) independently of their consequences. So, for example, one might consider it simply wrong to kill or torture another human being, no matter what beneficial consequences might be believed to flow from it.

So far, the most fully developed deontological view in the "animals" literature is Tom Regan's advocacy of animal rights. For Regan, to have a right is to have a morally grounded claim, *vis-à-vis* some identifiable moral agent, who is both able to meet, and be responsible for meeting that claim. For example, human individuals would be generally accorded a right to respectful treatment from others with whom they have dealings. This would not, on the rights view, be conditional (as for utilitarians) on any calculation as to the beneficial or harmful *consequences* of according such respect. Rather, on Regan's view, the right follows from possession by human individuals of what he calls "inherent value." There is, clearly, a close family relationship between this idea and the Kantian view of humans as "ends-in-themselves" who should never be treated merely as means to ends.

However, where Regan differs sharply from Kant, and, indeed, most other advocates of deontological moral theory, is in his view of the *range* of individuals who are held to have "inherent value," and, so, to possess rights. Regan's criterion is that such individuals should be "subject of a life," in that they have enough psychological complexity to have some sense of themselves as beings with a past and future, some capacity for purposive action, and liability to suffer harm, or be benefited by the actions of others. Notice, this is a much stricter qualifying condition than the utilitarian requirement of "sentience," but, arguably, much stronger protection is accorded to such beings as are included. There are, it would seem, absolute, as distinct from merely contingent and conditional restraints on how such beings are to be treated.

Nevertheless, the qualifying condition is much looser than in other deontological theories, which generally assign "inherent value" to human moral agents only. If the full panoply of abilities to reflect upon the relationship between a contemplated action and one or more universal moral principles, and then to act in accordance with the outcome of such moral reflections, is required as a condition of "inherent value," then it would certainly seem that only *human* individuals could qualify as rights holders. However, as Peter Singer, Tom Regan and other advocates of the moral standing of animals argue, such a restrictive criterion would also rule out many human individuals who do not have full moral agency in the required sense: people in early infancy, and, possibly, sufferers from addictions, various sorts of compulsive psychological disorders, and so on.

The supporter of human rights who would reject animal rights is faced with a dilemma over these types of case. He or she must either deny that these individuals, though human, have moral rights, or concede some looser criterion of inherent value that allows into the charmed circle of rights holders, at least some nonhuman animals. Regan makes an important distinction here, between moral agents and moral patients. All individuals who are "subjects of a life" – i.e. who can be benefited or harmed by the actions of others – are "moral patients." This is to say, they are, on their own terms, entitled to moral consideration from those who are capable of giving such consideration. However, in so far as they lack (full) moral agency, moral patients cannot be held to have duties toward other moral agents or patients. Moral patients may be either human or nonhuman.

Moral agents, however, are (probably and contingently) all *human* individuals. They owe *direct* duties to other moral agents *and* to moral patients, to respect their rights. This distinction between moral agents and moral patients does seem to make sense of widely held moral intuitions, and undermines some popular and "populist" jibes against animal rights. It conforms to widely shared moral intuitions in that we commonly do, in everyday life, accept that we have moral obligations toward others without being able to expect reciprocal obligations from them. Bringing up children is a central life experience in which this generally holds. Where people take on responsibility for caring for animals as pets or for other purposes, similar asymmetrical moral relationships are quite generally recognized and powerfully felt. The distinction between moral agents and moral patients also helps to explain why support for animal rights does not entail, for example, defending the rights of prey animals against their predators, imprisoning lions for murder, and so on. The rights of moral patients can be claims against moral agents, only, and not against other moral patients.

However, the rights view does, I think, have some weaknesses. One, which is addressed by Regan himself, is that it is difficult to square absolute rights with other moral intuitions. In particular, there is a problem about what to do when rights conflict with each other – especially when the rights of human beings conflict with the rights of animals. No matter how careful we are to offer subsidiary principles to deal with such cases, it is hard to avoid acknowledging that in practice the protection afforded by rights will be diluted and rendered conditional by the sheer complexity and diversity of real-life situations in which decisions have to be taken. Another set of problems has to do with what content rights might have in the case of nonhuman animals. Many of the rights about which the liberal tradition has been most concerned – freedoms of conscience and speech, political rights, and so on, – could have no meaning for nonhuman animals. Yet another difficulty is that in many human contexts in which rights claims have been used as a means of liberation it has been crucially important that those who claimed rights (racial minorities, women, slaves) were able, at some point in the struggle, to articulate their own, self-defined status as rights holders.

Each of these considerations is, I think, important, but none is decisive against the rights view. It may well be – especially through the analogies with infants and severely mentally damaged humans – possible to define a meaningful and substantive set of rights that ought to be respected in the case of many nonhuman animals. My concern, here, is with two potentially much more serious limitations of the rights view. These are, first, that the "subject of a life" criterion acquires a more powerful moral protection for some animals at the cost of excluding far more of them from any kind of moral consideration at all. It is, more generally, difficult to link the rights view in this version with any broader environmental ethic. The second concern I have is a large gap between, on the one hand, the philosophical advocacy of a right and, on the other, the ability of those who are assigned such a right to live well and flourish unhindered.

I hope my first cause for concern is fairly self-evident, and Regan has himself begun to address such considerations. One promising way forward would be (as Regan does) to treat the "subject of a life" criterion as a sufficient condition for qualifying as inherently valuable, but not a necessary one. There might be a whole range of other reasons for valuing other kinds of nonhuman beings, and this might ground duties on the part of (human) moral agents to protect them, refrain from destroying them, nurture them, or whatever. If this path were chosen it would take us in the direction of an approach to morality which was both more inclusive than the rights approach and more pluralistic and contextually sensitive in the sources of moral concern which it recognized. We might, for example, want to value and protect nonhuman species (such as marine molluscs, or bats, or spiders) precisely because they represent such profoundly and awesomely different solutions to the problems of biological existence from our own rather than because, as on the rights view, they share certain psychological attributes with us.

These considerations do, of course, raise big questions about the nature and role of moral theory. They would imply abandoning the attempt to build formal systems of logically connected propositions on the model of scientific theory. However, this most certainly would not mean going to the opposite extreme of abandoning the attempt to give reasoned justifications for moral decisions, or to achieve whatever degree of coherence and consistency is attainable across different contexts of action.

## Moral theory and effective action

Comparable issues concerning the nature and role of moral theory, and, indeed, morality itself, are posed by my second set of reservations about the rights view of the moral status of nonhuman animals. Let us accept, for purposes of this argument, that at least some nonhuman animals do, as the rights view insists, deserve moral concern, as inherently valuable beings. An aim of moral theory, then, would be to secure the well-being of such animals, and to avoid harming them. But how well-placed is rights theory, as a particular kind of moral discourse, to achieve this? Even more worrying, how well-placed is any kind of moral discourse to achieve this aim? These two questions, it should be noted, are an attempt to shift the terms of the existing debate. Rather than ask "do animals have the appropriate attributes to count as proper objects of moral concern?" I am suggesting that we ask "Assuming that they are, how much good is it likely to do them?"

### The problem of motivation

In part this is a matter of what is sometimes called the problem of motivation. It is quite possible that individual human agents who have dealings with nonhuman animals – people who have pets, look after zoo animals, are involved in experimental work, and so on – might accept the validity of the arguments for animal rights, but still be inclined to neglect them, behave abusively towards them, and so on. This is even more likely to apply to the actions that are often advocated by animal rights activists, such as consumer boycotts, shifting to a vegetarian diet and so on. In these cases, people are asked to undergo what they experience as deprivation, or to undertake inconvenient alterations in their normal life style for the sake of a supposed benefit to animals with which they never come into contact. To differing extents in different cases, the question arises, how effective is a moral appeal, no matter how rationally convincing, as a motivator of significant changes in human conduct?

Once we pose the question in this way, it becomes apparent that we need to go beyond questions of individual psychology (although that is, of course, important in its own right) to consider the variable contexts of social, economic, and political life in which moral issues about the treatment of nonhuman animals arise. Whatever rational grounds are offered by a moral theory for treating animals well, the chances that they *will* be treated well may depend more on how they (both the animals and the humans concerned) are located in human social practices and networks of relationships, than on how well the moral theory is argued.

So, for example, it is often used as an argument against reliance on an appeal to rights in the human case, that talk of rights and justice becomes relevant only when mutual ties of benevolence and sympathy in human communities have started to break down. This sometimes figures as part of an argument about the emergence of capitalist modernity. As capitalist market relations break up traditional ties of community, new forms of competitive and antagonistic individuality emerge. Discourses and practices of rights and justice then become necessary as a way of maintaining social peace in the absence of spontaneous bonds of affection and fellow feeling. Of course, this brief summary is guilty of a rather "rose-tinted" account of the precapitalist community, but, even so, the idea that we only *need* to assert "rights" when we cannot rely on benevolence and sympathy in relationships has a lot to be said for it.

If we apply this to the question of how to secure the well-being of nonhuman animals whose lives are affected by human social practices, then we might have some chance of linking up the concerns of moral theory with effective action. This will involve some sociological and anthropological work to give us a clear understanding of the different practices and institutional structures that involve the use of nonhuman animals. It will also involve some ethological work to understand how these species of animals behave, relate to each other (especially in the case of social animals), and meet their needs when they are *not* under constraints imposed by human social purposes. On this basis, we can get some clearer sense of the distortions imposed on the mode of life of an animal species as a result of human domination, and, in the light of that, be sensitive to symptoms of pathology – behavioral, psychological, physiological – which result.

Much work of this kind has been done. We do have studies of behavioral pathologies induced by confinement, social separation, boredom, anxiety and so on suffered by caged animals, let alone the more extreme and unmistakable suffering of the victims of intensive farming regimes and many nonhuman experimental subjects. However, this is only a starting point. We still have to ask how and

where intervention, and, in particular, intervention inspired by moral considerations might be effective in remedying such a situation.

### Human agency and social structures

Again, there are two aspects of the problem. To make this clear, we can usefully
consider two contrasting cases. The first is the intensive "factory" farm in which
animals are bred and reared for meat products. Individual animals are confined
to pens that allow only minimal physical movement, and prevent any of the normal
foraging behavior or social interaction in which the animals would otherwise
engage. Nutrients are supplied in amounts and in proportions calculated to maximize efficiency and profitability of meat production, and appropriate antibiotics
are administered to counteract the increased vulnerability to infection of animals
kept under these conditions. Neither farm managers nor employees have any long-
term or close contact with the animals, but rather participate in a complex division
of labor in which each has some specialized task to perform. Under these conditions, no person ever develops close emotional or caring relationships with any
animal, no person who does have contact with the animals has any significant say
in how the farm regime is organized, and all human beings depend for their jobs
or economic well-being on the efficiency of the farm as a maximizer of meat output
per unit of input.

Now, contrast this with the situation of family pet-keeping. Typically, the point
of the activity is the pleasure of the relationship with the animal, there is no
calculation of "inputs" and "outputs" and no overriding extraneous purpose of the
activity, such as profit maximization. Though there may be some agreed division of
labor between the members of the family that allocates different tasks of pet care,
there will usually be long-term and quite intimate relationships, with associated
emotional ties and commitments, between the pet animal and one or more family
members.

Now, some animal rights activists would object on principle to both types of
practice involving animals. However, one might, with a less stringent notion of
the moral status of animals, accept that pet-keeping on the model I have just
sketched is morally justifiable – even desirable. Nevertheless, it is clear that, from
the point of view of the animal, things can still go badly wrong. Even in a context
where affection and a sense of responsibility for the welfare of the animal are
present, the animal may be harmed or caused to suffer because its human carers
are mistaken about the nature of its needs, or are for various reasons unable to
meet them. More obviously, it is common for children to lose interest in family
pets as they grow up, or for people's circumstances to change so that the pet

becomes a nuisance, or too expensive – under these conditions various kinds of neglect and/or deliberate cruelty can arise.

The general characteristics of the social practice of pet-keeping are, however, such that the well-being of the animal will normally be an aim of activities that involve the animal, as a matter of course: benevolence and a voluntarily accepted responsibility for the animal will be the typical affective tone of the relationships. When things go wrong, a moral appeal can be made against the background of such normative expectations and well-established sentiments. Moreover, the human caregivers to whom the moral appeal is made are generally in a position to rectify any defects in the treatment of the animal by deciding to alter their conduct.

If we compare this with the intensive farming regime, the situation looks very different. A moral appeal to workers in such a unit has to persuade them to take action against their economic interests, which, indeed, might cost them their livelihood. Moreover, the moral appeal in this case is not made against a prevailing normative order or sentimental climate in which the well-being of the animals is an acknowledged priority. Whatever the personal feelings of the workers, they have to operate in an *institutional* climate in which the overriding concern is efficient meat production and profitability. But, and this is the key point, the situation is quite unlike the pet-keeping case in that, even if against the odds a worker in the intensive unit decides to act in accordance with the animal rights case, this can make little or no difference to the situation of the animals.

This is because their suffering comes from their location in a complex social institution, involving hierarchy and power relations among the human actors involved, the physical design of the plant, and the location of the whole within the constraints of a market economy. Although some people will be able to do more or less than others, depending how much power they have in the institutional framework, no individual, acting alone, may be able to do very much. Even if, in a case like this, animal welfare legislation is enacted, and regulation imposed, there are powerful pressures at work to dilute its requirements, negotiate its meaning, minimize its enforcement, or outright evade it. This is a familiar story, offering many parallels with attempts to legislate into being a recognition of the rights of oppressed or disadvantaged human groups.

The key point I want to draw from these illustrations is that moral discourses that seek to protect the interests of nonhuman animals are more likely to be effective in practice in the following circumstances:

1. The moral appeal they make complements rather than contradicts the already-existing sentiments and attitudes of the human agents involved.
2. The agents to whom the moral appeal is made are engaged in insti-

tutionalized social practices to which their decisions can make a difference.

So both individual openness to moral persuasion *and* institutional structures are likely to determine how effective moral theories can be in protecting the interests of animals who figure in human social life. A direct implication of this is that effective protection of these animals will require not just an adequate and persuasive moral theory, but also some sociological insights and a political program to transform a wide range of human social institutions. Such a program would need to have ways of taking into account and balancing the interests and needs of both human and nonhuman animals.

### Animals in medical experimentation

Turning now to the specific case of the use of animals in medical experimentation, it might seem that, if animals have rights, then any kind of experimentation is ruled out. The case is open and shut. However, even the strongest proponents of animal rights, who hold that animals, if they have rights, have *equal* rights, recognize that there arise situations in which rights conflict, so there can be, in practice, no such absolute and universal prohibitions. From a utilitarian perspective it might well be argued that the relief of suffering (both human and animal) that might come from medical experimentation could morally justify the suffering caused to the experimental subjects.

A more context-sensitive approach, such as I have been advocating, might recognize a number of, often competing, morally relevant considerations that have to be weighed against one another. So, with respect to any *particular* contemplated experimental use of animals, the following considerations might be relevant:

- Are there any feasible substitutes for live animals (e.g. use of tissue cultures)?
- How psychologically complex, and vulnerable to various sorts of suffering is the animal species concerned?
- What are the likely effects on the animal of the proposed intervention – intense but temporary pain?, permanent damage?, death?
- Could the objectives of the experiment be achieved without causing suffering, or by using a procedure which reduces suffering?
- Assuming there are no substitutes for the use of animals in the experiment, how important is it for the experiment to be done at all? (What part might the outcome be expected to play in finding a cure? How serious is the disease itself as a source of suffering?)

So far as a wider *policy* on medical experimentation is concerned, there might be other relevant considerations.

- What prospects are there for *developing* substitutes for live animals in experiments if sufficient resources are devoted to it?
- How can it be ensured that institutions that are allowed to conduct experiments on animals minimize this activity, and optimize the well-being of such animals as are used?
- How far can unnecessary duplication of experiments of this kind be avoided?
- How can it be ensured that experimental use of animals is not made for inessential or frivolous purposes?

Obviously, it is well beyond the scope of this short commentary, as well as beyond the reach of my own competence, to say how each of these questions should be answered. But simply listing the questions does, I think, serve to illustrate two important points. The first is that neither of the prevailing philosophical theories that accord positive moral status to animals is sufficiently attuned to the complexity and diversity of the relevant considerations in a real-life situation. I should emphasize, however, that *both* theories have had a significant role in sensitizing researchers and policy makers to the issue; my arguments are not so much arguments against them, as for including their insights in a more open-textured and sociologically informed debate.

The second major point is that all of the above questions presuppose one or more decision-making contexts – a university or medical school laboratory, a private sector research-and-development department, a grant-awarding body such as a UK research council, an independent regulatory body, a legislature, and so on. As these institutions are currently established, we can fairly readily see that the conditions I outlined above, which favor the *effectiveness* of moral appeals on behalf of animals, are quite unlikely to be satisfied. Most of these institutions would, in fact, be spread out along a spectrum between the pet-keeping and intensive-farming cases which I contrasted earlier, but probably with a bias toward the intensive-farming end. Why might this be so? In the case of private sector research and development the demand for product innovation and profitability in a competitive market situation is likely to override questions about the welfare of experimental animals, and employees who "blow the whistle" are unlikely to be tolerated easily. But in the case of university-based research, even where publicly funded, there will also be competitive pressures on research teams – to justify the next round of grant allocations by results, to build careers by getting there first, to enhance the reputation of their own institution, and so on.

All of this suggests, with our current pattern of institutions, the moral arguments may well point in one direction, but be overridden in the opposite direction by institutional pressures, career motivations, power structures, economic self-interest, and so on. But, since institutions *can* be changed, there is no need to be fatalistic about this. One can imagine, for example, a wide range of ways (some of them already exist as "best practice") in which the situation could be improved:

- A shift in the moral basis of the education of scientific researchers, which emphasizes respect for and empathy with other species, and associated changes in the standards and values of professional associations.
- Opening up decision-making about both the conduct of experiments, and the funding of projects to wider democratic accountability.
- A public right to information about, and justifications for, any experiments that are allowed.
- Legal protection for "whistle blowers" who speak out about abuses they witness.

These and other reforms might, we may hope, take us in the direction of a culture and a society in which the well-being both of humans and of nonhuman creatures and their environments become the constant aims of public policy, and in which these aims were not systematically obstructed by hierarchies of power, economic vested interests, and narrowly self-interested moral perspectives.

## Suggestion for further reading

Benton, T. *Natural Relations: Ecology, Animal Rights and Social Justice.* London: Verso, 1993.

Carruthers, P. *The Animals Issue: Moral Theory in Practice*, Cambridge: Cambridge University Press, 1992.

Leahy, M. P. T. *Against Liberation: Putting Animals in Perspective.* London and New York: Routledge, 1991.

Regan, T. *The Case for Animal Rights*, 2nd edn. London and New York: Routledge, 1988.

Ryder, R. D. *Victims of Science.* London: Anti-Vivisection Society, 1983.

Singer, P. *Animal Liberation*, 2nd edn. London: Cape, 1990.

# The science of the environment

ANDREW PULLIN

EDITORS' SUMMARY

Andrew Pullin, PhD, an environmental scientist, sketches the almost frightening devastation to the world's ecosystem caused mainly by a value-judgment by human populations that the environment must be tamed and tailored to industrial and economic ends. This attitude, Pullin argues, must be changed into one that extracts from the incredibly complex ecosystem only that which it can sustain, so that we can pass on to the next generation the same or even better resources than we ourselves have enjoyed. The chapter explores especially the horrible results on human health and well-being caused by environmental disorders we ourselves have created.

"Our environment is hostile, posing many threats to our health and welfare, and as such should be subdued, made safe, and turned to good use for better standards of living and wealth creation."

This may have been a typical and understandable view of a hypothetical observer of the birth of the industrial revolution, foreseeing rapid advances in technology by exploiting cheap energy and natural resources. Few would have disagreed with this view at that time. The alarming truth is that, with all the changes that have subsequently transformed our society, it is still only a few who disagree, at least who disagree strongly enough to turn words into actions.

I begin the subject of the environment in this way because it is essentially the way we view the environment that dictates the way we treat it. Humankind has an attitude problem. This is not a problem of political or economic theory, but a problem of our personal relationship with our environment. It is consequently the values within ourselves that we must question.

In this chapter I review the recent challenges to our environment that have rendered the opening words so dangerously out of step with reality today. But first I want to consider what the scientific study of our environment has revealed about its nature.

## Scientific study of the environment

There are many things we take for granted about our environment, its provision of oxygen and water for example, but most significantly its stability. We may think that the weather is changeable and unpredictable, but this is true only within certain limits. The evolution of complex organisms such as human beings has been possible only because of the relative stability of the earth's atmosphere and climate. Over geological time major changes have occurred, continents have drifted from the tropics to the poles and ice ages have come and gone, but these changes have usually been gradual, allowing most species to adapt to new conditions and evolve new relationships within their environment. On the few occasions that the climate has changed rapidly, mass extinctions have been written in the fossil record, the most famous of which is the demise of the dinosaurs some 60 million years ago.

Herein lies a lesson for our species. The complexity of our environment, both living and nonliving, buffers it against rapid change within limits, but those limits can be exceeded at all levels of organization, from the individual through the habitat and the ecosystem to the biosphere. We have clearly exceeded those limits at the lower levels, with many recorded species extinctions and destruction of habitats. We are clearly now testing the limits at higher levels, with degradation and mass destruction of entire ecosystems and projected perturbations of climate that have measurable impact upon the biosphere.

In order to further examine the problem, I want to briefly consider the major features of ecosystems, the factors that are the important agents in their function and then examine the challenges they face from humankind.

## Ecosystems

All ecosystems require an input of energy and almost without exception this energy comes ultimately from the sun. Green plants (the primary producers) then convert this energy into biomass by the process of photosynthesis. Plants are subsequently eaten by herbivores (secondary consumers) and herbivores by carnivores (tertiary consumers). At each stage a conversion of energy takes place and the majority is lost as waste or heat. Therefore, far less energy is stored at any one time in tertiary consumers than in primary producers and this partly explains why predators, and in particular large predators, are much fewer in number than herbivores or plants. The passage of energy is not, however, a linear process as implied in the concept of a food chain. There are many complex interrelationships that make the concept of a food web much more appropriate.

If ecosystems are not particularly efficient in the transfer of energy they are much more frugal when it comes to recycling of nutrients. An example is the important role bacteria and fungi play in the breakdown and recycling of dead organic materials. Little is wasted and most nutrients, such as nitrogen, phosphorus and sulfur are recycled by biological and chemical processes (biogeochemical cycles).

The integrity of ecosystems results from the diversity of its components and the different roles that component species play in the whole. The stability and buffering capacity derives from the complex interactions that exist among species and between species and their environment (living and nonliving). Small perturbations are absorbed by the complexity of the web's structure. The more strands there are, the larger the perturbation needs to be before the web collapses.

Ever since human beings evolved the use of tools and weapons we have had the capacity to perturb ecosystems. This was enhanced with the development of agriculture. Large areas of woodland were denuded of game by hunting and felled, producing a much more open landscape, often free from large herbivores and their predators. Two related factors have, however, over the last few centuries, transformed the relationship between human beings and the environment; the rise of the industrialized global economy and exponential population growth. The resulting impact on our environment has been profound, with many directly visible results but also many complex interrelated changes that only detailed scientific research has been able to bring to our attention.

The industrial economy capitalizes on advances in technology and the exploitation of fossil fuel (mainly coal) and mineral resources. These advances have enabled human beings to change their role in the food web, to dominate the system rather than being a cooperative part of it and inevitably to perturb it to the point of collapse. Initially this was not accepted as a problem; technology brought wealth to a few and they dominated the course of society. The belief was, and still is, that technology will answer all our problems.

A major dilemma is that human morality has had no difficulty in coping with many of the results of greater use of science and technology. For example, the increased use of resources per capita in industrialized countries means that each individual is having a much greater impact on the environment (and not only his or her own immediate environment). This thirst for resources has resulted in the exploitation of Third World countries, ensnaring them in a poverty trap of high interest loans and falling commodity prices.

At the same time, advances in Western medicine have radically reduced death rates and increased longevity without relieving the poverty that is an underlying reason for continued high birth rates. The consequent rapid population growth in Third World countries gives them little hope of sustainable development.

The shift in the position of humankind in the global environment has resulted in two categories of stress with which our biosphere has had to cope, which I will now look at in detail.

## *Pollution and ecosystem degradation*

Inevitably unsustainable and wasteful use of resources leads to the production and accumulation of unwanted materials and chemicals. The short-term answer in a market economy is to dump it cheaply. This habit was formed during the time when it was assumed that the capacity of the environment to absorb waste products was more than sufficient. Even though scientific research has shown that this capacity is very limited and has already been exceeded in many cases, a mixture of economic pressures and ignorance has ensured that this habit has continued throughout the period of rapid industrial and population growth, even in the richest countries.

Over the same period, research has shown how many of these waste products have had profound effects on our environment that we term pollution. There are four key components of ecosystems that suffer pollution: air, water, soil and food webs. Of course these exchange pollutants and none can be viewed entirely in isolation from the others.

### *Air*

Air pollutants arise primarily from heavy industry, fossil fuel power stations, vehicle exhaust emissions, and secondarily from the manufacture and use of consumer products (e.g. chlorofluorocarbons (CFCs) from aerosols and refrigerators). Four major consequences for our environment are currently of concern:

*Acid rain*  This is caused by normal rainfall washing out oxides of sulfur and nitrogen from the atmosphere and in the process forming sulphuric and nitric acids. The result is the acidification of soil and water, causing death of trees and of entire lake ecosystems.

*Ozone depletion*  Release of certain chemicals into the atmosphere, most prominently CFCs results in the depletion of the ozone layer in our atmosphere that absorbs much of the harmful ultraviolet radiation produced by the sun. Ozone is very reactive and is readily destroyed by CFCs causing depletion of this layer and increased levels of ultraviolet radiation at the earth's surface. Research has shown that this is likely to damage plant life as well as increasing the incidence of skin cancer in humans.

*Respiratory disease* There are specific examples where increased levels of pollutants in the air we breath has caused increased suffering and death from respiratory disorders. Ironically one pollutant is ozone, the same chemical that protects us when found higher up in the atmosphere, but toxic when we breath it in, because of its high reactivity. Evidence is particularly strong that poor air quality increases the incidence of asthma attacks in the population. A recent period of poor air quality in the UK in June 1994 saw the number of people reporting to hospitals with asthma problems increase up to eight-fold above the norm.

*Climate change* Evidence is accumulating that our climate is changing at an ever-increasing pace. A number of pollutants released by human activity, most significantly carbon dioxide and methane, act as a thermal blanket around the earth, allowing heat from the sun to penetrate but trapping more and more of the heat that would otherwise be lost as infrared radiation from the earth's surface as their concentration in the atmosphere increases.

We rely on the stability of our climate above all else for our survival. Rapid changes will cause disruption of our agricultural systems and extremes of temperature and rainfall could lead to widespread collapse of our natural ecosystems.

### Water

Water is a commodity that many of us take for granted, we expect clean drinkable water to come out of our taps. But this is becoming increasingly difficult and expensive to achieve. All of our water courses are polluted to some extent; it is just a question of acceptable levels. The pollutants come again from heavy industry but also from domestic sewage and run-off from fields under intensive cultivation. Each pollutant has its own effect and dangers, but I have listed three types as examples (see the next three sections).

*Eutrophication* This is a term used to describe increasing nutrient loads in the water. Key nutrients are nitrates and phosphates and they come mainly from domestic sewage, particularly high phosphate washing powders, and from run-off of fertilizer applications from agricultural fields. One result of eutrophication is an increase in algal growth in water, often to very high concentrations, referred to as "algal blooms." Some of these algae release toxins into the water, making it unsafe to drink. Of more long-term importance is the fact that increases in the nutrient load seriously disrupt the aquatic ecosystem, permanently changing its character and eliminating many characteristic species.

*Pesticides*   Chemicals used in agriculture and horticulture to kill insects are frequently washed out into the water courses and some more persistent types are present in our drinking water, for example DDT can still be detected at low levels in some drinking water supplies, despite a ban on its use in the USA since 1972. There is increasing concern over possible links between organochlorine and organophosphorus pesticides and certain types of cancer in humans.

*Toxic heavy metals*   Solutions containing heavy metals such as cadmium, are routinely dumped in our rivers by heavy industry, despite their known and suspected health risks to humans. Equally they cause profound changes in the river flora and fauna.

### Soil

The soil is intimately associated with both the water and the air and therefore shares some of their pollution problems such as acid rain and eutrophication. It does, however, suffer two further problems resulting from intensive agricultural use.

*Soil erosion*   Continued breakdown and exposure of the top soil resulting from intensive cultivation has resulted in massive soil erosion in some areas. When the vegetation is removed from some soils, the topsoil dries out in the sun; combined with the disturbance of plowing it can break down into small particles that are easily blown away, leaving infertile subsoil exposed.

*Salinization*   In tropical countries where irrigation systems are used to maintain the water supply to crops, rapid evaporation in high temperatures leaves behind salts and minerals that accumulate in the soil and quickly reach levels inhibitory to plant growth.

### Food webs

When pollution occurs in air, soil, and water it inevitably contaminates living organisms and is spread around the food web in any ecosystem. A particular problem then occurs with those pollutants such as pesticides and heavy metals that tend to accumulate within the organism rather than being excreted. Since a herbivore will eat many times its body weight in plants, any persistent pollutants in the plants will accumulate at a higher concentration in the herbivore. The same goes for carnivores eating herbivores, so that the higher the position of an individual in the food web the greater the concentration of pollutant that will accumulate in its body. The classic illustration of this was the discovery that the rapid decline of birds of prey during the 1950s and 1960s in North America and the UK was due to accumulation of organochlorine pesticides in their fatty tissue. At high concentrations this caused death, but at lower concentrations also caused females to lay thin-shelled eggs and consequently fail to rear chicks.

In summary, a wide range of pollutants perturb and degrade our ecosystems. In small quantities the robust nature of most ecosystems enables them to cope and recover. But the scale of pollutant production and release today has resulted in widespread and in some cases irreversible degradation of our environment.

## Habitat destruction and the biodiversity crisis

A more serious and immediate problem than the indirect environmental degradation caused by pollution is the environmental destruction directly caused by human activities. In the past the conversion of grassland, forest, or wetlands into croplands has been viewed as universally good, and understandably so, but today the problems are largely ones of the scale of these activities and the speed at which they are taking place. Many natural ecosystems have become so altered and fragmented that the small pockets that remain are not sufficient to support the species that characterize those systems and large-scale extinction is the result. The reasons behind habitat destruction are complex and beyond the scope of this chapter but they usually fall into two broad categories. Destruction for development and profit by governments and multinational companies or destruction for survival by a rapidly increasing dispossesed, Third World population. The result is the rapid disappearance of natural resources that play many important roles in the quality of human life. A typical figure for the proportion of natural habitat lost in African and Asian countries is 75%, but can be as high as 97% in the case of Bangladesh.

Scientific research has been most concerned with the biodiversity crisis, the rapid extinction of species before humankind even has a chance to record their existence. Tropical rain forests cover 7% of the earth's surface but are tremendously rich in species, containing approximately 50% of the global total. To date about half of all tropical rain forest has been destroyed and an area larger than the state of Florida is destroyed or degraded each year. At this rate our richest ecosystem may be almost totally lost within one generation; our generation. The statistics make devastating reading but the consequences are more important. The diversity of species is one of our major sustainable resources for use in agriculture, medicine, industry and recreation. They also form the integrated network that is so important to the stability of ecosystems and the welfare of communities that depend on them.

Perhaps of most concern is that we do not know the value of what we are losing. For example, some of the key drugs used today to treat cancer and heart disease come from species that had no apparent value 20 years ago. But there are also key ethical questions arising from our squandering of these resources and depriving future generations of their inheritance.

## Conclusion

The scientific advances in understanding the interdependability of different components of our environment force us to question many of the activities that past generations viewed as morally and socially justified. Recent research on climate and global pollutants have shown society that we are reaching the limits of perturbation that our global environment can sustain without undergoing unpredictable and almost certainly catastrophic changes. Now humanity must strive to retain many things that were previously taken for granted.

Scientific and technological advances have provided some solutions in the fields of pollution control, clean technology, ecosystem management and ecological restoration; but if they are to be implemented effectively, values within society must shift away from the exploitative practices of the industrial period to a truly sustainable economy that functions within the limits of our environment, leaving the next generation with the same wealth of resources that we inherited.

### Suggestions for further reading

Carson, R. *Silent Spring*. Harmondsworth: Penguin 1962 (reprinted 1982).

Chiras, D. *Lessons from Nature: Learning to Live Sustainably on the Earth*. Washington, DC: Island Press, 1992.

Devall, B. and Sessions, G. *Deep Ecology*. Salt Lake City, UT: Gibbs Smith Publisher, 1985.

Ehrlich, P. R. and Ehrlich, A. H. *Extinction: The Causes and Consequences of the Disappearance of Species*. New York: Random House, 1981.

Gates, D. M. *Climate Change and its Biological Consequences*. Sunderland, MA: Sinauer Associates, 1993.

Gore, A. *Earth in the Balance: Ecology and the Human Spirit*. Boston, MA: Houghton Mifflin, 1992.

Myers, N. *A Wealth of Wild Species.* Boulder, CO: Westview Press, 1983.

Wilson, E. O. *The Diversity of Life.* New York: Penguin, 1992.

World Conservation Monitoring Centre. *Global Biodiversity: The Status of the Earth's Living Resources.* London: Chapman & Hall 1992.

# Environmental ethics

ANDREW DOBSON

EDITORS' SUMMARY

Arguing that the critical leap for ethics is across the species divide, Andrew Dobson, PhD, is able to demonstrate how expanding our instinctive concern for the environment creates fundamental problems for philosophy because we want to know *why* that instinct is justified. Should animals and the environment be protected because they have intrinsic moral worth? But how could we claim that a hill has moral worth on an equal footing with the rabbit that makes its warren there? If this is not the case, then should animals and the environment be protected only because of their extrinsic value, their usefulness to human beings? And if this is the approach we should use, how would that differ at all from the wholesale manipulation and devastation of the ecosystem in favor of industrial and economic development aims? It is clear that some newer awareness of communal and ecological values is required to justify our concerns.

## Preamble

All ethical systems have two features – ideas about the standards of morality that are to apply to any given group or community, and ideas about who or what that group or community is to be. Of the two, most by far has been written about the first, and in this sense the history of this vast branch of philosophical endeavor is the history of debates regarding the nature of moral behavior: what makes an act or a person moral? Most of these debates have been untroubled by thoughts regarding the second question: who or what constitutes a moral community? With one or two vocal but subordinate exceptions (of which more later) it has just seemed obvious to everyone that the *human* community is the community in question and that the issue of ethical standards arises only in this context. Of course, it has not always been the case that the *whole* human community has been viewed in this way. Standards of morality have often not been applied universally across the human community because not all human beings have been regarded as equally worthy – women and "other races" (as the philosopher John Stuart Mill delicately puts it) have been obvious historical exceptions. These exceptions aside, the human community has traditionally been thought of as the proper context for talk of moral behavior.

Environmental ethics, however, as the name itself suggests, has a much wider frame of reference as far as the "moral community" is concerned. From this point of view the environment itself (including human beings, of course) is an object of moral concern and should be taken into account when one considers the nature of moral standards and to whom or what they should apply. "The environment" is in many senses an unwieldy term as it means different things to different people, and it is indeed possible to argue in favor of an ethics for the environment without thinking that the whole of the environment should be taken into account. I structure what follows, then, in the manner of breaking this term down into some recognizable (but by no means indisputable) parts. I begin with animals.

## Ethics and animals

Animals no more constitute the whole environment, of course, than do human beings. They do not even constitute a particularly large part of the biotic environment (in terms of biomass rather than species), let alone the abiotic environment that would have to be included in a fully environmental ethic. And when we realize that by no means all those who argue for the inclusion of animals in the moral community argue for the inclusion of all of them, we see that, as far as extending the boundaries of moral concern goes, we may not be progressing very far.

Nevertheless, any leap across the species divide is a big one for moral philosophers, even if once we get there we only put up a temporary camp staffed by a ragbag collection of renegade frontierspeople. The historical weight of arguments in favor of restricting the moral community to humans is heavy, and resisting it even a little amounts to making possible much more ambitious forays into territory "beyond the human." For this reason it makes sense to spend a little time considering the arguments surrounding the admission of (some) animals to the realms of moral considerability.

The belief that animals can and should be the objects of moral concern is by no means a new one. In the third and fourth centuries BC Epicurus and Aristotle set up battle lines in this regard that have been handed down to us practically unchanged. Aristotle believed that what distinguishes humans from animals is the former's capacity for rational thought, and that while desire is a feature of the human condition it should remain subordinate to the rational faculty. Moreover, value lies primarily in this rational faculty and those beings not possessing it could properly be seen as means to the ends of those that do: "Plants exist for [the sake of animals, and] animals exist for the sake of man," writes Aristotle in his *Politics*. Epicurus, on the other hand, held less store by the rational faculty and argued that the greatest good was pleasure. While he acknowledged that the possession of a rational faculty brought with it the possibility of mental pleasures and pains, and that these were (in his opinion) more serious than physical ones, the fact that animals can experience pleasure and pain brought them into the fold of moral consideration.

Ever since these arguments between the followers of Aristotle and Epicurus were established they have, respectively, been used either to illustrate the existence of the species divide or to cross it. What makes us (as humans) different is that we possess a rational faculty, and what makes us the same (as animals) is that we can experience pleasure and pain. The problem is that our perceptions can become skewed by excessive anthropomorphism. In slightly modified form, indeed, we can jump with Aristotle and Epicurus into the twentieth century and see how they informed the two most influential approaches to modern animal rights or animal liberation: those associated with Peter Singer and Tom Regan.

Peter Singer argues (most fully in his *Animal Liberation*, 1976) that, if (as most of us accept) animals are sentient, then they can experience pleasure and pain. It follows, he suggests, that we are morally constrained in our behavior towards them. In this he agrees not only with Epicurus but also with jurist Jeremy Bentham who famously wrote of animals in his *Introduction to the Principles and Morals of Legislation* of 1789 that the question should not be, "Can they reason?" nor "Can they talk?" but, "Can they suffer?" This apparently simple question, and the affirmative answer most of us would give to it, has opened a can of complex philosophical worms and while this is not the place to open it too far, some worries ought to be mentioned. Granted that both humans and animals are pain- and pleasure-experiencing creatures, does it follow, for example, that they have equal moral weight? This matters, for if animals have less moral weight than humans then we can legitimately kill them for food as long as they are killed painlessly. An argument for equal moral weight, though, would suggest that they can only legitimately be killed for food if we are prepared to do the same to human beings.

Further, some utilitarian theories of morality (of which Singer's is one example), argue that the morality of an action is to be judged by its consequences, and that the consequences are to be determined by a calculation of the aggregate utility (presence of pleasure, avoidance of pain) produced by the act. Apart from the difficulties of calculation always associated with utilitarianism, the upshot of this is that pain to individuals can be justified by the pleasure experienced by the wider community. Experiments on animals (or, indeed, on humans) can be justified by the wider benefits derived from them. So, while making sentience the quality a being must possess for it to be morally considerable enables the species divide to be crossed, it is by no means a conclusive solution to the problems that exercise animal liberationists.

Tom Regan's approach, however, is designed to provide a solution that avoids these problems, but it does so at the cost of raising one or two doubts of its own.

One of the difficulties with Singer's theory is that some of those who cannot swallow the "equal moral weight" (of humans and animals) argument will admit that humans and animals can both experience pain but that humans are due an additional – and "trumping" – consideration by virtue of their mental complexity. Regan takes this position at face value and makes *his* leap over the species divide by arguing that mental complexity crosses the divide rather than being confined to one side of it. He suggests that some animals can be seen as "subjects of a life" in the same way as human beings, and that this makes them individual holders of a right to life that, crucially, cannot be trumped by utilitarian calculations of aggregate utility.

We have already seen that an ethic for animals is by no means the same thing as an environmental ethic, largely because of what it (the former) leaves out. Regan's notion of "moral patients" is extremely restrictive, as he feels that only "normal mammalian animals aged one or more" are owed moral consideration in the same way as humans. Further, rights-based theories of this sort are subject to standard objections: do unacquired rights exist, for example? Bentham notoriously suggested that rights that are not enshrined in positive law are as insecure as "nonsense on stilts." Others argue that only moral *agents* can be holders of rights, and that animals do not qualify because they lack the qualities of agency required. Then there are those who suggest that the granting of formal rights does nothing to guarantee the substantive enjoyment of them, and that therefore political action needs to be taken to provide contexts within which formal rights can become substantive ones. Finally, the refusal by rights theorists of utilitarian calculations of the greater good means that the killing of an animal to find a cure for cancer cannot be sanctioned.

Neither of these theories of animal liberation or animal rights, then, does a watertight job in terms of producing an irrefutable ethic for animals. What they do manage, though, is to raise the spectre of "speciesism" – discrimination on the basis of species alone – and to ask us whether such discrimination can be rationally justified. To the extent that we conclude that it cannot, we have crossed the species divide and are consorting – however temporarily – with the ragbag of like-minded renegades I mentioned earlier. How far we go from here, though, can only be determined by a survey of the territory occupied by environmental ethics.

## Environmental ethics

The name that most often crops up in this territory (perhaps appropriately thought of as a wilderness) is that of Aldo Leopold. Leopold worked for the United States government's Forest Service from 1909 and came to represent those who argued for the preservation of wilderness rather than its "development" even for recreational purposes. In 1948, a week before he died, Leopold sent to press *A Sand County Almanac* in which he developed the notion of a "land ethic," "All ethics so far evolved rest upon a single premise: that the individual is a member of a community of individual parts . . . The land ethic simply enlarges the boundaries of the community to include soils, waters, plants, and animals, or collectively: the land."

At the beginning I pointed out that a common feature of all ethical systems is ideas about who or what are to be members of the ethical community. It seemed to stretch a point somewhat to claim that even animals could be members of that community, but this claim pales into insignificance alongside Leopold's "simple" enlargement of the community to include plants, soils, and so on. Not the least of the problems raised is that of working out just why plants and soils might have a moral claim upon us. In the rather easier case of animals (as we saw), theorists point to the similarities between them and us and suggest that it is irrational to base *our* moral considerateness on these characteristics and then deny it to them. What is it that plants and soils have in common with us? Life? For plants, certainly, but soils? Probably not.

A favourite concept invoked by environmental ethicists at this point is that of "interests." It is said that just as human beings have interests, so do other parts of nature – the nonhuman parts. But once again, although we might agree that the well-being of a creature rests in part with it staying alive, and that it therefore has an interest in staying alive, and that this interest is morally considerable, it is hard to conceive of the hill in which the rabbit creates its warren (for example) as having a well-being interest in the same way that the rabbit itself does. These problems have given rise to the deployment of another characteristic that beings and systems are said to have, and which makes them morally considerable: "autopoiesis," or "self-production." Any autopoietic entity – many of which are, of course, collections of individual organisms – is owed moral consideration by virtue of being ends in themselves.

This stress on "collections" or "groups" is essential to an understanding of environmental ethics. Up to now, the ethical focus has been largely upon individual organisms and the extent to which they can be held to have moral value. Environmental ethicists want to push the boat out farther and suggest that "environmental wholes" can have value – wholes such as species and ecosystems. This is clearly the import of Leopold's land ethic which refers "collectively" to "the land." If is accepted that ecosystems are morally considerable, then the parts that contribute to the whole are morally considerable – to some extent – as well. In this way, soils have a well-being interest to the extent that the ecosystems of which they are a part have a well-being interest.

Naturally enough, the arguments surrounding these far-reaching claims are fiercely contested. Apart from the problem of defining the environmental whole in question – do species or ecosystems exist in any morally meaningful sense? – opponents of environmental ethics in this sort of form worry about the future of the parts if the whole is the original source of moral standing. Leopold himself suggests that, "A thing is right when it tends to preserve the integrity, stability, and beauty of the biotic community. It is wrong when it tends otherwise." This can be read in such a way as to sanction the elimination of a part of the biotic community provided that the elimination results in preserving the latter's integrity, etc. This is the classic conservationist's dilemma. As is clear from Andrew Pullin's contribution to this volume, environmental scientists very rarely talk about individuals, since they are more concerned with species and the ecosystems of which individuals are a part. From an ecosystemic point of view the periodic culling of elephants may be justified, but if each elephant is held to have intrinsic value (like a human being) then the cull amounts to a form of "environmental fascism." Complicated calculations regarding the relative worth of the parts and the whole seem to be the only way out of this conundrum, but the nature of such calculations is such that they are unresolvably contested.

It is the belief of all conservationists that the environment should be protected, but environmental ethicists wrestle with the question of why. Often it is argued that it is a human interest to protect the environment for, as Andrew Pullin points out, "the diversity of species is one of our major sustainable resources for use in agriculture, medicine, industry and recreation." Thus, any one of the species that are disappearing at an alarming rate from the world's rain forests may harbor cures for presently untreatable diseases. From another point of view, though, it can be argued that species (and/or their diversity, and/or the stability/integrity of the ecosystem of which they are a part) have an intrinsic value that suggests their protection whether they are useful to human beings or not. These two views have given rise to the distinction between "anthropocentric" and "ecocentric" reasons for protecting the environment. Some will say that the former has nothing to do with ethics at all since it views conservation (e.g. of the podacarpus tree) as a means to an end (e.g. preventing landslides in cloudforests) rather than as an end in itself.

Extending the moral community, however, is not only a question of including ever more exotic life forms (or collections of them) within it. Andrew Pullin also points to the claims of future generations of human beings. There is no question that resource depletion (for example) affects the ability of future generations to meet their needs satisfactorily and to live fulfilled lives. This may also be true, of course, for those parts of the present generation whose share in current resource use and the goods derived from them is miserly. The difference, though, is that the present generation is alive and future generations are not (yet), so what is the source of our moral duty towards future generations? Some will say that we have none, and indeed the principles of much standard moral theorizing lead to that conclusion. Others will say that an imagined contractual situation in which we develop principles of justice without knowing which generation we are to be born into will produce a rational basis for inter-generational justice. Still others will say that our intuitions incline us in such a direction. Whatever the merits or otherwise of these arguments, it is plain that environmental ethics has as its principal intention the widening of the moral frame of reference.

## Deep ecology

Without minimizing the differences between all the positions outlined up to now and the difficulties associated with them, the one thing they all have in common is an attempt to expand the moral community beyond the human community. We can caricature two ways of doing this. The first is to shuffle rather timidly over the bridge that spans the species divide, dragging bits of our humanity with us, and seek out life forms that we can recognize as similar to us. When we find them we grant them moral consideration on the grounds of their similarity to us in some or other relevant respect. This is the approach of many animal rights theorists. The second gambit is to take a giant leap across the divide and open up as generously as possible to the claims of the beings and objects with which we find ourselves surrounded. This is the preferred tactic of environmental ethicists as I have described them above, but they still err on the side of caution to the extent that the moral qualification of ecosystems is based on their possession of a feature common to both them and to human being – "well-being interest" or even "auto-poeisis," for example.

Deep ecologists make arguably the biggest leap of all by taking what they find at face value, and suggesting that the working hypothesis regarding value should be "biospherical egalitarianism." This means what it says. As expressed by the man most widely associated with deep ecology, the Norwegian Arne Naess, biospherical egalitarianism "intuitively" means "the equal right to live and blossom" for (all) forms and ways of life. Of course, such a principle cannot be made to work in practice because as Naess himself says: "Any realistic praxis necessitates some killing, exploitation and suppression." This raises not only the hoary question of "how much?" (killing etc.) but also "Of what?" Like the environmental ethicists above, deep ecologists seem inevitably driven to constructing a hierarchy of valued beings and objects – despite their professed biospherical egalitarianism. Indeed it is uncanny how often the hierarchy turns out to look very much like the one with which most of us operate, with human beings at the top and the smallpox virus or human immunodeficiency virus (HIV) somewhere near the bottom. These issues aside, the very real importance of the principle is that if it were followed the onus of justification would be on those who sought to interfere with this equal right, rather than on those who seek to defend it. There can be little doubt that our relationship with the nonhuman natural world would be fundamentally altered in such a moral environment.

## Conclusion

It is just this – the altering of our relationship with the nonhuman natural world – that environmental ethicists are seeking to achieve. The American Thomas Paine wrote of the rights of man, and the English writer Mary Wollstonecraft responded with the rights of woman, in the belief that naming the holder of a right made it more secure. The right of the nonhuman world to live and blossom is but the latest in a long line of moral namings that seek to ground just treatment in its necessary condition: moral recognition.

### Suggestions for further reading

Brennan, A. *Thinking About Nature: An Investigation of Nature Value and Ecology*. London: Routledge, 1988.

Dobson, A. *Green Political Thought*, 2nd edn. London: Routledge, 1995.

Gruen, L. and Jamieson, D. (eds.) *Reflecting on Nature: Readings in Environmental Philosophy*. New York: Oxford University Press, 1994.

Leopold, A. *A Sand County Almanac, and Sketches Here and There*. New York: Oxford University Press, 1949.

Midgley, M. *Beast and Man: The Roots of Human Nature*. London: Routledge, 1995.

Naess, A. *Ecology, Community and Lifestyle*. Cambridge: Cambridge University Press, 1989.

Regan, T. *The Case for Animal Rights*. Berkeley, CA: University of California Press, 1983.

Sessions, G. (ed.) *Deep Ecology for the 21st Century: Readings in the Philosophy and Practice of the New Environmentalism*. Boston and London: Shambala Press, 1995.

Singer, P. *Animal Liberation: A New Ethics for our Treatment of Animals*. London: Jonathan Cape, 1986.

Wheale, P. and McNally, R. *Animal Genetic Engineering: Of Pigs and Men*. London: Pluto Press, 1995.

# Human activity
# and environmental ethics

ANDREW JAMETON

EDITORS' SUMMARY

Andrew Jameton, PhD, issues a clarion call for the future of human society based on the human and technological interventions in the world environment. He clearly establishes the necessity, not option, of both thinking and living differently from the way we have in the past. The disparities created by technology in general, and medical technology in particular, he argues, can no longer be tolerated. This argument proceeds not only from justice, but also from self-interest: we are all linked together throughout the world. The collapse of any one part will seriously damage the whole. This awareness of both the need to live more simply (to downsize) and the need to think more complexly (interactively) is the challenge for the next century's ethics and political systems.

In the last two centuries, human activity has made great changes in the global environment of humans and other species. Although specific details are controversial, the major outlines of an emerging environmental crisis are clear:

· The human population has become very large and is growing, adding over 90 million new people per year.
· Other species are becoming extinct at a rate so rapid as to demarcate an evolutionary epoch.
· World-wide industrial production has increased 50-fold in the last century,

and a billion people live at levels of wealth and consumption inconceivable in human history.

- Inequality in human welfare has been increasing rapidly. About a billion people live in great poverty.
- The welfare of the next generations is threatened, since population level, scarcity, and pollution are increasing together at exponential rates.
- Weaponry and conflict, augmented by the great destructive potential of nuclear weapons, is wrecking habitats.

This is a perilous situation and worsening exponentially. To cope, human societies must change the direction of their development. The developed world must reduce its consumption; the developing world must improve the condition of the absolute poor; the world must protect and restore natural habitats; this generation must provide for the next generations; humans must stop population growth; and nations must resolve conflicts with less violence.

It may be fantasy to consider achieving even one of these interdependent ethical and political goals, but if humans are to accept these goals, societies will need to set different priorities in their cultural values, and individuals will need to take into account factors they have not considered before. An agenda for ethics and values reflecting the new human situation could follow these outlines:

- In the last 200 years in the West, policy makers have regarded expansion and material growth as key to human progress; now, conservation and modesty are required. Moreover, justice and harmony will need to take a stronger position in relationship to individual liberty and achievement.
- Most people think of events happening on the other side of the globe as unconnected to local decisions in daily life; now, all resource use has global consequences; everyday decisions must consider the needs of those living far away.
- For the last thousand years, most Western thinkers have founded ethical principles on the unique characteristics of human beings; now, philosophers need to rearticulate those principles that unify human beings and other natural creatures. Moreover, past thinkers have regarded nature as an infinitely bountiful servant of humankind; now, nature must be respected for itself.
- Creation and harboring of wealth for this generation meant preparing a firm foundation for the next; now, the wealth of this generation spells the deprivation of the next.
- The right to reproduce and to maintain a family is fundamental to many Western societies. Now, the need to limit population must shape individual reproductive decisions.

· War and production for it are essential features of modern national economies and the use of force and violence characterize the culture of many males;
now, environmental limitations put this last resort even further out of
consideration.

Although the rough outlines of change are clear; the specific implications for
daily life are controversial. In the next sections, I discuss these value changes in
a little more detail. Then, I touch on the implications of these value considerations
for health care practice and biological research.

## Ethical implications of environmental challenges

### *Limiting wealth*

Reducing consumption and limiting material wealth have become the primary
ethical duties of the developed nations, and their most difficult challenge. Pro-
digious modern industrial economies are diminishing forests, wetlands, fisheries,
rivers, aquifers, and the atmosphere, and they are spreading toxic materials
throughout the biosphere. The widespread production of packaging, waste, power,
automobiles, meat-based diets, and the other consumables of the developed world,
if continued, will destroy the natural foundations of human survival. The limitations of the planet permit only a few people to live at this level for long; probably
not the current billion, and certainly not billions more. As the world's developing
middle class begins to acquire refrigerators, televisions, and washing machines,
the limits of the earth will become even more apparent, even though it is ungener-
ous of those who already possess these things to begrudge such amenities to others.

Although efficiency in manufacture can do much to reduce the environmental
burden of the developed world, it cannot do enough. Economic ground-rules and
social policies will need to limit income, luxury consumption, resource-intensive
technologies, shopping, junk foods, automobiles, packaged goods, soft drinks, and
advertising, much as growing policies now limit consumption of alcohol and
tobacco.

To achieve such limits, people need to revise their concept of human liberty,
which is now articulated most strikingly in making individual consumer choices.
Such choices express the philosopher John Stuart Mill's theory that liberty belongs
most to those private personal actions that have little or no effect on others. In
the new environmental situation, however, every action involving consumption of
natural resources has an impact on others. Thus, the concept of human liberty
must change from a focus on individual consumption to the liberty of communities
to plan and the capacity to sustain the material foundation for liberty in the long
term.

Moreover, concepts of individual aspiration must change. For generations, people have looked upward on the economic scale in order to visualize achievement. But it has now become important to look downward and learn to live more simply: to walk, ride bicycles and streetcars (trams), eat vegetarian meals, have water instead of fizzy drinks, travel less, conserve, reuse, recycle, and repair.

## Haves and have-nots

Anyone concerned for the poor must be awed and dismayed by the billion or so people now living in absolute poverty. Contrary to the utilitarian principle that the amount of good can outweigh the bad, increasing neither the wealth of the rich nor the numbers of the moderately well off can remedy the moral wrong of this massive human suffering. Nor can any principle of justice rationalize the degree of economic inequality characterizing the present world. Why not?

First, the fate of everyone on earth has become interconnected. The effects of individual actions ripple throughout the world. Trade and communication now connect almost everyone in a global network of consumption, pollution, and destruction of nature. Drugs to benefit developed countries are tested on people in less-developed ones. Thus, the wealth of an average individual in a developed nation cannot be separated from the poverty of those in distant nations.

Second, fostering the wealth of a few can no longer be expected to ameliorate the condition of the poor. Many economists believe that as the rich get richer, they bring the poor up with them. For this scheme to work, the world must have the capacity for indefinite expansion; but the world is reaching the limits of its ability to support humans, and even many animals, and so the vision of indefinite growth and expansion no longer serves humanity. The pie cannot be made much bigger; in the interests of justice, nations are going to have to slice the pie differently.

Third, there is no way that the absolute poor of the world can be held to be undeserving. In the USA, discussion of the welfare of the poor has long distinguished the "deserving" from the "undeserving" poor. Communities want to help those who are trying to improve themselves and who are victims of misfortune; they are less anxious to assist lazy or antisocial individuals. But, the world today contains whole nations on the brink of starvation; it is impossible to think that these are whole nations of "undeserving" people. The distinction itself is worthless.

Fourth, the developed world cannot be held blameless. The wealthy of the world use land, oil, forests, minerals, capital, money, and other resources throughout the world, which if available for the poor, would do much to reduce poverty and inequality. To many in the developed nations, their wealth is evidence that their nation is blessed by God. But wealthy nations initially acquired their resources by colonialism and violence and continue to use arms to protect their wealth; to many of the poor in the world, the wealth of the developed world is a curse.

Finally, individual wealth is only partly the product of individual achievement. It arises mainly from the prior work of generations who have respected their duties to future generations and the common heritage of nature, which ultimately should never be the property of single individuals. Thus, what people now possess individually should be open to the demands of stewardship and equality. In the last 30 years, the income ratio of the world's richest fifth of people has risen from 30 times to 60 times that of the world's poorest fifth. Although justice permits merit, need, and circumstances to justify differences in wealth, justice is primarily committed to equality. Sixty-fold differences in global economic income cannot be rationalized.

### Humans and nature

Many religions and philosophies have characterized the agenda of humanity as separate from the laws of nature. As Andrew Dobson (this volume) pointed out, theories often begin by distinguishing humans sharply from animals (i.e. non-human animals) and then deriving ethical principles from humanity's unique features. But, this approach to ethics distorts the relationship of humans to nature. Humans are totally dependent biologically on the earth. The progress of evolutionary theory, cell biology, biochemistry, and genetics in the last century has been to characterize humans as very close to other animals. For instance, human DNA differs from chimpanzee DNA by less than 2%. Just as human connectedness to nature is strong, so should responsibilities to nature and the long-term ability to live with it be strong.

Ethical and religious theories should thus first articulate the duties of humans as animals and biological beings: to recycle, to limit use according to environment, to hold health in high esteem, not to dirty our nests, not to exhaust resources, to stay safe, to respect our biological community, to be concerned about the welfare of future generations, etc. It is still also important to respect those duties that distinguish humans from other animals: to choose responsibly, to be reasonable, and to respect other humans. But, these duties must be understood in the context of other biological duties and not override them.

For those who love and respect animals and plants, it is easy to grieve over the decline of songbirds, turtles, frogs, forests, newts, and large mammals. Even those creatures people often dislike, such as some insects and predators, and those normally invisible, are full of significant information and interest to some. Yet, whether plants and animals deserve respect in themselves or not, they are a vital part of human community and provide much of the beauty and joy of life. Even in major cities the experience of fresh air, greenery, trees, and plants woven into these cities are needed for urban well-being. No more dismal picture of the world can be conceived than that robbed of the beauty of the natural environment. However, much fiction has made the interiors of space-ships seem bearable, metal, plastic, and television cannot compare to what nature has to offer. If one is looking for dismal, one can imagine a wealthy world made entirely of suburban malls, streets, and television, or a poor one of dusty streets, cinder block, and no shade. An economics of respect for nature can be sustained for generations; an economics of exploitation in a limited, integrated biosphere is inevitably self-consuming and self-destroying.

### Future generations

Providing for children and grandchildren has been long a fundamental responsibility and preoccupation of humans. The meaning of individual lives involves an integrated sense of self as connected with ancestors and descendants, if not one's own children, the children of others.

The main way to provide for children in the industrial West has been for the present generation to aggregate wealth; present wealth meant wealth for the future. But now, if "wealth" means material goods, then future generations are more likely to suffer from the wealth of this generation rather than to benefit.

Careful studies of natural resources for the next 50 years predicts decline rather than improvement. Indeed, a recent Worldwatch Institute report predicts famine for half of humanity in 40 years. Thus, "future generations" means our present children and grandchildren in their maturity; indeed, many millions of adults living now are likely to be alive 50 years on; it is prudent, not speculative, to think of responsibilities to future generations. The wealth that now must be sought and protected for future generations is harmony with nature, and equality and peace among humans.

### Population

Population and resource use are linked. If there were few humans, people could squander nature harmlessly, but because people are many, the sins of overconsumption magnify immensely. A large population must live modestly and limit its increase. When Thomas Malthus and others introduced the modern political and economic controversy over population 200 years ago, it was unclear whether births could be limited. Economists believed that if food were shared equally in an exponentially growing population, inevitably an equal portion of food for each would mean that all would eventually starve. So, population growth initially seemed to justify inequality.

But, births can be limited; in some countries, population growth is currently at a standstill. Although halting growth has involved methods that stir controversy – birth control and abortion – population growth must be stopped if nature is not to be destroyed and if humans are generally to enjoy an adequate standard of material welfare.

This situation has implications for basic and long-held assumptions about the ethics of family planning. Having children is a decision, like the use of property and resources, which has consequences for others, nature, and future generations. Every nation should have a population limitation policy that limits growth while being as respectful of individual family decisions as possible.

Population control has imposed undue burdens, especially on women, whose health and welfare are most at risk in planning families and controlling births. But, there are ways to reduce these burdens. Poor women around the world need basic economic goods and health services. If the entire focus of health services is birth control, as it is in some places, women are not well served. Thus, it is important that births be limited by providing, first, an adequate economic environment for families to thrive, and, second, basic public health and medical care so children and adults feel safe in having few children. Family planning services should be part of these basic services.

### Weapons and war

The easing of East–West tensions has eased the immediate danger of nuclear Armageddon, and nuclear stock-piles are now thankfully being reduced. But control over the spread of nuclear weapons has weakened, and risks of nuclear conflict are spreading geographically. Nuclear weapons, with their long-term radioactive damage and global environmental destruction, are the worst among many new weapons that increase the environmental costs and health destructiveness of war. Moreover, the costliness of weapons, about equal to the income of the bottom half of humanity, limits the availability of resources for more beneficial activities.

Abstaining from war will prove culturally, as well as economically, challenging. From war a male culture emerges that is destructive to the environment: the ability to kill, to be armed, and to destroy have long been among the tests of male adulthood and legitimacy. In its extremes, this male culture fosters domestic and street violence, disregard for the environment, consumption of harmful substances, self-seeking notions of achievement, excessive competitiveness, and accumulation of wealth. Thus, for the agenda of response to environmental change to become real, culture will need to change at many levels and in complex ways.

## *The role of medicine*

Poverty, scarcity, war, overcrowding, and environmental pollution create health problems. Indeed, there is good reason to worry that in the next 50 years, a worldwide public health disaster of major proportions is shaping up. The goals of health care and biological research should respond to this worry.

## Health care resource use

As one of the major sectors of any national economy, health care deserves more attention as a consumer of resources. Any brief visit to the medical world suggests areas of potential conservation, reuse, and recycling: hospital lighting and heating, disposable gowns, drapes (covers), and needles; helicopter, plane, and automotive transportation; records, documents, and paper; disposal of toxic and radioactive materials; and so on. Although some highly technological health care equipment is neither massive nor a large user of energy at the point of delivery, environmental costs of manufacture and transportation of such equipment is extremely high.

Health care is also dependent on the environment for many of its resources – some important pharmaceuticals are derived from plant species threatened by environmental change. Environmental change thus threatens the long-term capacity for medicine to continue to provide and improve its present technologies. Moreover, health care tends not to recognize its dependency on native civilizations and species, part of the common heritage of humankind.

Although there is a tendency to exempt health care from its environmental responsibilities because of its strong life-saving potential, many of health care's life-saving services could be provided with greater environmental efficiency. Moreover, if health care is simply viewed as saving lives in this generation to the detriment of the next, it is not providing health care in a sustainable fashion.

## Exploitation of suffering

United States health care uses roughly 40% of the world's health service resources on only 6% of the world's population. Much of this is luxury health care, driven by consumerist economy. Wasteful procedures often mentioned include: diagnostic testing to avoid malpractice; computer-aided tomography, magnetic resonance imaging and positron emission tomography when X-rays could substitute; coronary artery bypass operations for weak indications; radial keratotomy for near-sightedness when glasses would do; and fetal monitoring when regular checks by stethoscope are adequate. Medicine has often been called to task for wastefulness, but in the context of a global health crisis and resource scarcity, United States consumerist styles of medical practice join other economic excesses as an obscenity in the global health picture.

## *Public health goals*

Despite the tremendous investment noted above, the USA has some of the worst public health statistics in the developed world. Part of this can be laid squarely at the door of lack of a commitment to public health. Since environment is the key to health, no amount of health care can make up for lack of a healthy environment. Three main contributors to mortality are smoking, overeating combined with lack of exercise, and alcohol consumption. Traffic accidents, exposure to toxic materials, and violence are also major contributors to morbidity and mortality. All of these problems are partly results of excessive consumption. In a world of open resources, the wealthy patient who overindulges and overuses medical resources would be merely comic or pitiful, but in a world of scarce resources and desperate public health problems, public policy tolerant of this patient is criminally tragic. The United States health care system has long been criticized for its public health failings, but the burgeoning worldwide public health crisis of the coming century greatly increases the weight of arguments on behalf of prevention as more efficient and humane.

## *Contributions of acute care*

As health problems worsen due to environmental change, and cancers, epidemics, and new diseases spread, acute care will have an active role. Medicine is also important for control of population. By helping people to feel that their children will survive, parents can feel safer in bearing fewer children. And, medicine is needed for effective birth control and safe abortion procedures without which population growth cannot be stopped responsibly.

## Research and appropriate health care technology

Although many recent scientific advances in biology and medicine can be criticized for being expensive and lacking in public health impact, scientific developments will be essential to addressing environmentally caused health crises. Some critics attribute the world's problems to science and technology, but scientific developments have permitted detection and analysis of many environmental problems. For instance, we would not know of the ozone and greenhouse gas problems without the capacity for precise atmospheric measurement and analysis. And, the green revolution saved much of the world from the population and food apocalypse predicted by environmentalists in the early 1970s (unfortunately, agriculture cannot count on a second green revolution in the next 50 years). Scientific study and technological development are absolutely essential to make manufacture more efficient, improve recycling, address pollution, increase the availability of resources, reduce the level of violence in conflicts, and adapt acute care to the resources and conditions of the world as a whole.

Right now, consumerist, high-tech medicine seems to be the model toward which the whole world strives, but much medical technology is inappropriate for the conditions to which it is applied. For example, high-tech chemical cleaning materials used in the USA to keep hospitals aseptic cannot be handled safely by China's sewage systems. Similar cases abound. As scientists develop new health care technology, they should be thinking how it can be used for a much wider world than merely the wealthy countries.

## Conclusion

Environmental change brings human civilization two great tragedies. The first great tragedy is growing out of an irreconcilable conflict between the needs of humans and respect for nature. On one hand, since humans and nature share limited resources, humans can only obtain more for themselves if they take more from nature. More wealth for humans, where the population is large, can only mean losses to other animals, plants, and their habitats.

On the other hand, if humans strive to save nature and neglect equity and poverty among humans, then humans must suffer. For example, the Florida panther requires a large habitat in order to survive, some thousands of acres per individual, and hundreds of individuals are needed for the panther to survive in the wild. Meanwhile, Florida is growing rapidly in population and developing economically. It is hard to imagine that this rush of development will find some way to respect the panther's habitat.

The second great tragedy arises from losses to modern philosophical and moral aspirations. Much of the philosophical effort of the Enlightenment has been to develop an aspiring picture of humans as rational, individual, separate from nature, capable of great achievement, and ultimately free. Current events are drawing attention away from these notions of ethics; humans increasingly need to attend to the business of safety, health, and survival. These values, from an Enlightenment point of view, are much lower on the philosophical scale.

There is no easy way out. Philosophers should attempt a synthesis that takes growing global needs into account. Philosophical thinking should seek to build cultural values that can help human societies to respond to the new environment. But as population, production, injustice, and pollution grow rapidly, time to make cultural changes is short, and philosophy changes slowly.

## Suggestions for further reading

Boyden, S. V. *Biohistory: The Interplay between Human Society and the Biosphere: Past and Present*. Park Ridge, NJ: Parthenon Publishing Group, 1992.

Brown, L. R., Denniston, D., Flavin, C., French, H., Lenssen, N. and Lowe, A. M. *State of the World 1994: A Worldwatch Institute Report on Progress toward a Sustainable Society*. New York: W. W. Norton & Company, 1994

Chivian, E., McCally, M., Hu, H. and Haines, A. (eds.) *Critical Condition: Human Health and the Environment: A Report by Physicians for Social Responsibility*. Cambridge, MA: MIT Press, 1993.

Durning, A. T. *How Much is Enough?: The Consumer Society and the Future of the Earth*. New York: W. W. Norton & Company, 1992.

Katz, M. B. *The Undeserving Poor: From the War on Poverty to the War on Welfare*. New York: Pantheon Books, 1989.

McMichael, A. J. *Planetary Overload: Global Environmental Change and the Health of the Human Species*. New York: Cambridge University Press, 1993.

Meadows, D. H., Meadows, D. L. and Randers, J. *Beyond the Limits: Confronting Global Collapse, Envisioning a Sustainable Future*. Post Mills, VT: Chelsea Green Publishing Company, 1992.

Pointing, C. *A Green History of the World: The Environment and the Collapse of Great Civilizations*. New York: St Martin's Press, 1991.

Rolston, H, III. Environmental protection and an equitable international order: Ethics after the Earth Summit. In *Proceedings on The Ethical Dimensions of the United Nations Program on Environment and Development, Agenda 21*, ed. D. A. Brown. Conference held at United Nations, New York, 13–14 January 1994.

Schrader-Frechette, K. S. and McCoy, E. D. *Method in Ecology: Strategies for Conservation*. New York: Cambridge University Press, 1993.

Seager, J. *Earth Follies: Coming to Feminist Terms with the Global Environmental Crisis*. New York: Routledge, 1993.

Weiner, J. *The Next One Hundred Years: Shaping the Fate of Our Earth*. New York: Bantam Books, 1990.

World Bank *World Development Report 1993: Investing in Health*. New York: Oxford University Press, 1993.

# Postscript

DAVID C. THOMASMA and THOMASINE KUSHNER

To paraphrase the philosopher Bertrand Russell, if a scientific civilization is to be a good civilization, it must aim at good human ends. But good human ends are precisely what science and technology on their own cannot provide. The problems encountered in this book are not problems of science and technology, as Eric Cassell illuminated in his chapter, but rather, problems of human agency. We cannot blame blameless objects or processes. We are responsible.

Technology has become the stage on which the lives of health providers and patients have been played out. Thus, it provides the particular dramas that occupy both our consultative and educational strategies. The needs of a technological society run exactly counter to those stressed by the humanities. We are experiencing even now a collision course with medical technology, and technology in general. There is a struggle between the needs of a technological society for passive acquiescence, perhaps most strongly characterized by "watching" television, passive entertainment, virtual reality, and other forms of video entertainment, that must overemphasize emotional content to draw us in, and diminish, correspondingly, our intellectual and analytic capacities, and the humanities, that have traditionally stressed the goals of self-determination and a blend of reason and emotional harmony.

Caution must be exercised especially with regard to the goals of science. Science itself cannot suggest the proper goals of a human civilization. The issue of how to develop such goals is joined at the difference between what *can* be done and what *should* be done. The theologian Joseph Fletcher asks about genetic engineering whether there is a socially responsible science today. He ruminates:

> Whenever I read news stories about "activists" who demonstrate
> against nuclear weapons or nuclear power installations, or express
> outrage about toxic waste hazards, or allege the dangers to society
> and human life of research and development on frontiers such as
> biotechnology or ABC warfare (atomic, bacterial, and chemical), I
> cannot just dismiss it all as emotional imbalance. Instead a web
> of wonder starts to spin itself.

Fletcher concludes that there are moral limits to knowledge, that values ought to outweigh truth, when that truth can be put to destructive use.

Gaining control over those goals of science and technology is the most important political and moral task of the twenty-first century. As E. G. Mesthene observes:

> New technologies by their very nature must challenge existing social
> values and institutions. The opening up of new options for human
> action must call the old ones into question. As man gains control of
> the process of change, he is forced to decisions on the ends to which
> he will direct his own future.

As our power over nature increases, we increase our power over ourselves as well. To seek to control the interlocking processes is to seek to control our own biology, ecosphere, society, and culture. Physician Leon Kass starts his book on biology and human affairs with this observation:

> We belong to nature naturally, we place ourselves outside of nature
> to study it scientifically, in part so that we may be able to alter and
> control it technologically. Yet because we belong to the nature we
> study and seek to control, our power over nature eventually means
> power also over ourselves. We are not only agents but also and
> increasingly *patients* of our scientific project for the mastery of
> nature [author's emphasis].

Although this understanding of technology and science provided the impetus for the book, the chapters themselves focused on those issues that are explicitly created by new technologies and scientific discoveries in biology and medicine. The chapters of the book proceeded from the earliest forms of life through the stages of human development, growth, decline, and death, until the final chapters about issues surrounding the use of animals and concerns for the environment. Through the issues explored by our authors, the vastness of the field of bioethics is apparent. Its impact is felt on almost every stage of human life, on relationships, family, reproduction, care of the family members when aging, to requesting interventions that bring about a loved one's death, and possibly donating organs. There can be no question that scientific advancements create ethical dilemmas.

Looking ahead to the future, what are the major ethical tasks for the twenty-first century in bioethics?

## Technological restraint

Most remarkably the first conclusion one can draw from all the chapters is that medical and other technology, originally designed to assist and expand human capacities, is now almost "out of control." Not only does it create fundamental problems for us in thinking about the nature of the human person (e.g. treating the defective neonate), the value and objectification of body parts (e.g. in providing organs for transplant) and the values of families and society (e.g. controlling end-of-life decisions), but it also creates unjust imbalances among rich and poor in the health care system, not only in one's own country, but also world wide, between rich and poor countries.

In the final essay, Andrew Jameton proposed an ethic sensitive to the environment, but that ethic is actually one for the twenty-first century. The centerpiece is restraining and downsizing our capacities rather than promoting and advancing them. Without curtailing scientific advance, such restraint and sensitivity to complex outcomes will challenge us even more than did the ethic of domination that so characterized the Industrial Revolution.

## Dominion and ownership

A new ethic for the twenty-first century must involve responsibility for all interventions in human life and in the environment. This means that we have extra responsibility when we intervene in natural processes, say in genetics or in the reproductive cycle, or by contrast, in the process of dying. That "extra" responsibility revolves around "objectifying" a process that before was not the object of our direct manipulation.

With respect to genetics or reproduction, in the past we simply tried to fulfil the biblical injunction "Be fruitful and multiply," without major attention to the qualities of our offspring. Now, because of manipulation of the ovum, sperm, and embryo, we encounter moral problems such as surrogate motherhood, transplanting ovaries from aborted fetuses, paying young women for their eggs to be used in either research of fertility clinics, cloning or twinning from early embryos, and the status of frozen embryos that might no longer be wanted.

Also, in the past we allowed persons to die of their diseases. We then invented techniques and machinery that would assist others in dire straits to live. Later as the inventions were applied (some would say misapplied) to diseases for which they were not intended, moral problems about end-of-life decisions arose. As we saw, some people would argue that individuals deserve "everything possible" to save their lives. There are no boundaries to the possible it seems.

Ahead of time, we need to assess these new technologies so that more preparation for the moral impact can be made. Just as environmental impact statements are required by United States law for plans to build new freeways or other projects, so too we need "human values impact statements" for new biological and medical technologies.

## Use of knowledge

To what ends should we put our new scientific advances? As we noted above in the quotation from Leon Kass, human beings studying their biological nature and intervening to alter it become at once both the problem and the solution. If our inclination is to alter nature, then where do we turn for guidance about the goals of that alteration? Simply observing that technology and science must serve good human ends does not bring such ends into being. It is a common understanding that a scientific civilization creates alienation of human beings from one another, from the natural environment, and from their own nature as well. In the midst of that alienation, infected by it as it were, how can a vision of good human ends emerge? In the past, this was largely provided by nature itself. Now, as that nature is altered, as is the social and ecological environment, can it still stand as a benchmark for our ethics?

Perhaps the most explicit appeal to the natural order of things occurred in Jonsen's chapter, when he argued that reproductive technologies should be guided by interventions that promote the stable relationships embodied in the family. This notion is open to broad interpretation, but nonetheless appeals to the natural order of things, the way in which the family is the basis of society. But then, in today's world, what is a family or even a community?

## The goals of medicine

Many of the chapters focused on our difficulties in establishing an end or goal for medicine. Is the goal to be preserving life, or should it rather be preserving life within a personal, familial, social context that establishes boundaries for our scientific and technological interventions? The former has been, traditionally, the interpretation of medicine's ethic of not doing harm. The latter is now required by our science. We saw this dynamic in each section of the book, from birth to death. The science has not so much run amok; rather what has outstripped us is our own ability to understand the implications of scientific changes to our values. Thereby, a greater chance exists than in the past that such science and technology will either be misapplied, or its use unwarranted.

## Eternal life?

What exactly are we pursuing? By pushing forward on longevity, and perhaps providing, in the future, a clone of ourselves for autologous transplantation of our own bone marrow, blood, and organs, are we aiming at the goal of eternal life on earth, a reverse of the myth of the Garden of Eden? According to that myth, we were created to live in a lush garden, perpetually happy, without the specter of death to destroy our relationships. Through our own pride, we exceeded our capacities and ate from the tree of knowledge, participating in divine cognition that was forbidden. As a result, by exceeding our capacities, we introduced death into creation. Perhaps the myth was developed from our experience of doing just that!

If we were to create an eternal life on earth, death would come to mean release in a way we only glimpse today, through the lens of euthanasia and physician-assisted suicide. In those events, patients find that they have had enough of life, a suffering life to be sure. With increased longevity, unless quality of life is also enhanced, persons would tire of the limitations on their freedom brought on by chronic illness and decline. Since they might not *have* to die, perhaps they then would *wish* to die. Death as release, not only from suffering but also from the trials and tribulations of life itself, would then be pursued, as the heroes of the science fiction movie *Agent Orange* pursue it. Thus, a strange inversion would occur. As longevity increases, so too would longing for death. Control of living leads to a conjoined need to control the circumstances of our own dying.

## Educating health professionals toward persons and ambiguity

Instead of pressing on to do everything possible to save biological life, a crystal-clear goal compared with adjusting one's treatment plan to individual situations, the patient's values and family commitments, and social standards, health professionals must don the cap of practical wisdom to treat the whole person. This is much harder to do than the exhortations to do so would have it appear. Reading the chapters of this book offers details about why acquiring such practical wisdom is so hard. While biological science and the corresponding objectification of the body has exploded in the past two centuries, similar advances in value studies and human sciences have not been incorporated into medical school education. In one generation the scientific knowledge-base has advanced four-fold. There is already not enough time in the curriculum. How could this increase in knowledge be incorporated as well? Medicine should turn to exploration and development of values so that health professionals will be better able to cope with the dilemmas they encounter.

## Justice and distribution of health care

Though many nations have in place a system of health care that covers all their citizens, a breaking point is reached in attempting to cover the growing needs of some with the technological possibilities detailed in this book. Just how much intervention is just, and how much is ethically optional? Should everyone be entitled to kidney dialysis, as is the case in the USA, regardless of age? Or should age be used as cut-off for technology, as Daniel Callahan proposed. Similarly, how long should a neonate with bilateral strokes be treated in an intensive care unit before we withdraw treatment? By and large, medical indications are sought to help us to make these judgments, but how can the ever-expanding standards of care, being pushed by our discoveries, be used as a current measure of treatment choices? Most often, coupled with these medical indications, then, are community standards that are subject to the vagaries of political fortune.

## Less focus on autonomy-based ethics, more on community-based ethics

It is obvious that a communitarian ethics is required by our advancing knowledge. If our ethics is based just on rights, then why should not a person receive "all there is" to prolong and enhance his or her life? After all, they have a "right" to it, don't they? But if, instead, our ethics is based on a balance of persons and environment, and our commitments are to live simply by controlling our technology and directing it to good human and environmental ends, then the good of the whole takes precedence over individual rights, however strongly they must be measured and protected. The good of the individual can, and must, be meshed with the good of all.

## Alternative ethics to rights and utility

Throughout the chapters there has been a consistent call for a different kind of ethics. Most of medical ethics has developed from the concept of autonomy, which in turn requires a "right" to exercise one's freedom in the biological or medical environment. Then, too, some of our authors pointed out how the principle of utility has been utilized in medical ethics. Despite the considerable success of these methodologies, a new ethics is especially called for by the myriad challenges biological and medical developments have created.

## Expanding the envelope of concern to the most vulnerable, from the most obvious to the less obvious

When we "cross the boundaries" between species and between living and inanimate matter, the ethical challenge is to develop the reasons why our instincts tell us that we must "care for" animals and the environment, life at its earliest beginnings as fertilized ova, and life near its end, when it is played out in a vegetative person who cannot respond to our queries.

All vulnerable forms of life share one quality: they cannot speak for themselves. They must rely on us to form our social conscience about them. So far we have a poor track record. The widespread manipulation of vulnerable forms of life and our habitat does not auger well for the future unless humanity changes its way of caring for the most vulnerable forms of life and the nonliving environment.

But why should we be concerned? Should it be only that, if we are not, we will not survive ourselves? That is self-interest at its peak. While a forceful argument can be made that self-survival is at stake, it is not the best ethical principle on which to base our concerns. Instead we would argue that an experiential *a priori* requires these concerns. The experiential *a priori* simply means that, as we dialogue across cultures and within our own societies to try to solve some of the major problems bequeathed us by our science and technology, we must bring to the table our own past experience, with all its telling destructiveness, so that we commit ourselves to fundamental concern for the vulnerable as a first principle. Nothing else will do.

### Suggestions for further reading

Fletcher, J. The moral limits of knowledge. *Virginia Quarterly Review* (1988), **64**(4), 565–84.

Mesthene, E. G. How technology will shape the future. *Science* (1968), **161**, 135–43.

Kass, L. *Toward a More Natural Science*. New York: The Free Press, 1985.

# Index

abortion 15, 16, 324
  debated 58–69
  and genetic disease 27, 28–9, 67–8
acid rain 342
Ackerman, T. 274
acquired immunodeficiency syndrome (AIDS)
    166, 173–4, 221–2
  research 266, 285, 292–3
Addison's disease 137
aging/old age 133–62
  and autonomy 142–5, 148–9
  and care 136, 140, 146–52
  and depression 135, 137–8
  ethical issues 142–53
  frailty 137–8
  hormone therapy 133–5
  hospitalization 140–1
  and nutrition 137, 138–9
  personal meaning of 142, 143–6, 152
  and personality change 146–9
  polypharmacy 136, 137
  scientific advances 133–40
  see also dementia
Alexander, L. 262
alkaptonuria 9
altruism 110, 120, 128–9
Alzheimer's disease 142, 154, 160–2, 196, 197
  advances in 135–6
  and euthanasia 223, 228
  experiences of 155–9
  perception of 146
  research 293–4
amyloid beta-protein 135
anencephaly 80, 101, 173
Animal Liberation Front (ALF) 309–10
animal research 301–37
  alternatives to 306–7, 308
  animal testing 306–7

animals in education 307–8
criticisms 305–6
duties to animals 319–22, 326–8, 330, 333
justification of 304–5
pain and distress 302–3
policy and protocol 335–7
regulation of 303–4
utilitarianism 319–21, 327–8, 335, 351, 374
animal rights 305, 314–24, 349–52
  groups 303–4, 308–10, 311
  and social practices 325–37
Animal Welfare Act (USA) 303–4
Annas, G. 24, 262
apolipoprotein E 135
Aries, P. 222
assent, in children in research 274, 276, 277,
    280–1
assentist theory and abortion 62–8
assisted reproduction technology (ART) see
    reproductive technologies
autonomy 374
  in the aged 142–5, 148–9
  in animals 321, 322
  in children 89–92, 94, 108, 270–4, 276–8,
    280–1
  and euthanasia 172, 199–200, 201–5, 211,
    233–4, 245, 253–4, 256
  in medical research 265, 289, 293, 294, 295–6,
    322–3
  and transplantation 107–8, 115–16, 123,
    128
autopoiesis 352–3, 354
azothioprine (Imuran) 102

barbiturates 214–15
Beauchamp, T.L. 106–7
Beecher, H. 264
Belmont Report 265–7, 271–2, 286, 288–9

beneficence 265, 288–9, 290
  in medical research 271–2, 274–6
  in organ donation 109–13
Bentham, J. 319, 350–1
Bernard, C. 259
Bibace, R. 92
Binding, K. 210
birth rates, multiple 46–7
Bluebond-Langer, M. 93–4
brain death 100, 111–13, 195–6
breast cancer screening 28
Brock, D.W. 228, 229
Brown, Louise 36, 52
Budin, P. 72

cadaveric organs, financial incentives for 119–31;
  *see also* transplantation
Calne, R. 102
cancer 11, 28, 44–5, 139, 166
Carey, S. 92–3
caring
  for children 85–98
  for dementia patients 136, 146–9
  for the dying 163–75, 191–206, 243, 244–5,
    256, 257
  for the elderly 136, 140, 146–52
Carruthers, P. 327
children
  caring for 85–98
  concepts of illness/death 91–7
  and medical technology 79–84
  as moral agents 89–91, 92; *see also* autonomy
  as organ donors 108, 111
  in research 270–82
Childress, J.F. 106–7
chromosomal abnormalities 6–9, 11
cloning 37, 51
Cognex (tacrine) 135–6
cognitive disorders, researching of 293–4, 296;
  *see also* Alzheimer's disease; dementia
consanguinuity 46
contraception 48, 66–7
Crick, F. 9–10
criminality 22, 30–1
Cruzan, Nancy 196, 226
curare 215

death
  acceptance 152, 203, 204
  demonizing 16–18
  ending futile treatment 169–75
  in institutions 166–8
  as medical failure 243–4, 245
  "personal" and "medical" death 163–8,
    221–2
  *see also* dying, care of the; euthanasia
Declaration of Helsinki 263–4, 272, 286
dementia 135–6, 145–9, 154–62, 166, 196;
  *see also* Alzheimer's disease
depression 135, 137–8
deviancy, medicalization of 18–19
diethylstilbestrol (DES) 286, 294
DNA, discovery of 9–10
Doe, baby, controversy 76, 86
"double effect" doctrine 219–20
  case histories 225–7
Down syndrome 6, 135
dying, care of the 243, 244–5, 256, 257
  advances in 193–7
  ethical issues 163–8, 169–75, 198–205
  historical perspectives 191–4
  and technology 191–7
  *see also* death; euthanasia

ecosystems 340–6, 353
education
  student screening 29
  use of animals in 307–8
Edwards, R. 52
elders *see* aging
embryo transfer (ET) 36, 38–9
embryos 35–49
  research 298–9
employment and genetics 14, 29
Engelhart, H.T. 159
environment 339–67
  ecosystems 340–6, 353
  ethical issues 348–55, 357–67
  and genetics 24
  habitat destruction 345–6
  and pollution 342–5, 359
  and population 357, 358, 362–3
  role of medicine 364–6

and technology 339, 341, 365–6
and war 363
and wealth 359–60
ethics
    committees *see* Institutional Review Boards
    four principles approach 106–18; *see also*
        beneficence, nonmaleficence, justice,
        autonomy
eugenics 13–14, 21–5, 32–3, 55
euthanasia 165, 197, 199, 207–58
    arguments against 220–1, 231–46
    arguments for 222–7, 247–58
    and autonomy 253–4, 256
    case histories 225–7, 249–51
    "double effect" doctrine 219–20, 225–7
    Dutch experiences 213–15, 216, 238–9
    history of 207–15
    public policy on 210, 212–16, 229, 232, 238,
        254–7
    and religion 207–8, 211, 213, 248
    societies 209, 210, 213
eutrophication 343–4

farming, intensive 333–4
fertility drugs 37, 44–5
fetus
    rights of 62–5, 66
    tissue research 58, 65–6
    tissue transplants 299
    *see also* abortion; prenatal diagnosis
Fletcher, J. 50–1, 53, 211, 369–70
Food and Drug Administration (FDA) 265, 275,
    294
Fox, R. 184, 185
fragile-X syndrome 11
Freedman, B. 274
freedom of choice *see* autonomy
Friedman, M. 88
futility, medical 81–2, 170, 171, 173–4

Garrod, A. 9
Gaylin, W. 274
genetics 5–34
    chromosome abnormalities 6–9
    counselling 25
    ethical use of knowledge 14–20, 21–33

gene therapy 26–7, 285
genetic code 9–10
    history 6, 9
    probabilities in 25–6
    research 296–8
    screening 26–33
    *see also* Human Genome Project; prenatal
        diagnosis
geriatrics *see* aging
germline therapy 48
Gilligan, C. 86–8
Gorovitz, S. 184

heavy metal pollution 345
Hess, J.H. 73
Hippocrates/Hippocratic oath
    and death 166
    and euthanasia 219, 224–5, 240–2
    and harm 110, 265
    and medical futility 171–2
Hoche, A. 210
hormone replacement therapy 133–5
hospices 165, 220–1, 243
hospitalization
    and dying 166–8
    in old age 140–1
Human Genome Project 10–11
    use of knowledge from 14–20, 23–4, 26–33,
        266
humans as research subjects 259–99
Hume, D. 209
Huntingdon's disease 27
Huxley, A. 52–3, 178
hyaline membrane disease 74, 75

immigration 21–2, 32
Imuran (azothioprine) 102
infant mortality 17, 72–4
infant welfare movement 72–3
infertility treatments 35–49
    ethics of 53–7
    history of 51–3
    recent advances 39–42
informed consent 202, 289, 295–6
    and children 94, 270, 271
    *see also* autonomy

inherited diseases 6–9
  genetic basis for 10–11
  screening for 26–8
  *see also* genetics; prenatal diagnosis
insemination techniques 37, 54
Institutional Review Boards (IRBs) 265–7,
  276–9, 283–9, 304
insurance, health 14, 31
intensive care, futile treatment in 170, 171
*in vitro* fertilization (IVF) 36, 38–9, 42–4, 52,
  54–5
  research 266, 298–9
Ivy, A. 260, 261, 262

Jackson, M. 51
justice
  and aging 149–52
  and distribution of health care 374
  in medical research 265, 271–2, 274–7,
  288–90
  and transplantation 113–17

Kamisar, Y. 236, 238
Kantian philosophy 323, 329
Kevorkian, J. 216, 223, 233
Kleegman, S. 51
Kleinfelter's syndrome 8
Kübler-Ross, E. 212

law
  and buying/selling organs 115–17, 120,
  122–31
  and eugenics 22
  and euthanasia 210, 212–16, 229, 232, 238,
  254–7
  and genetic information 23
  and medical death 165
LD$_{50}$ tests 303, 309
Leahy, M.P.T. 326–7
Leopold, A. 352–3
life expectancies 139–40, 191
life, prolonging 133, 163–90
living wills 148, 213

MacIntyre, A. 184
male factor infertility 37, 40–2, 46
malnutrition in old age 137, 138–9

market approach and transplantation 119–31
Marx, K. 209
Matthews, G.B. 93
McCormick, R. 273–4
medical review boards *see* Institutional Review
  Boards
Mendel, G. 6, 13
menopause 133, 134
mental disorders
  and detention 30
  and human rights 330
  laws affecting people with 13–14, 22
  researching of 293–4, 296
Mesthene, E.G. 370
Mill, J. Stuart 171, 255, 349, 359
minority groups
  and mortality 137
  and research 290, 294–5
moral agency/status
  and animals 328–30, 350–2
  and children 89–91, 92
  in dementia 145–9
  in the dying 202
  *see also* autonomy

Naess, A. 355
National Commission
  Belmont Report 265–7, 271–2, 286, 288–9
  and regulation of research 276, 277–9
National Institutes of Health (NIH)
  and animal research 304, 308, 311
  and research on humans 264–6, 268, 284–5,
  288, 291–2, 294–8
Nazism
  and eugenics 13–14, 15, 16, 22
  and euthanasia 210–11, 213
  and human experiments 261–3
neonatal brain hemhorrage 74, 75
neonatal intensive care 71–8
  benefits and costs 80–3
  development of 72–6
  use of technology 79–83, 185
neonatal mortality rates 72–4
Netherlands
  animal research data 303
  euthanasia 213–15, 216, 238–9
neurotransmitters 135

Noddings, N. 86–8
nonmaleficence, in organ donation 107, 109–13
Nuremberg code 262–4, 286
nursing homes 145–6, 167
nutrition in old age 137, 138–9

old age *see* aging
organs *see* transplantation
"orphaning clauses" 275–6
orthostasis 137
ovarian cancer 44–5
ovarian stimulation 37
ozone depletion 342–3

palliative care 204, 256, 257
Pancoast, W. 51
Parkinson's disease 139, 197
paternity cases 32
penicillin 193
persistent vegetative state
    and euthanasia 212–13, 226, 237
    and futile treatment 170–1
    and organ donation 101
    and technology 195–6, 200
pesticides 344
pets 333–4
phenylketonuria 26
physician-assisted suicide (PAS) 197, 218–29, 248
    arguments against 231–46
    arguments for 222–7
    case histories 225–7
    helping patients to die 227–9
    *see also* euthanasia
physicians, professional ethics 231–46;
    *see also* Hippocratic oath
Piaget, J. 91–2
pollution 342–5, 359
polygenic diseases 11
polypharmacy in elders 136, 137
polyploidy 40–1, 45
Popper, K. 171
population
    screening 27
    world levels 357, 358, 362–3
Power of Attorney laws 148, 213

pregnancy
    and drug trials 275
    *see also* assisted reproduction; prenatal diagnosis
premature babies 44, 72–7; *see also* neonatal intensive care
prenatal diagnosis 6, 14, 26, 27–8
    consequences of 67–8
    parental choice over 15–16
proxy consent 272, 273, 277, 280

Quill, T.E. 223–4, 228, 233
Quinlan, Karen Ann 196, 212, 244

Ramsey, P. 273–4
rationality
    and dementia 154–60
    in reproduction 50–7
Regan, T. 323, 327, 328–30, 350–1
religion, and euthanasia/suicide 207–8, 211, 213, 248
reproductive technologies 35–70
    assisted (ART) 35–49, 50–7, 67–8
    history 51–3
    problems with 45–7
    recent advances 39–42, 54
    rights/ethical issues 53–7, 58–70
research on humans 259–99
    history of 259–69
    regulation of 262–9, 276–9, 283–99
    unethical research 260–5, 270–1, 284, 286–7
    use of children 270–82
    *see also* animal research
respect for persons 288–9, 290; *see also* autonomy
respirators 170, 193
resuscitation techniques 194
"right to die" 199–200, 201–5, 211, 213, 226;
    *see also* euthanasia
rights
    and abortion 58–70
    of fetus 62–5, 66
    and justice 113–15
    and reproduction 36, 58–70
    *see also* animal rights
risks
    in medical research 273–4, 278–81, 283, 287, 290–1
    in organ donation 109–11, 116–17
Russell, B. 369

severe combined immunodeficiency disease (SCID) 27
Shirkey, H. 275
Singer, P. 329, 350
Smith, D.H. 159
social pressure/control 15–16, 17, 18–19, 143–4
soil erosion 344
Steptoe, P. 52
sterilization 22
strokes 134, 135
suicide, assisted *see* physician-assisted suicide
surfactant 75, 80
surrogacy 56, 68
Swazey, J. 268

tacrine (Cognex) 135–6
Tarnier, E. 72
technology 369–71
    and care of the dying 191–7
    control and overuse of 177–89
    and the environment 339, 341, 365–6
    and treatment of children 79–84, 185
thalidomide 286, 294
Tomas, L. 222
transplantation 99–132
    allocation of resources 113–17, 125–7
    autonomy in donors 107, 108, 115–16, 123, 128

buying and selling organs 115–17, 119–31
    ethical issues 103–5, 106–18, 119–31
    policy on 115–17, 120, 122–31
    risks to donors 109–11, 116–17
    scientific advances 99–105
tuberculosis 179
Turner's syndrome 8
Tuskegee Syphilis Study 264
twinning 37, 51

utilitarianism 319–21, 327–8, 335, 351, 374

ventilators 170, 193

Walsh, M.E. 92
war 363
Watson, J. 9–10
Whitehead, A.N. 181, 184
Williams, G. 211
Willowbrook School study 270–1, 272
Wittgenstein, L. 326–7
women, as research subjects 290, 294–5
World Medical Association
    policy on euthanasia 212, 248
    and research on humans 263, 272, 286

Yesavage Geriatric Depression Scale 138